Business and Technology of the Global Polyethylene Industry

Scrivener Publishing
100 Cummings Center, Suite 541J
Beverly, MA 01915-6106

Publishers at Scrivener
Martin Scrivener (martin@scrivenerpublishing.com)
Phillip Carmical (pcarmical@scrivenerpublishing.com)

Business and Technology of the Global Polyethylene Industry

An In-depth Look at the History,
Technology, Catalysts, and
Modern Commercial Manufacture of
Polyethylene and Its Products

Thomas E. Nowlin

Scrivener
Publishing

WILEY

Co-published by John Wiley & Sons, Inc. Hoboken, New Jersey, and Scrivener Publishing LLC, Salem, Massachusetts.
Published simultaneously in Canada.

For general information on our other products and services or for technical support, please contact our Customer Care Department within the United States at (800) 762-2974, outside the United States at (317) 572-3993 or fax (317) 572-4002.

Wiley also publishes its books in a variety of electronic formats. Some content that appears in print may not be available in electronic formats. For more information about Wiley products, visit our web site at www.wiley.com.

For more information about Scrivener products please visit www.scrivenerpublishing.com.

Cover design by Russell Richardson

Library of Congress Cataloging-in-Publication Data:

ISBN 978-1-118-94598-8

Printed in the United States of America

10 9 8 7 6 5 4 3 2 1

Contents

Preface   xviii

1. **Global Polyethylene Business Overview**   1
  1.1   Introduction   1
  1.2   The Business of Polyethylene   2
  1.3   Cyclical Nature of the Polyethylene Business   2
     1.3.1   Global Feedstock Cost Variations   3
     1.3.2   Change in Middle East Feedstock Cost Advantage   4
        1.3.2.1.   2012 Capital Requirements for an Integrated Ethylene/Polyethylene Facility   5
  1.4   Early History of Ethylene and Polyethylene Manufacturing   6
     1.4.1   Discovery of Polyethylene   8
        1.4.1.1   Branched High-Pressure Polyethylene   8
        1.4.1.2   Linear Polyethylene   9
     1.4.2   Linear Low-Pressure Polyethylene – High-Density Polyethylene   9
     1.4.3   Manufacture of HDPE (1955–1975)   10
        1.4.3.1   Phillips Catalyst Produced HDPE with Higher Product Demand   11
        1.4.3.2   Phillips Catalyst Manufacturing Advantages   11
        1.4.3.3   Second Generation Ziegler Catalysts   11
     1.4.4   Single-Site Ethylene Polymerization Catalysts   12
     1.4.5   Status of the Polyethylene Industry as of 2010   13
     1.4.6   Global Demand for Polyethylene in 2010   14
     1.4.7   Polyethylene Product Lifecycle   14
        1.4.7.1   North American Polyethylene Market   18
     1.4.8   Comparison of Other Global Polyethylene Markets with the North American Market   19

1.4.9   Growth of the Global Consumer Class   20

      1.4.9.1   Quantitative Forecast for the Growth of the Global Middle Class   23

1.4.10   Global Economic Freedom   25

1.4.11   Future Economic Growth in India and China   26

1.4.12   Long-Term Global Polyethylene Capacity Expansion (2010–2050)   27

      1.4.12.1   Location of New Ethylene/Polyethylene Capacity (2010–2050)   27

1.4.13   Ethylene Feedstock Costs   28

      1.4.13.1   Manufacture of Ethylene   28

      1.4.13.2   Crude Oil and Natural Gas Prices   30

1.4.14   Impact of the Shale Natural Gas Revolution on Global Polyethylene Business   33

1.4.15   Natural Gas Liquids   34

      1.4.15.1   North American Natural Gas Supply   36

      1.4.15.2   Globalization of Natural Gas   36

      1.4.15.3   Status of Ethylene Costs as of 2010 and Future Trends   37

      1.4.15.4   Additional Feedstock for Ethylene Production   38

      1.4.15.5   U.S. Ethylene Forecast by Burns & McDonnell (2013)   38

      1.4.15.6   Ethylene Based on New Feedstock   39

      1.4.15.7   Methane/Methanol as Feedstock for Ethylene   39

      1.4.15.8   Biomass as Feedstock   40

      1.4.15.9   Governmental Policy and Regulation   40

1.4.16   Environmental Factors   42

      1.4.16.1   Polyethylene Recycling – Waste-to-Energy   43

      1.4.16.2   Biodegradation   43

1.4.17   Biobased Ethylene   44

References   44

2. Titanium-Based Ziegler Catalysts for the Production of Polyethylene   **47**

2.1   Introduction   47

2.2   Titanium-Based Catalyst Developments   47

2.2.1   Historical Developments 47
2.2.2   The Role of Professor G. Natta 50
2.2.3   Historical Controversy – Isotactic Polypropylene
with Cr-Based Catalyst 51
2.3   Titanium-Based Catalysts for the Manufacture
of Polyethylene 52
2.3.1   First Generation Ziegler Catalysts for the
Manufacture of Polyethylene 53
2.3.2   Types of Metal Alkyls Investigated 54
2.3.3   Soluble Titanium-Based Complexes for
Ethylene Polymerization (1955–1960) 55
2.3.4   Mechanism of Polymerization 58
2.4   Second Generation Ziegler Catalyst for the
Manufacture of Polyethylene 62
2.4.1   Early History of Ti/Mg-Based Catalysts – Solvay
& Cie Catalyst 62
2.4.1.1   Solvay & Cie Catalyst Details 62
2.4.2   Gas-Phase Fluidized-Bed Polymerization 68
2.4.3   Impact of High-Activity Mg/Ti Ziegler Catalysts
on the Polyethylene Industry 69
2.4.4   Overview of Particle-Form Technology 69
2.4.4.1   Historical Introduction 69
2.4.5   Growth of the Polymer Particle 71
2.4.6   Catalyst Polymerization Kinetics and
Polyethylene Particle Morphology 73
2.4.7   Magnesium-Containing Compounds that
Provide High-Activity Ziegler Catalysts 74
2.4.8   Additional Preparation Methods for
Catalyst Precursors 75
2.4.8.1   Reduction of $TiCl_4$ with
Organomagnesium Compounds **75**
2.5   Catalysts Prepared on Silica 76
2.5.1   Physical Impregnation of a Soluble
Mg/Ti Precursor into the Silica Pores 76
2.5.2   Chemical Impregnation of Silica 77
2.6   Characterization of Catalysts Prepared with Calcined Silica,
Dibutylmagnesium or Triethylaluminum and $TiCl_4$ 82
2.6.1   Spray-Drying Techniques 87

	2.6.2	Ball-Milling Techniques	89
	2.6.3	Characterization of High-Activity Ti/Mg-based Ziegler Catalyst Precursors	90
	2.6.4	Additional Electron Donor Complexes	91
	2.6.5	Catalysts Based on Magnesium Diethoxide and TiCl$_4$	92
	2.6.6	Spherical Magnesium-Supported Catalyst Particles	93
	2.6.7	Catalysts Prepared with Grignard Reagent/TiCl$_4$ with and without Silica	94
	2.6.8	Polyethylene Structure	94
	2.6.9	Characterization of Reactivity Ratios in Multi-site Mg/Ti Catalysts	95
2.7		Kinetic Mechanism in the Multi-site Mg/Ti High-Activity Catalysts	96
	2.7.1	Introduction	96
	2.7.2	Multi-center Sites	99
	References		104
	Appendix 2.1		107
3.	**Chromium-Based Catalysts**		**109**
3.1		Part I – The Phillips Catalyst	109
	3.1.1	Early History of the Phillips Catalyst	109
	3.1.2	Preparation of the Phillips Catalyst	112
	3.1.3	Unique Features of the Phillips Catalyst	115
	3.1.3.1	Control of Polyethylene Molecular Weight	115
	3.1.3.2	Initiation of Polymerization at the Active Center	116
	3.1.3.3	Possible Initiation Steps for Cr-Based Catalyst	116
	3.1.4	Characterization of Polyethylene Produced with the Phillips Catalyst	117
	3.1.5	Improvements to the Phillips Catalyst	119
	3.1.6	Review Articles for the Phillips Catalyst	125
3.2		Part II – Chromium-Based Catalysts Developed by Union Carbide	126
	3.2.1	Bis(triphenylsilyl)chromate Catalyst	127
	3.2.2	Chromocene-Based Catalyst	132
	3.2.3	Hydrogen Response of the Chromocene-Based Catalyst	137

	3.2.4	Effect of Silica Dehydration Temperature on the Chromocene-Based Catalyst	138
	3.2.5	Bis(indenyl) and Bis(fluorenyl) Chromium(II) Catalysts Supported on Silica	141
	3.2.6	Organochromium Compounds for Ethylene Polymerization Based on (Me)₅CpCr(III) Alkyls	142
	3.2.7	Organochromium Complexes with Nitrogen-Containing Ligands for Ethylene Polymerization	145
	3.2.8	Catalysts for Ethylene Polymerization with *In-Situ* Formation of 1-Hexene	160
3.3		Next Generation Chromium-Based Ethylene Polymerization Catalysts for Commercial Operations	164
References			165
4.		**Single-Site Catalysts Based on Titanium or Zirconium for the Production of Polyethylene**	**167**
4.1		Overview of Single-Site Catalysts	167
	4.1.1	Expanded Polyethylene Product Mix	167
	4.1.2	Types of Single-Site Catalysts	168
4.2		Polyethylene Structure Attained with a Single-Site Catalyst	169
	4.2.1	Product Attributes of Polyethylene Manufactured with Single-Site Catalysts	171
	4.2.2	Processing Disadvantage of Polyethylene Manufactured with Single-Site Catalysts	171
4.3		Historical Background	172
	4.3.1	First Single-Site Catalyst Technology – Canadian Patent 849081	174
	4.3.2	Discovery of Highly Active Metallocene/ Methylalumoxane Catalysts	176
		4.3.2.1 Early Publications of Kaminsky	177
		4.3.2.2 Kinetic Parameters of the Homogeneous Cp₂ZrCl₂/MAO Catalyst	179
	4.3.3	Alkylalumoxanes – Preparation, Structure and Role in Single-Site Technology	180
		4.3.3.1 Background	180
		4.3.3.2 Preparation	180

4.3.4 Structure of Alumoxanes 181
    4.3.4.1 Role of Methylalumoxane in Single-Site Catalysts 183
    4.3.4.2 Supporting Evidence for Cationic Active Site for Ethylene Polymerization 184
4.3.5 Additional Methods for Activating Metallocene Single-Site Catalysts 187
4.3.6 Characterization Methods that Identify Polyethylene with a Homogeneous Branching Distribution Obtained with Single-Site Catalysts 189
4.3.7 Control of Polymer Molecular Weight 192

4.4 Single-Site Catalyst Based on (BuCp)$_2$ZrCl$_2$/MAO and Silica for the Gas-Phase Manufacture of Polyethylene 193

4.5 Activation of the Metallocenes Cp$_2$ZrCl$_2$ or (BuCp)$_2$ZrCl$_2$ by Solid Acid Supports 197
4.5.1 Activation of Bridged Metallocenes by Solid Acid Supports 200

4.6 Dow Chemical Company Constrained Geometry Single-Site Catalysts (CGC) 202
4.6.1 Cocatalyst Activation of Constrained Geometry Catalyst 203
4.6.2 Processability of Polyethylene Manufactured with Dow's CGC System 204

4.7 Novel Ethylene Copolymers Based on Single-Site Catalysts 205

4.8 Non-Metallocene Single-Site Catalysts 207
4.8.1 LyondellBasell Petrochemical 210

4.9 New Ethylene Copolymers Based on Single-Site Catalysts 211
4.9.1 Ethylene/Norbornene 211
4.9.2 Ethylene/Styrene Copolymers Using Nova Chemicals Catalyst 213

4.10 Compatible Metallocene/Ziegler Catalyst System 215

4.11 Next Generation Catalysts 217

References 219

Appendix 4.1 222

5. Commercial Manufacture of Polyethylene **223**
   5.1   Introduction   223
       5.1.1   First Manufacturing Facility   224
       5.1.2   Early Documentation of Manufacturing Processes   225
   5.2   Commercial Process Methods   226
   5.3   Global Polyethylene Consumption   228
   5.4   High-Pressure Polyethylene Manufacturing Process   229
       5.4.1   Historical Summary   229
       5.4.2   Details of the Discovery of the
              High-Pressure Process   229
       5.4.3   Developments during World War II (1940–1945)   231
       5.4.4   Post World War II Developments (1945–1956)   232
       5.4.5   Rapid Growth Period – Demand Exceeded Supply   233
       5.4.6   Polyethylene Growth (1952–1960)   235
            5.4.6.1   Polyethylene Product Attributes that
                     Resulted in Rapid Growth   237
            5.4.6.1.3   Property Comparison   238
                 5.4.6.1.4   Pipe Applications   239
       5.4.7   Worldwide High-Pressure LDPE Capacity
              Increases (1980–2010)   241
       5.4.8   Future of High-Pressure Manufacturing Process   243
   5.5   Free-Radical Polymerization Mechanism
       for High-Pressure Polyethylene   243
       5.5.1   Initiation Step   244
       5.5.2   Propagation Step   244
       5.5.3   Termination Step   246
   5.6   Organic Peroxides as Free-Radical Source for
       Initiation Process   246
       5.6.1   Types of Organic Peroxides   247
   5.7   Structure of High-Pressure LDPE   248
       5.7.1   General Features   248
       5.7.2   Ethylene at High Pressures   249
       5.7.3   Autoclave Reactor   251
       5.7.4   Characterization of Short-Chain Branching
              (SCB) in LDPE   253
            5.7.4.1   Structural Differences between
                     LDPE and LLDPE   254

5.8	Low-Pressure Process		255
	5.8.1	Early History	255
	5.8.2	Particle-Form Technology for Low-Pressure Process	256
		5.8.2.1 Historical Background	256
		5.8.2.3 Solution Mode Operation	260
		5.8.2.4 Slurry Polymerization Mode	260
		5.8.2.5 Pilot Plant Designs for Particle-Form Reactors – Development History	261
		5.8.2.6 Phillips Pilot PlantVertical Pipe-Loop Reactor Design	262
		5.8.2.7 Operation of the Phillips Pilot Plant Pipe-Loop Reactor	264
	5.8.3	First Ziegler Catalyst Commercial Process	265
	5.8.4	Chevron-Phillips Slurry Loop Process Status as of 2010	268
		5.8.4.1 Reactor Scale	268
	5.8.5	Reactor Start-Up	269
	5.8.6	Product Transitions	270
	5.8.7	Reactor Fouling	271
	5.8.8	New Vertical Loop Reactor Design	273
5.9	Gas-Phase Process		274
	5.9.1	Historical Introduction	274
	5.9.2	BASF Early Gas-Phase Reactor	279
	5.9.3	Horizontal Gas-Phase Process	279
	5.9.4	Union Carbide Gas-Phase Reactor	281
		5.9.4.1 Gas Distribution Plate	282
	5.9.5	Gas-Phase Univation Process (2012)	283
	5.9.6	Fluidized-Bed Gas-Phase Operation Overview	283
		5.9.6.1 Ethylene Partial Pressure	285
		5.9.6.2 Catalyst Feed Rate	286
		5.9.6.3 Product Discharge	286
		5.9.6.4 Condensing Agent	286
		5.9.6.5 Reactor Start-Up/Product Transitions	287
		5.9.6.6 Reactor Fouling	288
		5.9.6.7 Catalyst Requirements for Gas-Phase Fluid-bed Reactor	289

5.10	Gas-Phase Process Licensors	290
5.10.1	Background	291
5.10.2	Gas-Phase Process Company History	291
5.10.3	Reactor Size and Configuration	292
5.10.4	Gas Phase and Slurry Loop in Series	294
5.11	Solution Process	294
5.11.1	Historical Introduction	294
5.12	DuPont Sclair Process	295
5.12.1	Background	296
5.13	Solution Process (2012)	298
5.13.1	Overview	298
5.13.2	Bimodal MWD in Solution Reactors	299
5.13.3	Dowlex Solution Process	300
References		300

6 Fabrication of Polyethylene **303**

6.1	Introduction	303
6.1.1	Fabrication Business	304
6.1.2	Terms and Definitions Important in Polyethylene Fabrication	304
6.1.3	Development of Melt Index Instrument	306
6.1.4	Polyethyene Product Space for the Fabrication of Finished Products	307
6.2	Early History of Polyethylene Fabrication (1940–1953)	308
6.2.1	Post World War II	309
6.3	Stabilization of Polyethylene	310
6.3.1	Introduction	310
6.3.2	Thermal Oxidation Mechanism (1920–1960)	311
6.3.3	Polyethylene Melt Processing	314
6.3.3.1	Scission	315
6.3.3.2	Crosslinking	315
6.3.3.3	Temperature Dependence of Crosslinking and Scission	316
6.4	Historical Overview of Some Common Polyethylene Additives	316
6.4.1	Polyethylene Additives (1935–1955)	316
6.4.1.1	Carbon Black	317
6.4.1.2	Antioxidants	317

	6.4.1.3	Flame Retardants	317
	6.4.1.4	Lubricants	318
	6.4.1.5	Anti-Static Agents	318
	6.4.1.6	Calcium Carbonate as Filler	318
6.5	Examples of Additives Presently Used in the Polyethylene Industry (2012)		318
	6.5.1	Antioxidants	319
	6.5.2	Secondary Antioxidants	320
	6.5.3	UV-Light Stabilizers	321
	6.5.4	Mineral Fillers/Reinforcing Agents	321
	6.5.5	Lubricants	324
	6.5.6	Blowing Agents	324
	6.5.7	Flame Retardants	324
	6.5.8	Antiblock Agents	324
6.6	Rheological Properties of Polyethylene		326
6.7	Fabrication of Film		327
	6.7.1	Introduction	327
6.8	Blown Film Extrusion		328
	6.8.1	Description of Blown Film Extrusion	328
	6.8.2	History of Polyethylene Rapid Growth in Film Applications	329
		6.8.2.1 Cost	329
		6.8.2.2 Shelf-Life	329
	6.8.3	Blown Film Apparatus	329
	6.8.4	Multilayer Films	330
	6.8.5	Low-Density Polyethylene Films	332
	6.8.6	LLDPE with a Broad MWD	333
	6.8.7	Blown Film Process for HMW-HDPE Film	334
	6.8.8	High-Stalk Extrusion	336
	6.8.9	Cast Film Line	337
	6.8.10	Pipe Applications	339
6.9	Fabrication of Polyethylene with Molding Methods		341
	6.9.1	Blow Molding	341
		6.9.1.1 Brief History of Blow Molding (ca. 1850–1960)	341
		6.9.1.2 Environmental Stress Crack Resistance	343
		6.9.1.3 Types of Blow Molding Machines	344

		6.9.1.4	Method to Decrease Die Swell	346
		6.9.1.5	Milk Bottle Resin	347
	6.9.2	Injection Molding		348
		6.9.2.1	Introduction	348
		6.9.2.2	Polyethylene Shrinkage	349
		6.9.2.3	New Product Applications	350
		6.9.2.4	History of Injection Molding Process	350
		6.9.2.5	Some Aspects of the Machine Design	352
		6.9.2.6	Mold Design	353
6.10	Rotational Molding			355
	6.10.1	Background History		355
6.11	Thermoforming			357
	6.11.1	Thin- and Thick-Gauge Thermoforming		357
	6.11.2	Grades of Polyethylene for Thermoforming		359
References				359

7. Experimental Methods for Polyethylene Research Program   361
7.1	Introduction			361
	7.1.1	High Throughput Laboratory Equipment		363
7.2	Experimental Process			363
	7.2.1	Catalyst Preparation		364
		7.2.1.1	Some Catalyst Preparation Operation Guidelines	364
	7.2.2	Catalyst Evaluation Process		366
	7.2.3	Catalyst Performance Characteristics		367
7.3	Important Considerations for Laboratory Slurry (Suspension) Polymerization Reactors			368
	7.3.1	Background Information		368
	7.3.2	Basic Laboratory Polymerization Reactor Design		368
	7.3.3	Polymerization Rate/Total Polymer Yield		370
	7.3.4	Isolation of Polyethylene Product		371
	7.3.5	Steady-State Polymerization Conditions		371
		7.3.5.1	Determination of Steady-State Conditions During Polymerization	373
		7.3.5.2	Operation Guidelines for a Slurry Polymerization Reactor under Steady-State Condition	375

7.3.6   Polymer Characterization   375

     7.3.6.1   Laboratory Characterization Equipment   375

     7.3.6.2   Melt Index and Density Data   375

     7.3.6.3   Infrared Method   378

     7.3.6.4   Differential Scanning Calorimetry   380

     7.3.6.5   Gel Permeation Chromatography   382

     7.3.6.6   Temperature Rising Elution Fractionation   385

     7.3.6.7   CRYSTAF Method   388

     7.3.6.8   Carbon-13 Nuclear Magnetic Resonance   388

7.3.7   Catalyst Process Attributes   389

     7.3.7.1   Catalyst Activity (g PE/g cat) and Polymerization Kinetics   389

     7.3.7.2   Reactivity with Higher 1-Olefins such as 1-Butene, 1-Hexene and 1-Octene   391

7.3.8   Additional Features of Commercial Catalysts   391

7.4   Polymerization Reactor Design for High-Throughput Methods   391

7.5   Polymer Characterization   393

7.6   Process Models   393

References   394

Index   **397**

Preface

This book, *Business and Technology of the Global Polyethylene Industry*, is an in-depth look at the history, technology, catalysts, and modern commercial manufacture of polyethylene and its products.

Primary emphasis has been placed on documenting the history of the important technical developments over the past 80 years, and the scientists and engineers responsible for them, that have led to continued growth in the polyethylene industry—from the manufacture of a few hundred tons of polyethylene in 1940 to the manufacture of over 160 billion pounds of polyethylene in 2012.

The book is dedicated to the thousands of scientists and engineers that have been involved in research, development, manufacturing and fabrication of polyethylene since its discovery in 1933. The role of these industrial scientists was to translate scientific results into practical innovations for commercial purposes which have greatly contributed to an improvement in the standard of living of people around the globe.

Each of the seven chapters is written to appeal to a wide range of individuals that are involved in the polyethylene industry today.

Chapter 1 is an overview of the polyethylene business, written as an executive summary, which may appeal to anyone interested in understanding the status of the global polyethylene industry today and the factors that may lead to the continued growth of the industry over the next half-century.

Chapters 2, 3 and 4 discuss each of the three catalyst types used today to manufacture polyethylene, with Chapter 2 devoted to titanium-based catalysts; Chapter 3 to chromium-based catalysts and Chapter 4 to single-site catalysts. These chapters would primarily appeal to scientists that are involved in developing ethylene polymerization catalysts.

Chapter 5 is a summary of the processes used to manufacture polyethylene, which includes the tubular and autoclave high-pressure process and the three methods that operate at relatively low-pressure—the slurry process, gas-phase process and solution process.

Chapter 6 is an overview of the methods used to fabricate polyethylene into end-use products and includes a brief discussion of the additives required to stabilize and enhance the properties of polyethylene in these end-use applications.

Chapter 7 discusses the design and operation of a research laboratory dedicated to developing new polymerization catalysts for the polyethylene industry.

Because no single book could cover all of the complex subjects associated with the global polyethylene industry, several recently published books are recommended. There have been many excellent books published between 2000 and 2010 that discuss the significant new technology that has been commercialized for the manufacture of polyethylene and other polyolefins around the globe. Much of the recent discussion has dealt with the introduction of single-site catalyst technology for the polymerization of 1-olefins. However, the books by Andrew J. Peacock and Dennis B. Malpass are directed more specifically at the polyethylene industry and cover some additional topics in more detail than this book. Peacock's book covers the testing methods required in the industry and the book by Malpass includes a chapter on metal alkyls that are used extensively in the polyethylene and polypropylene industries, and importantly, includes comments on the safety and handling of metal alkyls.

The following books are recommended:

Metallocene-Based Polyolefins, edited by John Scheirs and Walter Kaminsky, John Wiley & Sons, Ltd., Volumes 1 and 2, ISBN 0-471-9886-2, 2000.

Handbook of Polyethylene, Structure, Properties, and Applications, by Andrew J. Peacock, Marcel Dekker, Inc., New York, ISNB 0-8247-9546-6, 2000.

Tailor-Made Polymers, via Immobilization of Alpha-Olefin Catalysts, edited by John R. Severn and John C. Chadwick, Wiley-VCH, ISBN: 978-3-527-31782-0, 2008.

Stereoselective Polymerization with Single-Site Catalysts, edited by Lisa S. Baugh and Jo Ann M. Canich, CRC Press Taylor & Francis Group, ISBN: 13: 978-1-57444-579-4, 2008.

Handbook of Transition Metal Polymerization Catalysts, edited by Ray Hoff and Robert T. Mathers, John Wiley & Sons, Inc., ISBN: 978-0-470-13798-7, 2010.

Introduction to Industrial Polyethylene, by Dennis B. Malpass, Wiley-and Scrivener Publishing; ISBN: 978-0-470-62598-9, 2010.

One important aspect of industrial research that is critical in the commercialization of new technology is the opportunity to work alongside other scientists and engineers with a wide variety of other skills necessary to move new technology from the laboratory stage through the development process and into a commercial facility.

I was privileged in my career to work with a large number of gifted people at both Union Carbide Corporation Research Center in Bound Brook, New Jersey, and Mobil Chemical Company in Edison, New Jersey, that made the pursuit of new technology a rewarding and satisfactory career. This book is especially dedicated to these people. However, I would like to acknowledge and thank Dr. Robert I. Mink and Dr. Yury V. Kissin, who I worked with on a daily basis at the Mobil Technology Center in Edison, New Jersey. Working with these individuals, we were able to develop two new catalyst systems in use today for the commercial manufacture of polyethylene.

Finally, I would like to thank Dr. Max P. McDaniel of Chevron Phillips Chemical Company in Bartlesville, Oklahoma, for many helpful discussions on the development of the Phillips loop process and catalyst systems developed by Phillips Petroleum Company, and later by Chevron Phillips Chemical Company for the slurry process.

1

Global Polyethylene Business Overview

1.1 Introduction

Over the past 80 years the organic material designated as polyethylene has evolved from a laboratory curiosity into a global polyethylene business with a global demand in 2010 of more than 68 million metric tons (150 billion pounds), accounting for the creation of a large number of worldwide jobs involving the manufacture, fabrication and distribution of polyethylene and end-use products. These products involve the transformation of the polymeric material into thousands of necessary applications that have improved the standard of living of people around the globe.

The purpose of this chapter is to summarize the status of the global polyethylene business as of 2012 and to briefly discuss some historical aspects of this business which demonstrate how the technology of this seemingly simple material has continually improved over the course of eight decades as the result of the efforts of thousands of scientists and engineers worldwide. Other chapters in this book will discuss the technical development of the polyethylene business in more detail.

1.2 The Business of Polyethylene

The manufacture of polyethylene is a capital-intensive, complex business involving three distinct business categories:

1. Utilization of hydrocarbon-based feedstock such as naphtha or ethane to manufacture ethylene.
2. Polymerization of ethylene into an intermediate polyethylene material designated as low-density polyethylene with a material density very broadly defined from about 0.89 g/cc to 0.93 g/cc, or high-density polyethylene also very broadly defined with a density from about 0.93 g/cc to 0.97 g/cc. Each of these two product types are manufactured over a wide range of molecular weights and range of molecular weight distributions from very narrow to very broad.
3. Fabrication of polyethylene into a commercial item. This is achieved by melting a specific grade of polyethylene, allowing the molten polyethylene to form the desired shape and then cooling the polyethylene to form the solid material into the desired shape. Because the polyethylene molecular structure is not changed during the fabrication process, commercial items based on polyethylene are easily recycled into new products

1.3 Cyclical Nature of the Polyethylene Business

Each of these three business segments are considered capital intensive businesses where the cost of equipment is relatively very large compared to the fixed cost of labor to operate each business. Therefore, annual revenue per employee is relatively large. For example, a polyethylene plant with an annual production of 1–2 billion pounds of polyethylene requires a few hundred employees so that annual revenue per employee is about 2–4 million US dollars.

For the polyethylene industry, companies involved in the production of ethylene and polyethylene usually target an average annual return on capital (ROC) of about 15%, which is usually based on the performance of a particular segment of this industry over a 5–10 year time period due to the cyclical nature of these businesses on a year-over-year basis. Over the course of the past 80 years, there have been years in which the profitability of these two separate businesses (ethylene and polyethylene) has been remarkably different. For example, there have been periods in which the manufacture of ethylene has experienced a very high (> 20%) ROC (usually due to a low feedstock cost and/or a high ethylene demand); while during this same period, the manufacture of polyethylene has performed relatively poorly (< 5%) due to a relatively high ethylene cost and/or over-capacity of available polyethylene in the marketplace. Consequently, the trend in the petrochemicals industry over the past several decades (since about 1980) has been for companies to integrate the manufacture of both ethylene and polyethylene into a single business and to build petrochemical facilities at the same location that provide both the manufacture of ethylene and polyethylene. This has been especially true in the Middle East where ethylene and polyethylene capacity is always located at the same facility because the infrastructure to transport ethylene across the region is not available. In the early 1980s both Mobil Oil Corporation and Exxon Corporation built large petrochemical facilities in Saudi Arabia in joint ventures with SABIC (Saudi Arabia Basic Industries Corporation). Mobil's facility was located in Yanbu on the Red Sea, while Exxon's facility was located on the Persian Sea near Jubail, Saudi Arabia. Therefore, companies in the polyethylene business are also involved in the manufacture of ethylene and the financial results of these two businesses are usually combined to provide a ROC for both businesses.

In addition, the ROC of the ethylene and polyethylene businesses has varied over a wide range depending on (1) the global location of the petrochemical complex (i.e., the Middle East vs North America or Europe) and (2) the variations in economic growth as measured by annual changes in gross domestic product (GDP). Consequently, there have been times in which the business environment was extremely difficult in some global regions and therefore the ROC for the integrated businesses has been very poor (ca. negative or low single-digit returns), while at the same time in another global region such as the Middle East, the business environment has been relatively better and ROC was high, i.e., in the range of 15–25% or even greater.

1.3.1   Global Feedstock Cost Variations

One aspect of the ethylene/polyethylene business that has been extremely important since the 1970s has been the enormous variation in feedstock costs around the world. In the early years of the polyethylene business, from about 1940–1970, feedstock costs did not vary a great deal around the world due to low-cost crude oil in which crude oil demand did not exceed crude oil supply. This situation changed dramatically in the 1970s with the creation of OPEC due to the continued growth in the demand for crude oil in the developed regions such as North America and Europe that could not be matched with continued increase in the supply of crude oil in these developed regions. For example, crude oil production in the United States peaked in 1970 at about 11 million barrels per day, while crude oil demand in the United States continued to grow beyond this production limit. This regional demand/supply imbalance led to the creation of the crude-oil cartel—Organization of Petroleum Exporting Countries (OPEC)—and the increase in crude-oil price by a factor of about 10 (i.e., crude oil cost increased from about $3/barrel to over $30/barrel). Hence, feedstock costs needed to produce ethylene became the dominant business concern for the manufacture of polyethylene.

Since about 1970 through 2010, the low-cost feedstock advantage found in the Middle East resulted in the investment of enormous amounts of capital in the Middle East for the construction of ethylene/polyethylene petrochemical complexes in this region for export to other regions. The high levels of return on capital resulting from the very low ethylene manufacturing costs for the manufacture of polyethylene in the Middle East could not be matched anywhere else in the world.

1.3.2   Change in Middle East Feedstock Cost Advantage

Since about 2010, the feedstock cost advantage in the Middle East has begun to erode. The two primary reasons for this change in feedstock cost advantage are: (1) the shortage of ethane in Saudi Arabia and (2) the discovery of enormous amounts of natural gas and crude oil in North America, which has been designated as the "shale gas revolution."

A report by John Richardson in 2010 titled "Saudi Feedstock Problems Worsen," [1] discussed the complex issues that have at the very least led to a shortage of ethane availability in Saudi Arabia. This ethane shortage was due to several factors such as: (a) lower crude oil production, (b) lower levels of ethane (dry crude) in the crude oil produced, and (c) the relatively high cost of replacing the ethane-based crude oil with ethane-based natural gas production. In addition, as Richardson noted, these ethane production problems in Saudi Arabia also coincide with the shale gas revolution in North America. This ethane shortage in Saudi Arabia has led to an ethylene cost of $150/ton and could rise to $300–350/ton on limited supply of ethane. On the other hand, due to the increase in ethane availability in North America, ethylene costs have been reduced from $700/ton in about 2008 to $400–450/ton in 2012, which eliminates most of the feedstock cost advantage in Saudi Arabia. This would have been viewed as almost impossible only a decade-ago.

Therefore, since this recent development of new sources of natural gas and oil in North America, the significant reduction or elimination in the feedstock cost advantage in the Middle East will most likely cause a significant decrease in the addition of new polyethylene capacity in the Middle East over the next several decades (i.e., 2015–2040). These developments have resulted in a new era for the polyethylene business that will most likely result in the increase in ethylene/polyethylene capacity in North America that seemed highly unlikely as recently as 2005. It should also be pointed out that ethane derived from natural gas has become the low-cost feedstock for the manufacture of ethylene compared to crude-oil-based naphtha. For example, for crude oil at $90/barrel, naphtha-based ethylene costs about $1250/ton, while ethane at about $0.40/gallon derived from natural gas and condensed into a liquid provides ethylene at about $370/ton. Such a feedstock cost advantage for ethylene production based on natural-gas-based ethane will most likely continue in future decades and will be a primary consideration for the location of ethylene and polyethylene capacity expansions in North America during the 21st century.

It should be noted that shale-rock deposits exist in most other regions of the world, i.e., Europe and Asia, and the exploration of these deposits to produce other sources of ethane in future decades may offer the potential for polyethylene capacity expansions in these regions; although as of 2014, these regions have not been explored to a significant level due to lack of the necessary infrastructure for the drilling to take place.

1.3.2.1.   2012 Capital Requirements for an Integrated Ethylene/Polyethylene Facility

A press release by Chevron-Phillips Chemical Company [2] for the construction of a world-class petrochemical facility in Baytown, Texas, USA, illustrates the capital requirements and capacity as of 2012.

This $5 billion (USD) petrochemical plant will have an annual capacity of 1.5 million metric tons (3.3 billion pounds) of ethylene and will manufacture 1.0 million metric tons (2.2 billion pounds) of polyethylene. Note that ethylene capacity in a world-class complex usually exceeds the volume of polyethylene production. This excess ethylene capacity is either used to manufacture other ethylene-based chemicals at the same complex or is sold as a commodity chemical when access to a distribution pipeline is available. This plant will utilize ethane as the feedstock for ethylene, which has a cost advantage over ethylene plants that use oil-derived naphtha as feedstock.

Table 1.1 Comparison of capital requirements for the manufacture of ethylene and polyethylene.

Ethylene	$1.2/lb. of capacity
Polyethylene	$0.4/lb. of capacity

Note: These figures should not be used for any planning purposes, as actual capital requirements of any ethylene/polyethylene manufacturing facility can vary significantly. This comparison is used to very qualitatively demonstrate approximate capital requirements as of 2012.

A final comment on the relative capital requirements for the manufacture of ethylene and polyethylene is necessary. Polyethylene manufacturing facilities require much less capital than the capital required for the manufacture of ethylene. A typical world-class polyethylene facility with a capacity of 0.750–1.5 billion pounds requires about $0.3–0.8 billion dollars depending on the type of manufacturing process and scale. Hence, for the type of petrochemical complex cited above, approximately 70–80% of the total capital of $5 billion dollars is for the construction of the ethylene portion of the complex (Table 1.1).

1.4　Early History of Ethylene and Polyethylene Manufacturing

In the early years of the polyethylene business in North America (ca. 1945–1975), there were many petrochemical companies that were involved in the manufacture of only ethylene or polyethylene. During this period, ethylene was distributed through dedicated pipelines along the United States Gulf Coast, which allowed companies to participate in either business. However, during this period polyethylene was in the very early part of the product life cycle where the annual growth of the business was extremely high and neither business was near an overcapacity situation. Once growth rates slowed in the 1970s, however, polyethylene became more susceptible to business cycles and petrochemical companies determined that an integrated business was needed to provide a less cyclical business.

Polyethylene is a semicrystalline thermoplastic material classified as a synthetic organic polymer which was developed in the early 1930s by British scientists at Imperial Chemicals Industries (ICI). Since that time, polyethylene material has undergone a wide variety of changes as scientists across the globe have been able to develop new polymeric structures by varying the type and amount of side-chain branching present in the material, as well as the ability to control the polymer molecular weight and molecular weight distribution. Such changes have created a wide variety of commercial applications for polyethylene in broadly defined areas such as:

- Packaging
- Consumer and industrial products
- Building/construction
- Transportation
- Healthcare

The early history of polyethylene is an especially important period because the first practical uses of polyethylene were in electrical applications that played a significant role in the outcome of World War II. Polyethylene exhibits low dielectric loss and great water resistance, which are important product attributes for wire and cable applications. Prior to 1940, polyethylene was produced in relatively large-scale pilot plant operations, providing a few hundred tons of material which were available for these specialty wire and cable applications in support of the war effort. Consequently, polyethylene was used by the British in submarine wire and cable, but perhaps more importantly, polyethylene-based wire and cable were used to help make extremely rapid progress in the introduction of radar [3]. Radar played a significant role in the Battle of Britain (July 10 – September 24, 1940), allowing the Royal Air Force to monitor enemy aircraft long before the aircraft reached airspace over Great Britain, thus providing the RAF with a significant tactical advantage.

Robert Buderi [4] published *The Invention that Changed the World,* in which the importance of radar in various applications during the course of World War II was deemed essential in the defeat of Germany and Japan. Without this technology, the outcome may have been delayed for years. The importance of radar was described by Robert Buderi as follows:

"The Royal Air Force Fighter Command knew an invasion of Britain (code named *Sea Lion* by Hitler's forces) loomed but did not have the fuel or planes to maintain standing patrols in anticipation of enemy raids. Nor did the country have the time to breed another crop of brave and intelligent young men if German bombers surprised planes and pilots on the ground. The Nation's best hope for hanging on rested on being able to spot the Luftwaffe far out over the English Channel and then deploying its thin resources to meet the threat at hand. On this vital front, everything depended on the Chain Home Radar Network."

1.4.1   Discovery of Polyethylene

Polyethylene with industrial useful properties was discovered in 1933 by the British chemists Dr. E. W. Fawcett and Dr. R. O. Gibson in the laboratories of ICI (Imperial Chemical Industries) utilizing a free-radical mechanism under high ethylene pressure that produced a branched polyethylene chain molecule. This was the first type of polyethylene that possessed a sufficiently high molecular weight that provided useful mechanical properties and a melting point (ca. 115°C) that allowed the polymeric material to maintain useful properties at relatively high temperatures, yet made the material easy to fabricate into useful shapes once the material was melted and passed through shape-forming equipment.

1.4.1.1   Branched High-Pressure Polyethylene

This early version of polyethylene can be very simply represented with the structure:

$$[-(CH_2-CH(R)_n-]$$

where R is primarily H, n-butyl, or a very long alkyl chain group. Because this was the only commercial type of polyethylene available from about 1935–1955, it was unnecessary to distinguish this material from any other type of polyethylene. Today, this high-pressure produced polyethylene is referred to as LDPE (low-density polyethylene). In addition, polyethylene characterization studies carried out in later years showed that the actual structure of LDPE is extremely complex and contains a wide variety of branching groups. A more complete discussion of polyethylene manufactured with high ethylene pressure can be found in Chapter 5.

Because of the free-radical polymerization mechanism involved in manufacturing this material, high ethylene pressure was required in order to provide a high ethylene propagation rate in order to achieve a relatively high molecular weight material so that the polyethylene possessed useful mechanical strength properties for practical applications.

Commercial production started at ICI in 1938, and in 1940 polyethylene production had reached 100 tons which were utilized in early wire and cable applications to build radar systems and other applications to support the war effort. One could argue that these first 100 tons of polyethylene may have been the most important polyethylene ever manufactured. Toward the end of the war, British annual production was about 1,500 tons. In 1943, the great military significance of polyethylene led both the Union Carbide and DuPont corporations to license the ICI process and begin the manufacture of polyethylene in the United States.

1.4.1.2   Linear Polyethylene

By the mid-1950s this high-pressure produced polyethylene was designated as low-density polyethylene (LDPE) in order to distinguish the free-radical produced polyethylene from a linear type of polyethylene discovered in 1953 in two independent laboratories. This linear type of polyethylene was prepared using a low-pressure process and a transition metal catalyst and was designated as high-density polyethylene (HDPE) because branched chains were mostly eliminated from the molecular structure due to a completely different ethylene polymerization mechanism. The elimination of most branching in the molecular structure of this new type of polyethylene provided a solid polymer with a higher density, i.e., 0.960 g/cc vs 0.930 g/cc. Consequently, this new type of polyethylene attained a higher degree of crystallinity as the molten polyethylene solidifies after the polymer is fabricated into a final product, thus accounting for the higher density of the material.

In order to understand the very different polymerization conditions required for the manufacture of both types of polyethylene, the process conditions utilized for the production of each type of polyethylene are illustrated in Table 1.2.

1.4.2   Linear Low-Pressure Polyethylene – High-Density Polyethylene

In 1953, in a rather remarkable coincidence, high-density polyethylene (HDPE) was discovered in two separate research laboratories. In one laboratory based in the United States, a group headed by J. P. Hogan and R. L. Banks of the Phillips Petroleum Company, and in a second laboratory, another group headed by Karl Ziegler of the Max Planck Institute in Mülheim, Germany, resulted in each group producing a mostly linear

Table 1.2 Process condition requirements for polyethylene.

	High-Pressure	Low-Pressure
Reactor Pressure (Bar)	1,000–3,000	5–25
Temperature (°C)	100–350	50–250
Catalyst	Free Radical	Transition Metal
Comonomer (optional)	Vinyl ester or Acid	3–8 carbon 1-olefin

Note: The polymerization conditions summarized above have been greatly simplified only to illustrate the significantly different process conditions required for the commercial manufacture of each type of polyethylene. The different process conditions necessary to produce each type of polyethylene have resulted in the use of enormously different equipment and reactor designs needed for the manufacture of LDPE and HDPE in the commercial plants.

polyethylene structure with two different catalyst types based on different transition metals.

Working in a Phillips Petroleum laboratory in Bartlesville, Oklahoma, Hogan and Banks utilized a chromium-based catalyst supported on porous silica to synthesize linear polyethylene, while Ziegler's group in Germany produced a similar (but not identical) linear polyethylene from the interaction of ethylene with an organometallic compound based on aluminum alkyls and a Group IVa compound with a preferred titanium chloride compound. Both of these catalysts offered a low-pressure route to linear polyethylene. Hence, not only was a new process for the polymerization of ethylene disclosed, but a new product (HDPE) emerged.

It is important to note that the Ziegler catalyst was later slightly modified by Natta and coworkers in Italy in order to provide isotactic polypropylene, which has also developed into an important global plastic. However, the composition of matter patent on polypropylene underwent a very long litigation process because Hogan and Banks at Phillips Petroleum were actually the first scientists to isolate and characterize crystalline polypropylene (isotactic polypropylene) using a chromium-based catalyst. After this long litigation process, Phillips Petroleum was eventually awarded the composition of matter patent on polypropylene. But Natta's catalyst was the preferred catalyst for the commercial polypropylene industry because the yields of isotactic polypropylene were much higher using a titanium-based catalyst with aluminum alkyls as activators.

1.4.3   Manufacture of HDPE (1955–1975)

High-density polyethylene was produced commercially during these early years with both the Phillips Cr-based catalyst and the Ziegler Ti-based catalyst; however, each catalyst type provided a different type of polymer structure that addressed different polyethylene markets and applications. For the most part, each product type was unique to these markets and consequently HDPE manufactured by each type of catalyst did not compete with each other.

The HDPE produced with the Cr-based catalyst possessed a relatively broad molecular weight distribution and a relatively high molecular weight, which was found to be a better type of polyethylene for bottles fabricated using the blow-molding technique. The HDPE produced with the Ti-based catalyst possessed a relatively narrow molecular weight distribution and could be produced over a much wider molecular weight range, which was found to be a better type of polyethylene for items fabricated using the injection-molding fabrication technique.

1.4.3.1 Phillips Catalyst Produced HDPE with Higher Product Demand

During the period from 1955–1975, the chromium-based, Phillips catalyst was the preferred catalyst for manufacturing HDPE because of relatively higher catalyst activity. But more importantly, the HDPE manufactured with this catalyst possessed better mechanical and physical properties for the introduction of HDPE, blow-molded containers for a wide variety of organic and non-organic liquids. Both the LDPE and HDPE produced with the Ti-based Ziegler catalyst were unable to produce a grade of polyethylene that could compete with the HDPE provided from the Phillips catalyst for this blow-molding market.

1.4.3.2 Phillips Catalyst Manufacturing Advantages

Because of the higher Cr-based catalyst activity, a catalyst removal step was not necessary in the manufacturing process based on Cr-based catalysts. In addition, the Phillips catalyst was supported on amorphous silica that provided a template for the growth of the polymer particle into a large granular polymer particle which was insoluble in the organic solvents (n-pentane, isopentane and isobutane) used in the polymerization process. These attributes allowed the early introduction of a slurry polymerization process which was developed by engineers at Phillips Chemical Company to commercialize the Phillips loop reactor in 1960, which provided lower polyethylene manufacturing costs. This vertical slurry-loop polymerization process was designated as "particle-form" technology, as one catalyst particle produced one granular polymer particle that was easily removed from the polymerization process by filtration.

1.4.3.3 Second Generation Ziegler Catalysts

Although the initial ethylene polymerization catalyst developed by Ziegler (later designated as a first-generation catalyst) was used commercially from 1955–1975 to produce polyethylene for injection-molding and film markets, relatively low catalyst activity and a more costly polymerization process limited the growth of this type of polyethylene in various markets. As stated above, a modified Ziegler catalyst developed by Natta and coworkers was used commercially to manufacture isotactic polypropylene, which was a much more important business during this period than polyethylene produced with a Ziegler catalyst.

However, in the early 1970s, Ti-based Ziegler catalysts were significantly improved after it was found that the incorporation of magnesium chlorides (and other magnesium-based compounds) into the catalyst preparation produced a second generation catalyst in which catalyst activity was increased by a factor of 10–50 when magnesium compounds were used in the catalyst preparation. These high activity (second-generation) Ziegler catalysts were used for the manufacture of HDPE and a new type of low-density polyethylene, designated as linear low-density polyethylene (LLDPE), with a relatively narrow molecular weight distribution (MWD). The LLDPE was produced in various types of polymerization reactors by utilizing a 1-olefin (propylene, 1-butene, 1-hexene, or 1-octene) in the process which incorporated these comonomers into the polymer chain at about 2–20 mol%. In the presence of low levels (ca. 0.2–1.0 mol%) of one of these comonomers, these second generation Ziegler catalysts provide HDPE with a relatively narrow MWD.

A solution polymerization process was especially useful to commercially produce LLDPE products for the market, while the Phillips loop reactor was somewhat limited in producing LLDPE with relatively lower levels of comonomer in order to avoid process problems, as the solubility of the LLDPE increased in the slurry solvent. The LLDPE became commercially more important in 1977 when Union Carbide developed a high-activity, silica-supported Ziegler catalyst that was utilized to manufacture LLDPE in a gas-phase, fluidized-bed process.

Hence, by the early 1980s, the global polyethylene business consisted of three basic polymer types: LDPE, HDPE and LLDPE, as summarized in Table 1.3.

Since about 1980, each type of polyethylene has undergone significant growth because high-pressure produced LDPE and low-pressure produced HDPE and LLDPE each address different markets because of the differences in polymer engineering properties due to the differences in the molecular structure associated with each type of polyethylene.

1.4.4 Single-Site Ethylene Polymerization Catalysts

By 1980, Professor Walter Kaminsky and coworkers at the University of Hamburg reported the development of a new class of ethylene

Table 1.3 Types of polyethylene.

PE Type	Molecular Weight Distribution	Process/Catalyst
LDPE	intermediate	high-pressure/free radical
HDPE	broad	low-pressure/ Cr-based catalyst
HDPE	narrow	low-pressure/Ti-based catalyst
LLDPE	narrow	low-pressure/Ti-based catalyst

polymerization catalysts, which produced polyethylene with a new molecular structure that involved the polymerization of ethylene in the presence of other 1-olefins such as 1-butene and 1-hexene. This type of polyethylene exhibited a very narrow molecular weight distribution as indicated by an M_w/M_n molecular weight ratio of 2.0, which showed that only one type of active site was active during the polymerization process in which a chain transfer mechanism was involved in which one active site produced a large number of polymer molecules. In addition, the short-chain branching that was provided by the additional 1-olefin used in the polymerization process was uniformly distributed along the polymer backbone. The commercialization of this new type of catalyst resulted in the addition of new types of polyethylene into the marketplace that identified new grades of polyethylene that were available for new applications for the polyethylene industry.

1.4.5 Status of the Polyethylene Industry as of 2010

In order to put the global polyethylene market as of 2010 into perspective, it is helpful to compare the demand of the five most common thermoplastic materials. These are polyethylene, crystalline polypropylene, polyvinyl chloride, polyethylene terephthalate and polystyrene.

Polyethylene is a thermoplastic material that had a 2010 global demand of about 154 billion pounds. Polyethylene has the largest market of the five most common thermoplastic materials, as summarized in Table 1.4.

Polyethylene is unique among these five thermoplastic materials because the polyethylene market consists of three distinct types of products which are high-density polyethylene (HDPE), low-density polyethylene (LDPE)

Table 1.4 The five largest thermoplastic[a] materials.

Thermoplastic Type	Global Demand (billions of pounds)	Market Share (%)
Polyethylene (PE)	154[b]	31.2
Polypropylene (PP)	112	22.7
Polyvinylchloride (PVC)	75	15.2
Polyethylene terephthalate (PET)	119	24.1
Polystyrene (PS)	34	6.9
Total	494	100

[a] A thermoplastic material may be melted, formed into a useful shape and then cooled without any significant change in the polymer structure.

[b] Approximately 22 lbs./capita assuming a global population of 7 billion people.

Source: SRI consulting website.

and linear low-density polyethylene (LLDPE), each with very different mechanical properties that address different market applications. In addition, since the mid-1990s the polyethylene product mix has expanded due to the introduction of the single-site polymerization catalysts to include very low-density materials which are elastomeric in properties and unique in mechanical properties compared to previous types of polyethylene materials. This has, consequently led to new markets and applications.

1.4.6   Global Demand for Polyethylene in 2010

The SRI Consulting Group reported that global consumption of polyethylene in 2010 was approximately 154 billion pounds (70 million metric tons) and required an 82% capacity utilization rate. The consumption for each of the three types was [5]:

- HDPE: 69 billion pounds (45%)
- LLDPE: 46 billion pounds (30%)
- LDPE: 39 billion pounds (25%)

In 2004, Chemical Marketing Associates in Houston, Texas, reported a global demand of 135 billion pounds for polyethylene which was subdivided into each of the three product types, as shown in Table 1.5.

The increase in global polyethylene consumption was 14% for the six year period 2004–2010, which corresponds to an annual increase of just over 2%. It is important to note that in late 2008, the global economy experienced a deep recession and the global economic growth slightly declined between 2008 and 2010.

1.4.7   Polyethylene Product Lifecycle

Before discussing the lifecycle of polyethylene, it is necessary to define "product life cycle" as the length of time a particular commercial product

Table 1.5 2004 Global polyethylene market by product type.

Product Type	Billions of lbs.	Market share
HDPE	59	44%
LLDPE	38	28%
LDPE	38	28%
Total	135	100%

is manufactured. Most products developed over the course of history have a life cycle during which the product is introduced and utilized in the marketplace.

For example, in the petrochemical industry, a gasoline additive such as tetraethyl lead (TEL), which was used to improve the performance of gasoline as a fuel in the internal combustion engine, became obsolete due to health concerns over the emission of lead-containing compounds into the environment. Consequently, TEL was developed and introduced into the marketplace, underwent a growth period and then underwent a steady decline as the substance was phased out of gasoline. The life cycle of TEL was approximately 50 years.

The life cycle of any commercial product has four distinct phases [6]. A brief summary of each phase and how each phase relates to the polyethylene industry is outlined below.

1. **Product Discovery and Market Introduction:** This is the period between the discovery of the product and the time it takes to introduce the product commercially. In the petrochemical industry, initial product introduction is often done from a pilot-plant-scale process. The introduction is usually described as a period of initial slow growth as the product is introduced into the market. However, for polyethylene, this period was most likely shortened significantly because of the urgency in the development of radar at the beginning of World War II. Somewhat arbitrarily, this phase for polyethylene took place from 1933–1943.

2. **Growth Phase:** This phase follows the successful introduction of the new product and is characterized by very rapid sales growth, the addition of new competitors and rising capacity. Economies of scale are an important part of this phase as substantial profit improvement takes place. Polyethylene is presently in the product growth phase of the product life cycle. Product consumption per capita is the best measure of consumption.

3. **Mature Phase:** The acceptance of this product has reached saturation and product growth has stopped.

4. **Product Decline/Termination:** Product volume is in decline and may or may not stabilize at lower levels. This period is usually the result of the introduction of new competing technology, or in some cases, due to new governmental regulations.

The concept of a product life cycle is illustrated in Figure 1.1 and may be applied to a product category (such as polyethylene) or any other aspect of the product. For example, the polyethylene product category may be classified in general terms as any type of polyethylene or as a specific product type such as LDPE, LLDPE or HDPE.

The four phases of a product life cycle are illustrated in Figure 1.1.

Polyethylene is presently in the growth stage of the product life cycle and has evolved from the discovery to the present growth period in 80 years.

In order to better understand the global polyethylene business today, we will examine the status of the global polyethylene business and then examine the polyethylene business in each of the five major global markets, which are: (1) North America, (2) Latin America, (3) Europe, (4) Middle East/Africa, and (5) Asia/Pacific.

The product discovery and market introduction phase for polyethylene are as follows:

- 1933: Initial discovery of polyethylene by ICI.
- 1939: First commercial plant (autoclave) built by ICI.
- 1941: ICI licenses technology to DuPont.
- 1941: Union Carbide builds tubular pilot plant.
- 1943: First commercial tubular plant built in US.
- 1943: DuPont builds first autoclave plant in US under ICI license.

The discovery and market initiation phase for polyethylene lasted from 1933–1943. The growth phase began immediately after 1943 and continues

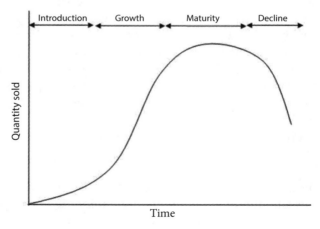

Figure 1.1 Product lifecycle.

today. After 80 years, the global polyethylene business remains in the growth phase of the product life cycle with little evidence that growth will slow significantly during the next 50 years. This continued growth is due to the remarkable discovery of new technology over this time period.

The present status of the product life cycle for global polyethylene business is shown in Figure 1.2. During the growth stage of the product life cycle new polyethylene manufacturing capacity has been added in all the above-mentioned five major global markets. Most market-share growth for any particular company involved in the manufacture of polyethylene is obtained by growing the business by adding new capacity. The global polyethylene market is presently in the early stage of the growth phase of the product life cycle. Figure 1.2 illustrates polyethylene consumption over the past 65 years, with anticipated growth shown for the next 50 years.

About 135 billion pounds of PE were produced in 2004 [7], which increased to 154 billion pounds by 2010, which is 14% growth over this 6-year period. Future growth (dashed line in Figure 1.2) was estimated assuming a 3% annual growth rate over 20 years (2005–2025) and then a 2.5% annual growth rate over the next 30 years (2025–2055), which would produce the global polyethylene demand shown in Table 1.6.

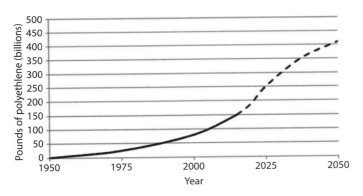

Figure 1.2 Global polyethylene product cycle.

Table 1.6 Estimated demand for the global polyethylene business[a].

Year	Demand	Annual Growth Rate
2004	135	N/A
2028	270	3.0%
2055	540	2.5%

[a] Data illustrated in Figure 1.2.

It should be noted that these estimates are not based on any long-term market forecast studies carried out by this author, but merely an attempt to demonstrate the enormous growth opportunities that may be experienced by the polyethylene industry if this business continues on a growth cycle and does not begin to enter the product mature stage, which is the next phase of a product life cycle.

Polyethylene Business Consulting Services are highly recommended for policy makers responsible for the polyethylene business. Business managers in the polyethylene industry have a large number of global business forecasting resources available that follow the polyethylene business on an annual basis. These consulting services are an excellent source of information on the polyethylene business, and future growth estimates are usually provided by such companies on a global regional basis.

In order to gain a better understanding of the global polyethylene market, it is important to examine the growth of the market based in the five individual regions of the globe. These are: North America, Latin America, Europe, Middle East/Africa and Asia/Pacific.

1.4.7.1   North American Polyethylene Market

The North American polyethylene market (considered as only the United States and Canada) was the first global region to undergo rapid growth, with most polyethylene capacity added from about 1945–1955 taking place in North America. Consequently, North America is ahead of the other global regions of the world in terms of per capita consumption. The North American polyethylene market is in a more advanced stage of the product growth phase. In fact, the North American polyethylene market may be approaching the mature stage of the product life cycle, as measured by the consumption of polyethylene/capita over the next 10–20 years. This forecast excludes the export of polyethylene produced in North America and shipped to other global regions. Table 1.7 shows the 2010 annual production of PE for the North American market for each type of polyethylene. The production totaled about 38 billion pounds of polyethylene, which is about 108 lbs PE/capita for a North American population of 350 million people. No other global region has developed anywhere near the North American region. From a marketing perspective, the North American market may be used to predict the future of the global polyethylene business over the next 25–50 years.

North American polyethylene production actually underwent a capacity decrease from about 2006–2010 [8], as approximately 2 billion pounds of capacity were either shut down or temporarily idled during 2008 [9] during the onset of the global recession.

Table 1.7 North American polyethylene market for 2010.

PE Type	Production (billion lbs)
HDPE	18
LLDPE	12
LDPE	8[a]
Total	38[b]

[a] A 2012 report estimates 6.8 billion lbs. in 2011[1].

[b] Approximately 108 lb /capita

[1] *Plastic News*, March 12, 2012, reported North American sales of LDPE as 6.7 billion pounds, with Formosa Plastics Corp. USA, Livingston, New Jersey, planning a LDPE capacity increase in Point Comfort, Texas, with a 2016 start-up date.

However, additional North American polyethylene capacity will most likely increase significantly, primarily for the export market, over the next several decades (2010–2050). This is due to the fact that feedstock costs (ethane) should be significantly reduced as the effects of the shale gas revolution begin to take effect over the next decade, resulting in a huge increase in ethane supply (and other natural gas liquids).

This market forecast is in sharp contrast to predictions made in the period from 1995–2005, in which business executives correctly assumed that polyethylene capacity would most likely not undergo much growth in North America due to the need to import natural gas and a continuing decrease in the production of crude oil. The likelihood of importing additional crude oil, natural gas and polyethylene from the Middle East region seemed inevitable.

1.4.8   Comparison of Other Global Polyethylene Markets with the North American Market

To better illustrate the relatively advanced growth stage of the North American polyethylene market compared to the total of the other four global regions, a comparison of the annual per capita consumption of polyethylene for 2006 in North America vs the remainder of the globe is shown in Table 1.8.

This data illustrate that the North American per capita annual consumption (108 lbs PE/capita/year) is a factor of 7.2 times greater than the combined per capita consumption in all other regions of the globe, supporting

Table 1.8 Comparison of global polyethylene market to the North American market.

Region	PE Consumption (billions of lbs.)	Population (billions)	Per Capita Consumption (lbs/capita)
North America	38	0.35	108
Other Regions	97	6.35	15
Global	135	6.70	20

Note: This data is approximate and should be viewed only to illustrate the annual per capita consumption of the North American region compared to the remainder of the globe.

the conclusion that the North American market is in a much later stage of the polyethylene growth phase (for internal consumption) compared to the remainder of the world. Moreover, this data shows that if the global per capita annual consumption of polyethylene increases from the present level of 20 lbs PE/capita/year to 54 lbs PE/capita/year the annual production of global polyethylene would be approximately 362 billion pounds, which is an increase of 170% over the present global production of 135 billion pounds in 2006. However, in order for such significant growth to take place over the next 50 years, the growth of the global consumer class needs to increase substantially.

1.4.9 Growth of the Global Consumer Class

The purpose of the discussion of the North American polyethylene market in terms of annual per capita consumption, compared to the annual per capita consumption of the global polyethylene market, was to illustrate the relatively advanced growth stage the North American market occupies compared to the remainder of the world.

Although Europe is also in a relatively advanced stage of the polyethylene growth cycle, the other regions in the world exhibit very low polyethylene consumption on a per capita basis. The reason for the large discrepancy in polyethylene per capita consumption in North America and Europe compared to the other regions is the size of the consumer middle class. North America and Europe have a relatively large consumer class, or middle class, which drives the consumption of polyethylene. A relatively large consumer middle class in the other regions of the world is required in order for the polyethylene market to develop in these other regions. In order to determine the global regions where there is potential growth in the consumer class, it is useful to examine the countries that make up the G20 (Group of Twenty Finance Ministers and Central Bank Governors) and divide these countries into two categories. The first category is countries that presently possess a relatively large consumer class, and the second one is the countries that presently have a small consumer class. These regions are referred to as the developed markets and the emerging markets, respectively.

As of 2014, the G20 economies consisted of 19 individual countries and the European Union (EU). The EU has 27 member countries; however, four of the EU countries are also individual members of the G20. Hence, a total of 42 countries are members of the G20. The G20 countries as of 2014 are shown in Table 1.9.

The European Union is the largest single-developed market in the world with more than 450 million people and a total gross domestic product (GDP) of 15.2 trillion US dollars. Table 1.10 shows the GDP, population and per capita GDP for each of the G20 regions. The table is arranged in order of the highest per capita GDP.

An increased demand for polyethylene will be due to the growth of the global consumer class primarily in the developing economies.

The G20 regions listed as data points 1–11 are considered well-developed economies with a substantial consumer class. These regions will most likely experience a relatively low increase in polyethylene consumption

Table 1.9 The G20 Countries as of 2014.

19 Individual Countries		European Union Countries	
Argentina	South Africa	Austria	Latvia
Australia	South Korea	Belgium	Lithuania
Brazil	Turkey	Bulgaria	Luxembourg
Canada	United Kingdom*	Cyprus	Malta
China	United States	Czech Republic	Netherlands
France*		Denmark	Poland
Germany*		Estonia	Portugal
India		Finland	Romania
Indonesia		France	Slovakia
Italy*		Germany	Slovenia
Japan		Greece	Spain
Mexico		Hungary	Sweden
Russia		Ireland	United Kingdom
Saudi Arabia		Italy	

*Also a member of the EU.

Table 1.10 2008 gross domestic product per capita for the G20 regions (IMF, 2008).

Rank	Region	GDP (Trillion US$)	Population (million)[a]	GDP/capita (US$/capita)
1	United States	14.2	300	47,300
2	Canada	1.3	34	38,200
3	Australia	0.8	22	36,400
4	United Kingdom	2.2	61	36,100
5	Germany	2.9	82	35,400
6	European Union	15.2	450	33,800
7	Japan	4.3	130	33,100
8	France	2.1	65	32,300
9	Italy	1.8	60	30,000
10	South Korea	1.3	48	27,100
11	Saudi Arabia	0.6	26	23,100
12	Russia	2.3	140	16,400
13	Mexico	1.5	110	13,600
14	Argentina	0.6	39	15,400
15	Turkey	0.9	72	12,500
16	Brazil	2.0	190	10,500
17	South Africa	0.5	49	10,200
18	China	7.9	1,340	5,900
19	Indonesia	0.9	230	3,900
20	India	3.3	1,160	2,800

[a] Population data is rounded to nearest million from 2008 or 2009 population estimates.

on a per capita basis when compared to global regions which are much less economically developed and, consequently, may exhibit a substantial increase in the population considered as the consumer class. The regions represented by data points 12–20 are emerging economies that must show a significant increase in the consumer class during the next several decades in order for the global polyethylene demand to continue to grow as postulated in Figure 1.2, which estimates a polyethylene demand of about 400 billion pounds by the year 2050.

These developing markets have a total population as of 2009 of about 3.33 billion people, which is approximately 50% of the global population.

However, it should also be noted that the regions representing the European Union presently possess well-developed economies as well as emerging markets. For example, the developed regions of Western Europe, including the Scandinavian countries and the United Kingdom, are also in a more advanced stage of the polyethylene growth cycle than some of the other regions presently part of the EU. Eastern Europe and the regions formed from the former Soviet Union may be considered emerging markets, which may also undergo a significant increase in their consumer class.

The global polyethylene market growth forecast shown in Figure 1.2 is based on an annual polyethylene per capita global consumption approaching 45 lb/capita. The global population as of 2009 is approximately 6.7 billion with a projected population of 9 billion by the year 2040. Hence, a global population of about 9 billion people may require approximately 400 billion lbs. of annual polyethylene production. Figure 1.2 shows that the present polyethylene market is only about 35% of that value.

Economic growth takes place most readily in countries where the Government is committed to free-market capitalism. Countries that are able to take advantage of a free market's ability of allocating capital and labor in an efficient manner do the best job of promoting prosperity for all citizens.

The countries of Brazil, Russia, India and China (designated as the BRIC emerging market countries) are considered the relatively more important global regions that will need to exhibit a sustained increase in economic growth over the next 40 years in order to create a substantial increase in the consumer class that will be required in order for the polyethylene industry to reach the growth levels shown in Figure 1.2 by the year 2050.

1.4.9.1   Quantitative Forecast for the Growth of the Global Middle Class

Homi Kharas and Geoffrey Gertz of the Wolfensohn Center for Development[1] at the Brookings Institution have provided a forecast in the population of the middle class by region and the purchasing power (in 2005 United States Dollars) of each region for the years 2020 and 2030 relative to the 2009 data for each global region. Their forecast is shown in Tables 1.11 and 1.12.

---

[1] The Wolfensohn Center for Development was a five year independent research analysis that investigated the topic of promoting people out of poverty. The program was under the leadership of former World Bank President James D. Wolfensohn. The results of this investigation were incorporated into the "Global Economy and Development" program at the Brookings Institute.

Table 1.11 Population of middle class in millions.

Region/Year	2009	2020	2030
North America	338	333	322
Europe	664	703	680
Central/South America	181	251	313
Asia Pacific	525	1,740	3,228
Sub-Saharan Africa	32	57	107
Middle East/North Africa	105	165	234
Total Middle Class	1,845	3,249	4,884

Source: Tweedy, Browne Fund, Inc.; 350 Park Avenue, New York, NY, (USA) Annual Report, March 31, 2011.

Table 1.12 Consumption of global middle class in US dollars.

Region/Year	2009	2020	2030
North America	5,602	5,863	5,837
Europe	8,138	10,301	11,337
Central/South America	1,534	2,315	3,117
Asia Pacific	4,952	14,798	32,596
Sub-Saharan Africa	256	448	827
Middle East/North Africa	796	1,321	1,966
Total Middle Class	21,278	35,045	55,680

Source: Tweedy, Browne Fund, Inc.; 350 Park Avenue, New York, NY, (USA) Annual Report, March 31, 2011.

The data summarized in the tables suggest that the global middle class and the consumption of the middle class will both increase significantly by the year 2030 relative to the 2009 statistics. However, this growth in the global middle class will come from regions other than North America and Europe. More specifically, the forecast projects that the total global middle class will increase from 1,845 million people in 2009 to 4,884 million people by 2030 (an increase of 3,039 million people) by the year 2030. The forecast for the Asia Pacific Region is particularly important, as this region is expected to undergo middle class growth from 525 million people in 2009 to 3,228 million people by 2030, or an astonishing 2,703 million people, accounting for 89% of the growth in the global middle class by the year 2030.

If such an increase in the global middle class were to be reached by 2030, the global annual consumption of polyethylene could reach approximately 400 billion pounds over the 2009 annual production of polyethylene of about 150 billion pounds.

1.4.10   Global Economic Freedom

Critical Question: Do the countries listed below possess sufficient economic freedom that will allow substantial economic growth to take place over the next four decades (2010–2050) to generate a large consumer class?

- Russia
- Mexico
- Argentina
- Turkey
- Brazil
- South Africa
- China
- Indonesia
- India

The countries listed above are the emerging markets and/or developing countries; however, it would also be accurate to refer to these nine countries as underdeveloped economies, which begs the question – Why have these particular economies underperformed many other regions? The answer has been categorized with an arbitrary quantitative analysis by Terry Miller and Kim R. Holmes in the Index of Economic Freedom [10].

According to this Economic Freedom analysis, which rated 179 out of 183 countries of the world in terms of ten measures of Economic Freedom, none of these nine underdeveloped countries can be considered as free or mostly free economies. The results of the analysis by Miller and Holmes for these nine countries are outlined in Table 1.13 in terms of decreasing economic freedom.

Examination of the above table shows that Mexico, South Africa and Turkey have moderately free economies which may undergo considerable economic growth. However, the remaining six countries fell under the category of mostly unfree economies, with Russia very nearly a repressed economy. In addition, with the exception of South Africa, the eight other underdeveloped countries listed above possessed Governments which were considered to have a relatively high level of internal dysfunction.

Table 1.13 Economic freedom around the world.

Country	World Rank[a]	Economic Freedom Value[b]
Mexico	49	65.8
South Africa	61	63.8
Turkey	75	61.6
Brazil	105	56.7
India	123	54.4
Indonesia	131	53.4
China	132	53.2
Argentina	138	52.3
Russia	146	50.8

[a] Out of 179 countries.
[b] Economic Freedom Values are based on the following scale:

80–100	Considered a free economy	7 out of 179
70–79.9	Mostly free economy	23 out of 179
60–69.9	Moderately free economy	53 out of 179
50–59.9	Mostly repressed economy	67 out of 179
< 50	Repressed economy	29 out of 179

1.4.11   Future Economic Growth in India and China

India, with an estimated 1.1 billion people, and China, with an estimated 1.3 billion people as of 2010, account for approximately 35% of the world's population. Hence, economic growth in these two particular countries is, therefore, especially significant in increasing global economic growth and has a significant impact in improving the standard of living for global citizens. If substantial economic growth does take place over the next several decades, the benefit of improved quality of life for citizens across the globe will be due to enormous growth in international trade.

Both India and China have undergone considerable economic growth over the past two decades. India is the world's most populated democracy, while China is an authoritarian state with the ruling Communist Party maintaining control of political expression, speech, assembly and religion. However, China has liberalized its economy significantly since the 1980s, accounting for the economic progress during this period. If China is able to avoid any long-term internal instability in the future, economic growth will continue. Culture will play an important role in the continued economic growth in China. Education is an important part of the culture in China. In addition, the people are hardworking and possess an entrepreneurial spirit. However, efforts by the Chinese population to accelerate the path toward a more democratic government may be a possible threat to political stability and economic growth. As of 2014, the fundamental change required in the role of government towards a less authoritarian format is highly unlikely, making it difficult to postulate that the present government in China will be able to respond to the needs and will of the people in the decades to follow. Consequently, the present consensus that Chinese economic growth will be the fastest growth region in the twenty-first century may be overstated.

India, as a democratic state, may be less susceptible to long-term internal instability. However, a culture that leads to a more class-based society will not provide a smoother path to improving the living standard for all its citizens. Such differences in Indian culture and the ineffectiveness of the Indian government over the past half century to respond to the needs of its citizens in terms of mass education to reduce poverty levels, will most likely result in slower economic growth in India relative to China during this century.

1.4.12   Long-Term Global Polyethylene Capacity Expansion (2010–2050)

It is not the purpose of this book to provide specific polyethylene global expansion forecasts for various regions. As stated previously, there are a large number of consulting firms worldwide that provide detailed ethylene/polyethylene global capacity estimates. The services of such consulting firms are highly recommended for global business managers involved in the petrochemicals industry.

As discussed above, future growth of the global polyethylene industry will be primarily determined by an increase in the global consumer class, which will be driven by an improvement in the economic freedom in global underdeveloped countries.

1.4.12.1   Location of New Ethylene/Polyethylene Capacity (2010–21050)

The location of new polyethylene capacity increases will be determined by primarily three factors, listed in decreasing importance:

1. Significantly lower feedstock and energy costs,
2. Low corporate tax rates and other governmental regulations, and
3. Availability of a skilled workforce.

1.4.13   Ethylene Feedstock Costs

Ethylene is an intermediate commodity chemical with four primary uses shown in Table 1.14. The specific end-use distribution for ethylene manufactured in the United States is shown for 2003 to illustrate the uses of commodity ethylene.

Examination of the table shows that ethylene is primarily used to manufacture polyethylene. For comparison purposes, the global utilization of ethylene for 2004 was approximately 225 billion pounds with about 130 billion pounds of polyethylene manufactured in 2004, which represents an ethylene utilization of 58% for the manufacture of polyethylene, demonstrating that the United States and global ethylene utilization for polyethylene are comparable.

The world ethylene production for 1994 was 50 million tons; 2000 was 93 million tons; 2004 was 110 million tons and the forecast for 2015 is approximately 160 million tons [11].

1.4.13.1   Manufacture of Ethylene

Ethylene is produced utilizing a steam-cracking process carried out at 750–950°C (steam cracking is the most energy-consuming process in the chemical industry), with primarily two feedstocks, naphtha and ethane, although other feedstock such as propane, butane and gas oil are also used.

The composition of naphtha varies, but cracking yields for ethylene are about 30%, with a wide variety of co-products such as methane, propylene, butadienes, butane and a mixture of other fuel oils accounting for the remaining 70%. With ethane as feedstock, ethylene yield is very high at about 80% with few byproducts.

The feedstock is passed through the pyrolysis or cracking section using a very short residence time of 0.1 to 0.6 seconds, with the relatively longer residence times required for a heavy feedstock such as naphtha. However,

Table 1.14 United States utilization of ethylene for 2003[a].

Polyethylene	54%
Ethylene dichloride	18%
Ethylene oxide	12%
1-Olefins	9%
Miscellaneous/other	7%

[a] 61 billion pounds total

Source: www.the-innovation-group.com

a lighter feedstock such as ethane or ethane/propane mixtures require a shorter residence time and provide higher conversion yields of ethylene.

Naphtha is derived from crude oil, so its cost is tied to the price of crude oil, while ethane is mostly a byproduct of natural gas extraction, so ethane cost is tied to the price of natural gas. Crude oil is priced as a global commodity, while natural gas is priced on a regional basis. In North America natural gas has become relatively inexpensive since about 2010 due to the development of the "fracking revolution." Because natural gas is difficult to transport to other regions, natural gas prices vary as much as a factor of five across the globe.

For worldwide polyethylene producers, feedstock costs necessary to manufacture ethylene are the single most important issue in achieving a profitable and competitive polyethylene business, requiring the polyethylene producers to back-integrate into the ethylene business. A comprehensive polyethylene business strategy involves (a) access to low-cost feedstock, (b) large world-class ethylene crackers to gain economies of scale in ethylene production, (c) high-capacity polyethylene reactors (the gas-phase and the slurry process for the manufacture of polyethylene are the lower cost processes) for low operating costs, (d) access to a skilled labor force, and (e) capacity location in low tax regions.

In Western Europe, approximately 95% of ethylene is produced from steam-cracking naphtha. In the United States, ethylene manufactured from ethane accounts for approximately 70% of ethylene production, while steam-cracking naphtha accounts for 30% of ethylene production [11]. In the future, ethane will almost certainly be used for the vast majority of ethylene manufactured. In the Middle East, ethylene production is based on ethane (or ethane/propane mixtures) derived from natural gas or extracted from crude oil. In addition, energy costs in the Middle East are about five times lower than in Europe.

World ethylene production [11] for 2004 is shown in the Table 1.15.

Forecasts for Middle East ethylene capacity indicate that the Middle East will overtake Europe by 2015. However, the ethane supply required for ethylene production has changed significantly from 1985–2010. Qatar has enormous supplies of natural gas and has become a global leader in supplying liquefied natural gas for transfer to global markets. Consequently, Qatar

Table 1.15 Distribution of world ethylene capacity in 2004 (in million tons).

	United States	Europe	Middle East[a]	Other
Ethylene Capacity	30	30	12	38

[a] Increased from 3 million tons in the 1990s.

has an excess supply of ethane, which will provide a low-cost feedstock for the manufacture of ethylene, and therefore has been a prime location for petrochemical plants that produce polyethylene. On the other hand, Saudi Arabia became a large center for petrochemical plants primarily starting in the early 1980s. By 2005 the demand for ethane as a feedstock for these petrochemical plants has become an increasingly limited resource in Saudi Arabia as natural gas production failed to increase at a sufficient rate to meet the ethane demand. Possible ethane supply shortages in Saudi Arabia in the early 2000s led to some speculation that short ethane supply may limit the fast-growing petrochemicals industry in the Middle East. Consequently, in 2005, Saudi Aramco took steps to double the ethane supplies by 2010 through an aggressive natural gas production program. This effort has been successful, so that additional polyethylene capacity is scheduled for startup in 2010–2015, but the feedstock cost advantage of the Middle East compared to North America has mostly eroded, favoring vast new capacity increases in North America starting in about 2015.

Prior to 1970, North America, Western Europe and Japan were the primary locations of petrochemical plants involved in the manufacture of ethylene and polyethylene. These regions were highly stable with a very low risk to loss of capital, were not at a feedstock cost disadvantage and had an ample supply of a skilled workforce required to operate these sophisticated facilities. However, after 1970, these regions became increasingly dependent on importing crude oil to meet the rising demand in energy. Consequently, the Organization Petroleum Exporting Countries (OPEC) were able to significantly increase the cost of crude oil to each of these regions. The historical price of crude oil in nominal and inflation adjusted terms from 1946–2009 is shown in Figure 1.3.

Examination of the price of crude oil clearly shows the oil price spike that occurred in 1973 from the OPEC oil embargo policy during this period. In inflation adjusted prices, this spike lasted until 1986, when oil prices gradually decreased until 1999. Post 1999, crude oil prices have undergone a rapid increase that continued until the global recession of 2008–2009. Continued economic growth in the developing regions after 2010 will most likely result in additional cost increases in crude oil. Consequently, regions such as Western Europe, where ethylene production from naphtha is the primary source of ethylene, will have a significant feedstock disadvantage over ethylene manufactured from ethane.

1.4.13.2   Crude Oil and Natural Gas Prices

Crude oil is a global commodity due to the relative ease with which the product can be transported to all global regions. The cost/barrel is

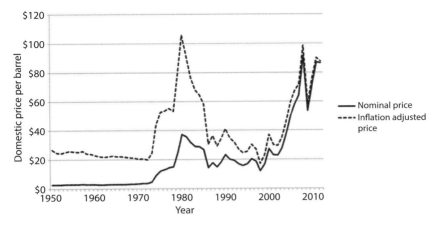

Figure 1.3 Crude oil prices, 1950–2013.

essentially the same in all regions, depending mostly on the quality of the particular grade. Because natural gas is not as easily transported as liquid crude oil, natural gas is best considered a regional commodity and not a global commodity. The price of natural gas is regional, depending on the supply and demand fundamentals in each region. Natural gas is primarily transported in pressurized pipelines across each global region. Although liquefied natural gas (LNG) may be utilized to ship natural gas around the globe, which involves cooling natural gas to −160°C to condense the gas to a liquid material which is then transported in specially designed shipping containers. Once received at a dedicated receiving terminal, the cooled liquid needs to be heated to transition back to a gas and then transferred to pressurized pipelines. Two additional methods of utilizing natural gas commercially are: (1) to convert the natural gas to a higher molecular weight material, such as diesel fuel, designated as a gas-to-liquids (GTL) process, or (2) directly using the natural gas as its recovery location to generate electricity, which is designated as the gas-to-wire (GTW) process.

It is sometimes stated that natural gas and crude oil prices are interrelated; however, the data in Figures 1.3 and 1.4 showing the cost of crude oil (global) and natural gas (United States) suggest that this may not be the case. The cost of natural gas from 1983–2000 decreased in nominal terms, and if adjusted for inflation decreased by about 50%. Meanwhile, the cost of crude oil increased from about $2/barrel in 1972 to $40/barrel in 1980, an increase by a factor of 20. From 1980 to 1999, the nominal price of crude oil fell from $40/barrel to about $15/barrel, or a decrease of about 62%.

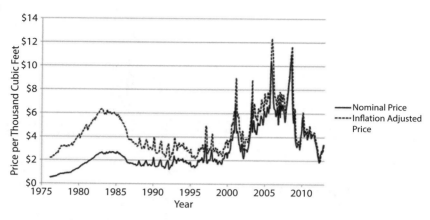

Figure 1.4 U.S. natural gas wellhead prices.

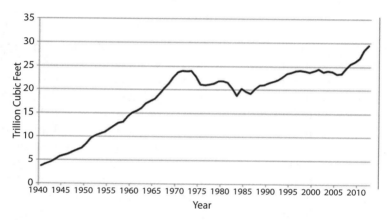

Figure 1.5 Annual U.S. natural gas gross withdrawals.

However, since 2000, the price of global crude oil has increased significantly, while the price of natural gas increased until about 2010 and since then has significantly decreased due to the shale gas revolution in the United States.

Before 2005, the consensus opinion was that natural-gas production in the United States was facing permanent decline and that it would require large-scale imports of liquefied natural gas (LNG) to meet US demand. Annual natural gas withdrawals peaked in the United States in the early 1970s and have remained relatively constant since that time. Figure 1.5 shows this data.

But, as mentioned previously, the drilling boom over the past five years from the new drilling techniques involving horizontal drilling, have created greatly increased estimates of US natural gas reserves. This new drilling technology involves hydraulic fracturing (termed fracking) to produce gas from hard shale-rock formations. Water, sand and drilling fluids are pumped into the shale under enormous pressure creating fractures in the rocks that allow the natural gas to escape from these previously encapsulated formations. The importance of this new technology was demonstrated in December, 2009, when ExxonMobil announced a $41 billion dollar bid to merge with XTO Energy, which was a leading producer of natural gas utilizing these new drilling techniques.

1.4.14   Impact of the Shale Natural Gas Revolution on Global Polyethylene Business

In 2010, Amy M. Jaffe, a Research Fellow for Energy Studies at Rice University in Houston, Texas, reviewed the future impact of shale gas around the world [12]. Ms. Jaffe argues that this new source of enormous quantities of natural gas will decrease the likelihood of the formation of a natural gas cartel (similar to OPEC for crude oil) that could control the future price of natural gas. Global regions that once looked as if they would become large natural gas exporters for supplying gas to other global regions with much less supply, may no longer be in a position to dominate the natural gas industry.

Utilization of these vast new natural gas reserves by utility companies to generate electricity, liquefied natural gas (LNG) exporters, natural gas-to-liquids plants (i.e., gas-to-diesel plants), and in applications such as in transportation using compressed natural gas could significantly increase the ethane supply in North America (United States and Canada) bringing down ethane feedstock costs for the manufacture of ethane-based ethylene.

Shale gas opportunities are also being pursued in Europe and Asia, although as of 2012 these developments are only in the very early stages of development. Estimates are that Europe has 639 trillion cubic feet of recoverable shale-based natural gas, which approaches the 862 trillion cubic feet of shale-based natural gas in the United States [13]. However, costs to produce shale-based natural gas in Europe may exceed North American costs [13] because the European shale gas is deeper underground and more difficult to extract. Also, Europe has a lack of drilling rigs and available pipelines for distribution of the gas. The process is also facing political pressure due to concerns over the safety of the hydraulic fracturing process. As of 2012, France has issued a moratorium on the process until the safety concerns can be more fully studied, while other countries such as Poland have been active in pursuing shale-gas opportunities. However, such infrastructure obstacles may be overcome in the decades ahead and could, therefore, have a significant effect on the location of new polyethylene capacity in Europe. Obviously, such developments would further erode the Middle East cost advantage that has already been eroded in North America.

1.4.15   Natural Gas Liquids

Natural gas is, in fact, a complex mixture of light hydrocarbons and the exact composition of the gas can vary over a wide range, as illustrated in the Table 1.16. Natural gas liquids refer to the amount of ethane, propane and butane that can be separated from the natural gas stream.

The typical composition of a natural gas source will vary over a relatively wide range as shown in Table 1.16.

The decision to determine the type of well to put into commercial production is a complex business decision determined from the composition of an individual well or field of wells. A gas discovery with a very low liquids component (referred to as a "dry" well) has a relatively lower heat index, making such a discovery more easily transported in a natural gas pipeline which has heat content specifications to prevent condensation of natural gas liquids during the transportation process. However, a commercial dry well does not benefit from revenue produced by selling natural gas liquids, which can be a lucrative market, improving the profitability of a particular natural gas production stream.

Gas streams with relatively higher natural gas liquid components require the removal of these liquids prior to entering the natural gas pipeline system, but these natural gas liquids require a distribution system in order for particular products to reach a market. For ethylene manufacture, ethane is the important feedstock. Distribution bottlenecks will affect the market price of natural gas liquids. In the United States, for example, the shale gas revolution has provided tremendous growth in the ethane business. Ethane is the preferred feedstock for the petrochemical industry to manufacture

Table 1.16 Typical composition of natural gas.

Component(s)	Weight (% of total)
Methane	70–90
Ethane, Propane, Butane	0–20
Carbon dioxide	0–8
O_2, N_2, H_2S	0–5

Source: www.NaturalGas.org

ethylene which is necessary for the manufacture of polyethylene as well as other plastics and chemicals.

In the second half of 2011, the United States ethane demand grew to about 933,000 barrels per day from 812,000 BPD in 2009 [14]. However, distribution bottlenecks have increased the ethane feedstock cost to 95 cents a gallon in late 2011, from 60 cents a gallon in early 2011.

For example, the Marcellus Shale natural gas deposits in the Appalachian basin on the East Coast of the United States are discoveries that are relatively dry (low natural gas liquids). However, Marcellus Shale natural gas deposits found in western Pennsylvania and West Virginia [15] have a relatively high natural gas liquids component which makes this particular region very attractive for the chemical industry, which utilizes the ethane component for the manufacture of ethylene.

The status of the North American polyethylene industry as of 2012 has changed dramatically since the discovery of shale-based natural gas. A summary report [16] was published by Shawn McCarthy in late 2011 on the impact of the shale gas revolution on North American polyethylene producers. Nova Chemical, based in the Calgary region of Canada, has gained access to the Marcellus region in the United States through Sunoco Logistics Partners LP in order to ship ethane to an ethylene plant near Sarnia, Ontario, Canada. Nova Chemical also announced long-term supply ethane contracts with Range Resources Corp. and Caiman Energy LLC. These ethane-related business plans allow Nova Chemical to replace higher-cost ethane derived from crude oil from Western Canada.

In late 2011, other polyethylene producers also announced plans to increase ethylene capacity in the United States. These include Dow Chemical, Royal Dutch Shell, Bayer AG, Chevron Phillips Chemical Company and LyondellBasell [16,17]. McCarthy's report [16] provided testimony from the American Chemistry Council to the United States Congress stating that "a sustained increase in domestic (United States) ethane supply from shale gas would result in 400,000 new jobs in the U.S.-based chemical industry and $132-billion (U.S. dollars) in added economic activity."

Dow will add about 5 billion pounds of ethylene capacity in the United States [16,18] through 2017, primarily on the Gulf Coast, with Range Resources supplying Dow's Louisiana operations the shale-gas-based ethane feedstock from the Marcellus region in Southwest Pennsylvania. The cost of ethylene produced from ethane is approximately $600 a ton as compared to $1,200 a ton for ethylene produced from crude-oil-based naphtha [17].

However, the addition of ethylene capacity by Royal Dutch Shell [19] in the Appalachia region of the United States (either Pennsylvania, West Virginia or Ohio), which was considered extremely unlikely less than a decade ago, may be the first of several such plants in a region of the United States labeled the "Rust Belt," as manufacturing jobs have left this region over the past four decades.

1.4.15.1   North American Natural Gas Supply

In 2009, the amount of natural gas available for production in the United States soared 58% since 2005 due to the discovery of enormous new gas fields in Texas, Louisiana and Pennsylvania. A report issued by the non-profit Potential Gas Committee estimated that the United States presently has reserves of about 2,100 trillion cubic feet, which is an astonishing 35% increase over the 2007 estimate. In addition, after 2010, future Federal government policy guidelines in the United States may encourage additional utilization of natural gas in other less traditional roles, such as a transportation fuel (compressed natural gas), to reduce carbon dioxide emissions from fossil fuels. Such policies will increase the annual consumption of natural gas significantly from the plateau of 18–20 trillion cubic feet level experienced from 1970–2005. In 2011, natural gas production began to increase to 23 trillion cubic feet and is forecast by the U.S. Energy Information Administration (EIA), a governmental agency of the United States, to continue over the next decade.

1.4.15.2   Globalization of Natural Gas

The onset of the industrial revolution in Britain in the 18th century was driven by the industrial use of coal and later crude oil. Natural gas was initially viewed as an unwanted byproduct during the search for crude oil or coal deposits. However, today natural gas will be the global growth fuel of the 21st century. Global consumption of natural gas is projected to increase by about 70% between 2002 and 2025 [11]. ExxonMobil has projected that natural gas consumption will increase from 290 billion cubic feet per day in 2009 to 450 billion cubic feet per day in 2030 [20].

As the global natural gas demand increases worldwide over the course of this century, it is possible that the natural gas price in each global region will change over the levels in 2010. As of 2010, a large number of intra- and intercontinental natural gas pipeline projects are either under construction or planned in order to move massive quantities of natural gas from the Middle East and Russia to European and Asian markets that may drive the natural gas industry towards a more global market. In addition, the increase in the global liquefied natural gas distribution system will also provide increased mobility of natural gas that will reduce, but not eliminate, wide differences in regional natural gas prices. It is important to understand that there are significant costs associated with the LNG business, so that large differences in the price of natural gas will most likely continue. The LNG is loaded into specially designed tankers for shipment across the globe. Table 1.17 illustrates the rapid growth of the global LNG system.

The hydrocarbon liquids (e.g., ethane and propane) associated with this massive increase in the LNG system will create a significant supply increase for these liquids. The export of ethane from North America to Europe may become an important factor in lowering ethylene manufacturing costs in Europe over the next several decades.

1.4.15.3 Status of Ethylene Costs as of 2010 and Future Trends

The price of ethylene from 2004–2010 represents an excellent example of the ethylene price volatility spanning relatively low feedstock costs to relatively high feedstock costs, with the price of ethylene in January 2009 illustrating the weakness in ethylene prices during a global economic recession. This data is summarized in Table 1.18.

Table 1.17 Global liquefied natural gas capacity and forecast [21].

Year	Capacity (Trillion cubic feet/year)
2008	10
2011	15
2013	18
2015	29

Note: Annual growth rates of LNG production from 2005–2015 is approximately 15%.

Table 1.18 Ethylene costs from January 2004 – January 2010.

Year	Cost (US $/lb)
2004	0.300
2005	0.445
2006	0.443
2007	0.554
2008	0.568
2009	0.275
2010	0.511

Source: Plastemart.com

As the highest cost producers, ethylene plants using oil-based feedstock such as naphtha determine the price of ethylene. This type of pricing heavily favors ethylene producers using natural gas-based ethane as feedstock. Costs as of January, 2010 are shown Table 1.19 [17].

1.4.15.4　Additional Feedstock for Ethylene Production

As outlined above, ethylene is primarily manufactured using an energy intensive steam-cracking process with either crude-oil-based naphtha or natural-gas-based ethane. Ethane became a significantly lower cost feedstock since about 2000, when crude-oil prices climbed from about $15–20/barrel to about $140/barrel. Global future demand for crude oil is expected to continue to increase due to relatively rapid growth in the developing regions. Crude oil demand may stabilize or decrease in North America over the next few decades due to higher fuel mileage standards and improvements in automotive technology, this decrease will not be sufficient to offset the expected growth in crude oil demand in other global regions. Hence, ethylene manufactured from naphtha will most likely remain at a significant cost disadvantage for polyethylene producers using naphtha-based ethylene.

Ethylene derived from ethane will remain the preferred source with lower- cost ethane in the Middle East and North America responsible for the enormous increase in the capacity of polyethylene in the future. For example, the Middle East added about 20 billion pounds of new ethane-based polyethylene capacity between 2008 and 2012 [22].

1.4.15.5　U.S. Ethylene Forecast by Burns & McDonnell (2013)

The Burns & McDonnell company headquartered in Kansas City, Missouri, provides a wide variety of consulting services. A recent report by M. W. Lockhart and E. Robertson, "Shale Gas Boom – A New Era of Technology for Ethylene and Ethylene Derivatives," presents an excellent discussion on the impact of new sources of natural gas-based ethane in North America on the cost and location of future ethylene and polyethylene capacity expansions in North America [23].

Table 1.19 Cost comparison of natural gas- and oil-based ethylene.

	Ethane/Natural gas	Naphtha/Oil Based
Ethylene Cost (US $/lb)	0.28–0.30	0.51
Ethylene Price (US $/lb)	0.55	0.55

This report forecasts the addition of about 40 million metric tons/year of ethylene capacity expansions in North America by 2018, and an additional 45 million metric tons of ethylene capacity by 2030, due to low-cost ethane supplied from shale-based natural gas liquids.

Because as much as 50–70% of this new ethylene capacity will be used for the manufacture of polyethylene, this new ethylene capacity will make North America a significant polyethylene export region similar to the Middle East. If the 2030 ethylene forecast is reached, then about 60 million metric tons of additional polyethylene capacity will be available by 2030, mostly for the export market.

This North American ethylene expansion will be driven by the cost advantage of ethylene produced from ethane compared to ethylene manufactured from naphtha, which will remain the primary source of ethylene in other regions of the globe. The Burns & McDonnell report estimates the cost of ethylene in the Middle East at 5–15 cents/lb, North American ethane-based ethylene at 20–30 cents/lb, and naphtha-based ethylene at 50–60 cents/lb. Finally, the Burns & McDonnell forecast estimates an investment of about 120 billion (USD) in new ethylene capacity in North America from 2013–2030.

1.4.15.6   Ethylene Based on New Feedstock

The discovery of new technology to manufacture ethylene from other raw materials could also change the feedstock economics for the manufacture of polyethylene. Two such possibilities are manufacturing ethylene from (a) methane or methanol, or (b) biomass material. New scientific discoveries would be needed to facilitate such a change.

1.4.15.7   Methane/Methanol as Feedstock for Ethylene

Since the early 1970s, scientists in the petrochemical industry have understood that the manufacture of ethylene from methane or methanol could significantly lower capital and operating costs relative to ethylene production based on today's use of naphtha or ethane.

The direct oxidative addition of methane to ethylene shown in the equation below is a thermodynamically favorable reaction (Gibbs free energy of −69 Kcal/mole).

$$2\ CH_4 + O_2 \rightarrow CH_2{=}CH_2 + 2\ H_2O \tag{1}$$

Early research into this methane-based route to ethylene was reported by Keller and Bhasin in 1982 [24]. Efforts in the oxidative coupling of methane to ethylene have increased since 2008, as Dow Chemical Company awarded research grants to research groups at Northwestern University and Cardiff University in January 2008 as part of its Alternative Feedstock Program [25].

In addition, the conversion of methanol-to-light olefins (MTO) process has received a significant level of industrial interest since the 1970s. Light olefins are known to be intermediates in the methanol-to-gasoline process discovered by Mobil Oil Corporation scientists using zeolite-based catalysts designated as H-ZSM-5 [26].

A report [27] by Professors Thorsten M. Bernhardt of the University of Ulm, Germany, and Uzi Landman of the Georgia Institute of Technology, Atlanta, Georgia, USA [27], described a process, at low temperature and pressure, in which ethylene was produced from methane using a catalyst based on gold.

1.4.15.8   Biomass as Feedstock

Ethanol produced from biomass material such as sugarcane, algae or cellulose is another route to ethylene utilizing an ethanol dehydration step:

$$CH_3CH_2OH \rightarrow CH_2{=}CH_2 + H_2O$$

Unlike ethanol production from corn, ethanol manufactured from sugarcane may be relatively cost-effective as an ethylene feedstock. Dow Chemical announced a sugarcane-to-polyethylene project in Brazil in 2007 that will play a role in evaluating the process economics in the manufacture of polyethylene from a biomass material.

1.4.15.9   Governmental Policy and Regulation

Although the location of the majority of new ethylene and polyethylene capacity will be determined by low feedstock and energy costs, other factors such as governmental policies and regulation could play a role in the future. The location of additional ethylene and polyethylene capacity in the future could be determined by governmental factors such as regional political stability, tariffs, environmental restrictions, corporate income taxes and other future taxes, such as a carbon emissions tax.

1.4.15.9.1   Tax Policies

In the early 1980s corporate tax rates in most of the developed countries in the world were relatively high in the 48–50% range, and had little impact in determining the capital investment decision. However, during the 1980s a tax cut policy was put into place in the United States that lowered both the income and corporate tax rates, which lead to strong economic expansion. This strong economic expansion in the United States had a significant influence on many other developed countries across the globe, which resulted in very large corporate tax cuts in the countries that were members of the Organization for Economic Co-operation and Development (OECD). The OECD is an international organization of thirty countries that accept the principles of a representative democracy and free-market economy. By 2008 the average corporate tax rate of these countries, excluding the United States, had declined from about 48% in 1988 to about 26% in 2008, placing the United States at a competitive disadvantage in regard to tax policies [31]. The combined corporate tax rate in the United States has been essentially unchanged since about 1993–2008 at about 40%. This tax data is illustrated in the Figure 1.6.

The result of this decrease in corporate tax rates across the globe may affect the investment of capital around the world. The combined corporate tax rates for the five highest member countries in OECD are:

- Japan   39.5%
- United States   39.3
- France   34.3
- Belgium   34.0
- Canada   33.5

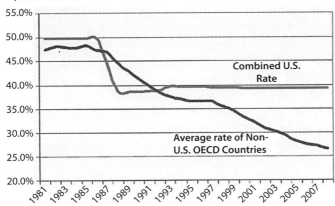

Figure 1.6 U.S. business tax rates compared to OECD average.

While the five countries in OECD with the lowest tax rates are:

- Turkey   20.0%
- Poland   19.0
- Slovak Republic   19.0
- Iceland   15.0
- Ireland   12.5

To demonstrate the effect of low tax rates on economic growth, Ireland experienced the highest economic growth rates over an 18 year period (1990–2008) compared to any other European country.

This tax competition across the globe has now made it increasingly necessary for many regions of the world to continue to lower corporate tax rates to attract the capital investments needed to stimulate economic growth. High economic growth will take place in global regions where tax rates continue to decline. If feedstock costs become less important in the future in determining the location of capacity increases for ethylene and polyethylene producers, then high economic growth regions will not only show a significant increase in polyethylene consumption, but these high growth regions may replace other regions as the leading centers for ethylene and polyethylene capacity expansion.

1.4.16   Environmental Factors

Climate change, recycling rates, waste to energy and non-biodegradation of polyethylene are all subjects that influence the public perception of polyethylene, in particular, and plastic materials in general. These topics may affect the growth of the polyethylene industry in the future, particularly some specific applications such as the merchandise bag that is perceived as causing environmental damage and has led government policy makers around the world to ban the use of such bags, causing the formation of advocacy groups in support of the plastic bag (e.g., www.SaveThePlasticBag.com created by Stephen Joseph).

However, there is little scientific evidence that support the claims of environmental damage caused by plastic bags and such issues need to be settled through the judicial system, where scientific evidence supporting such claims would be necessary. A report by the British Environmental Agency is especially important for the outcome of such cases, as this report states that a single-use polyethylene bag has a lower carbon footprint than alternative paper or reusable bags [28].

1.4.16.1   Polyethylene Recycling – Waste-to-Energy

One positive outcome from the debate on the utilization of plastics over the past 30 years has been the growth in the effort to collect and recycle polyethylene in order to fabricate new polyethylene products from the recycled material, or the utilization of organic waste as fuel to generate electricity, commonly referred to as waste-to-energy.

The polyethylene recycling effort will most likely continue to increase in the future and will have an effect on the amount of polyethylene capacity that may be needed in the future. Certainly this type of recycling program will result in the more efficient use of hydrocarbon feedstock and should be encouraged by the polyethylene industry.

1.4.16.2   Biodegradation

The public perception that plastic materials harm the environment has had a somewhat damaging effect on the growth of the polyethylene industry, especially in the area of polyethylene-based plastic merchandise bags. Certainly the littering of these items in the environment does contribute to the general negative opinion of this type of item, but littering is mostly a behavioral problem that applies to any item used in a commercial application regardless of the type of material utilized in the application (i.e., glass, metal, paper and plastic). The argument that biodegradable plastics will somehow cause less environmental damage is in contrast to the same environmentalists that have accepted the conclusion that carbon dioxide emissions contribute to global warming. In fact, plastics that do not biodegrade will reduce the carbon dioxide levels in the atmosphere due to the carbon remaining sequestered in a non-biodegradable item. Biodegradation produces carbon dioxide.

In fact, the non-biodegradability of polyethylene would reduce CO_2 emissions (referred to as a negative carbon footprint) if the ethylene used to produce the polyethylene was based on a biofuel such as ethanol produced from sugarcane. Sugarcane growth consumes CO_2 during the photosynthesis process. If ethanol is produced from the sugarcane and the ethanol is used to manufacture ethylene by dehydration, then the sequestered carbon in the polyethylene material would reduce atmospheric CO_2 emissions with the carbon that was removed from the atmosphere by photosynthesis being sequestered in the polyethylene molecule.

1.4.17   Biobased Ethylene

Commercial plans for a biobased ethylene source to manufacture polyethylene have been published by Dow Chemical Company for a process based in Brazil. In a report by Kevin Bullis published by *MIT Technology Review* on July 25, 2011, plans for the construction of a 240-million-liter ethanol plant using sugarcane as feedstock has been proposed by Dow as a joint venture with Mitsui. This ethanol-based ethylene would be used to manufacture 350,000 metric tons of polyethylene. This would be in addition to a 200,000-ton sugarcane-to-polyethylene plant operated by Brazil-based Braskem. However, for this ethanol-based ethylene and polyethylene facility to be successful, the ethylene costs must be competitive with ethylene produced from naphtha, which will mostly likely remain the high-cost ethylene source.

The polyethylene industry should not discount the future importance of biobased ethylene and such statements should be avoided. Technological innovations could even result in cost advantages of ethylene from ethanol as compared to ethane-cracking, which requires extremely large capital investments for construction of a world-class ethylene manufacturing plant.

One such possibility is ethanol from algae or other water-based bacteria that rely on water-based photosynthesis, which takes place at extremely fast reaction kinetics. Water-based carbon dioxide levels far exceed the carbon dioxide concentration in air, thus providing the possibility of much faster biofuel-based reaction rates. One example of this type of process was reported by Michael Totty [29] who described the biofuel approach taken by Joule Unlimited Technologies in Cambridge, MA, USA. Joule Unlimited Technologies has created genetically engineered micro-organisms that produce ethanol, diesel fuel and other hydrocarbons from a patented water-based photosynthesis described as a solar converter process.

Another example of algae-based research was announced in July 2009, when ExxonMobil announced a joint venture with Synthetic Genomics Inc. to investigate the synthesis of biofuel from natural and engineered strains of algae [30].

References

1. J. Richardson, Saudi feedstock problems worsen, 2010 Report, www.icis.com.
2. Chevron-Phillips Chemical Company press release, *Bloomberg News*, Dec.15, 2011.

3. T.O.J. Kresser, *Polyethylene*, Reinhold Publishing Corp., New York, p. 9, 1958.

4. R. Buderi, *The Invention that Changed the World*, Simon & Schuster, New York, NY, pp. 89-97, 1996.

5. www.sriconsulting.com/WP/Public/Reports/polyethylene.

6. P. Kotler, *Marketing Management*, Prentice-Hall, Inc., Englewood Cliffs, New Jersey, pp. 230-245, 1976.

7. Chemical Marketing Associates Inc., 11757 Katy Freeway Suite 700, Houston, Texas, 77079.

8. *Chemical & Engineering News*, July 4, 2011.

9. H. Rappaport, Chemical Market Associates Inc., Houston, Texas, in: *Plastics News*, p. 12, March 16, 2009.

10. T. Miller, and K.R. Holmes, "2009 Index of Economic Freedom," available from heritage.org/index.

11. T. Mokrani, and M. Scurrell, *Catalysis Reviews*, Vol. 51, p. 10, 2009.

12. *Wall Street Journal*, May 10, 2010.

13. The Economist, Nov. 26, 2011, www.economist.com.

14. Bentek Energy as reported in the *Wall Street Journal*, Dec. 19, 2011.

15. S. Martin, "US players mull ways to bring Marcellus Shale ethane to market," *ICIS News*, Oct. 19, 2010 (originally published Oct. 18, 2010), www.icis.com.

16. S. McCarthy, "Shale gas gives plastics sector new lease on life," *Globe and Mail*, Sept. 29, 2011, www.theglobeandmail.com.

17. L. Denning, *Wall Street Journal*, April 25, 2011.

18. J. Kaskey, *Bloomberg News*, April 21, 2011, www.bloomberg.com.

19. ICIS, additional reporting by S. Burns, B. Ford and N. Davis, "Shell's new ethylene cracker in Appalachia may be the first of several," *ICIS Chemical Business Magazine*, Aug. 29, 2011, www.icis.com.

20. *Wall Street Journal*, Vol. CCLIV, No. 128, Nov. 30, 2009.

21. *Wall Street Journal*, p. C12, June 26, 2009.

22. *Plastics News*, p. 12, March 16, 2009.

23. M.W. Lockhart, and E. Robertson, "Shale gas boom – a new era of technology for ethylene and ethylene derivatives," presented at ALCHE – Southwest Process Technology Conference in Galveston, Texas, Oct. 4, 2013, source: www.burnsmcd.com; accessed March, 2014.

24. G.E. Keller, and M.M. Bhasin, *J. Catalysis*, Vol. 73, p. 9, 1982.

25. Dow Chemical Company, Midland, Michigan, "Dow Chemical Awards Methane Challenge Grants to Cardiff and Northwestern Universities," Jan. 24, 2008.

26. T. Mokrani, and M. Scurrell, *Catalysis Reviews*, Vol. 51, p. 60, 2009.

27. T.M. Bernhardt et al., *Angew Chemie International Edition*, Vol. 49, No. 5, p 821, 2010.

28. *Plastic News*, Crain Communications, Vol. 22, No. 49m p. 3 March 7, 2011.

29. M. Totty, *Wall Street Journal*, Oct. 17, 2011.

30. www.exxonmobil.com; accessed Feb. 2012.

31. *Wall Street Journal*, Vol. CCLIV, No. 151, Dec. 28, 2009.

2

Titanium-Based Ziegler Catalysts for the Production of Polyethylene

2.1 Introduction

Ethylene polymerization catalysts based on titanium as the active center are primarily used in the low-pressure manufacture of polyethylene with a relatively narrow molecular weight distribution to include both high density polyethylene (HDPE) and linear low density polyethylene (LLDPE). Table 2.1 summarizes a few examples of end-use applications for titanium-based polyethylene in terms of the density, molecular weight distribution and molecular weight for various applications.

2.2 Titanium-Based Catalyst Developments

2.2.1 Historical Developments

In the early 1950s, Ziegler and coworkers [1] were attempting to produce high molecular weight polyethylene by repetitive insertion of ethylene between the aluminum-carbon bond in triethylaluminum, as shown in

Table 2.1 Examples of product applications for titanium-based catalysts.

Product Type	Melt Index (MI) $(I_{2.16})^a$	Flow Index (FI) $(I_{21.6})^b$	MFR $(FI/MI)^c$	Density (g/cc)
Film[d]	1.0	28	28	0.918
Cast Film	2.0	52	26	0.924
Injection molding	10	250	25	0.926
	30	720	24	0.953
Rotational molding	5.0	125	25	0.934

[a] Melt Index is an inverse relative measure of polymer MW.
[b] Flow Index is a measure of the polymer processability.
[c] Melt Flow Ratio of the FI and MI values is a relative measure of the polymer MWD where MFR values between 24–28 indicate that the MWD is relatively narrow. Melt Index, Flow Index, MFR and density are only approximate values for each application.
[d] Film products contain a comonomer (1-butene, 1-hexene or 1-octene) which is used to reduce polymer crystallinity. The type of comonomer used has a significant effect on film toughness properties. 1-butene is used for general purpose film applications and 1-hexene or 1-octene are used for more demanding applications.

the growth step (Equation 2.1), where R represents additional alkyl chains that are formed simultaneously.

$$AlEt_3 + CH_2=CH_2 \rightarrow R_2Al\text{-}CH_2\text{-}CH_2\text{-}(CH_2CH_2)n\text{-}CH_2CH_3 \qquad (2.1)$$

A termination step (Equation 2.2) involving beta-hydride transfer to an aluminum atom leads to the formation of a vinyl-terminated polymer molecule and an Al-H bond, which can insert an ethylene molecule to initiate the growth of additional polymer species.

$$R_2Al\text{-}CH_2\text{-}CH_2\text{-}(CH_2CH_2)n\text{-}CH_2CH_3 \rightarrow$$
$$R_2Al\text{-}H + CH_2=CH\text{-}(CH_2CH_2)n\text{-}CH_2CH_3 \qquad (2.2)$$

This reaction sequence was designated as the Aufbau process by Ziegler. Ziegler's data at the time indicated that the polymer growth step is favored at reaction temperatures between 100–120°C, so that polyethylene with molecular weights between 3,000 to 30,000 were produced depending on other reaction conditions such as ethylene pressure. However, polyethylene in this molecular weight range does not possess sufficient mechanical strength properties for most commercial applications. Higher molecular weight polymers were not formed because the termination reaction took place too rapidly at these high temperatures.

However, in late 1953 an anomalous finding was made by Ziegler and Holzkamp after one particular experiment carried out in a high-pressure autoclave at 100°C and 100 atm of ethylene. In this particular experiment, 1-butene was predominantly recovered, unexpectedly, from the reactor [2]. This result was traced to the accidental contamination of the reactor with a nickel compound that had been used in a previous experiment for a hydrogenation experiment. Ziegler and Holzkamp concluded that the nickel compound had modified the chemistry in such a way as to accelerate the termination reaction and result in the large quantity of 1-butene. This result led Ziegler, Holzkamp and Breil to search for another metal-containing compound that would modify the Aufbau reaction in the opposite manner to produce high-molecular weight polymer. Although Holzkamp examined zirconium acetylacetonate as a modifier in the Aufbau process first, it was Breil that repeated the experiment and obtained a white powdery substance, which was characterized as high molecular weight polyethylene [3].

A more active catalyst than the example above was found by mixing two liquids, titanium tetrachloride ($TiCl_4$) and triethylaluminum ($AlEt_3$), and this catalyst system was developed for large-scale production of high-density polyethylene and was designated by Ziegler as the Mülheim Atmospheric Polyethylene Process [4]. However, the interaction of these two liquids in hydrocarbon solvents such as heptane produces a solid catalyst that is insoluble in the reaction medium. Consequently, this catalyst is a heterogeneous catalyst system in which the ethylene polymerization takes place on the surface of the titanium-containing solid material. It is important to understand that the high-density polyethylene produced by Ziegler possessed a narrower molecular weight distribution than the high-density polyethylene prepared earlier by Hogan and Banks at Phillips Petroleum Company with their chromium-based catalyst (Phillips) system. Although both types of polyethylene offered significantly higher stiffness and higher heat distortion temperatures than high-pressure, low-density polyethylene, it was understood as early as 1955 that each variety of polyethylene would find its own markets based on the processability and properties of each variety.

At the 11th Annual National Technical Conference of the Society of Plastics Engineers in Atlantic City, New Jersey, in January of 1955, the attendees from Monsanto, Spencer Chemical and Union Carbide anticipated significant growth of both high-pressure and low-pressure polyethylene without appreciable competition with each other, based on the growth of separate markets and applications [5]. This turned out to be the case, as LDPE and the two types of HDPE each developed rapidly in the years to follow.

2.2.2   The Role of Professor G. Natta

An excellent account into the detailed role of Professor Natta and coworkers in the development of Ti-based catalysts was published in Boor's book [6,7]. A brief summary of these events is important to put the work of Ziegler and Natta (working with Ti-based catalysts) and Hogan and Banks (both of Phillips Petroleum and working with Cr-based catalysts) into historical perspective.

Professor Karl Ziegler was at the Max Planck Institute for Coal Research during this period working primarily with ethylene (Aufbau process) and much of his research that followed the initial discovery of the Mülheim catalyst (TiCl$_4$ + AlEt$_3$) dealt with ethylene homopolymerization and ethylene/propylene copolymerization. Before Ziegler disclosed his discovery in a technical publication, he disclosed his results to two industrial companies: Montecatini of Italy and Goodrich-Gulf of the United States.

Because Professor Natta was a technical consultant for Montecatini at the time, he undertook a research program as a result of a contractual agreement between Ziegler and Montecatini. Also at this time, Natta was active in carrying out investigations into the kinetics of the Aufbau reaction and was able to redirect his research program into studying the behavior of the Mülheim catalyst with other olefins such as propylene, higher 1-olefins and dienes. Hence, early in 1954, the Natta group obtained a mixture of amorphous and crystalline polypropylene and used solvent extraction methods to isolate the individual polymer fractions. Natta designated the crystalline form of polypropylene as "isotactic" because the polymer chain consisted of long sequences in which the propylene unit had the same configuration. Both Ziegler and Natta were able to document their research findings in scientific literature rather quickly and Natta was especially able to form large research teams of specialists in different areas to accelerate the rate of discovery mostly in stereo regular addition polymers. This work was summarized in a comprehensive book by two pioneers in polymer chemistry, N. G. Gaylord and H. F. Mark [8]. While at Goodrich-Gulf, S. E. Horne and coworkers used Ziegler's Mülheim catalyst to prepare cis-1,4 polyisoprene, which is the basic component of natural rubber [9].

Because of the numerous publications by Natta and Ziegler during the course of their research efforts in the mid-1950s, other scientists worldwide were soon able to participate in this type of research. One of the more important findings for obtaining excellent control of the molecular weight of various polymers prepared with Ziegler-Natta catalysts was reported by E. J. Vandenberg of the Hercules Research Center, who reported that molecular hydrogen was a chain transfer agent for these catalysts [9].

Chain transfer with hydrogen to ethylene coordinated to the active site produces a saturated polymer end group and a Ti-Et bond (or a Ti-H bond if chain transfer proceeds without an ethylene molecule coordinated to the active site) that is able to initiate the growth of another polymer molecule.

2.2.3   Historical Controversy – Isotactic Polypropylene with Cr-Based Catalyst

In order to clarify the historical events pertaining to the initial discovery of crystalline polypropylene, it is important to understand the original research of Hogan and Banks at Phillips Petroleum. The olefin polymerization research carried out by Hogan and Banks was initiated in late 1951, which was two years earlier than the developments of Ziegler and Natta [10]. As part of their research program at Philips Petroleum, Hogan and Banks examined the polymerization of numerous 1-olefins and mixtures of various 1-olefins with their Cr-based catalyst.

Consequently, the Phillips Petroleum scientists isolated crystalline polypropylene between October 9, 1951 and April 16, 1952. Although, initially, the United States Patent and Trademark Office awarded the composition of matter patent for isotactic polypropylene (prepared with a Ti-based catalyst composition) to Montecatini on February 6, 1973 (U.S. Patent 3,715,344), the Federal District Court of Delaware reversed the United States Patent and Trademark Office decision on January 11, 1980, and awarded the composition of matter patent to Phillips Petroleum based on the earlier research carried out by Hogan and Banks (See Chapter 3 of this book for additional details on the historical origins of polyethylene and polypropylene.)

However, this historical record should not diminish the important contributions made by Natta and coworkers. The Cr-based isotactic polypropylene was produced in low yields and was not an attractive method for the commercial manufacture of isotactic polypropylene. But examination of the research records by the court clearly showed that the Phillips scientists were the first group to prepare isotactic polypropylene, and correctly awarded the composition of matter patent to the proper owners.

The research by Natta's group demonstrated that significantly higher yields of isotactic polypropylene could be produced with modified Ti-based catalysts, so that the commercial development of this new type of polymer could take place. Hence, the Phillips scientists were able to show that their research produced the first samples of isotactic polypropylene, while the Natta group research resulted in the rapid commercialization of isotactic polypropylene.

2.3   Titanium-Based Catalysts for the Manufacture of Polyethylene

Catalyst developments for the manufacture of polyethylene with Ti-based catalysts can be best described as based on two different types of catalysts, which are broadly defined as first and second generation catalysts.

- The first generation Ziegler catalysts (ca. 1955–1970) were based on relatively low activity catalysts prepared with various titanium compounds and aluminum alkyls. Depending on polymerization process conditions, polyethylene manufactured with these low activity catalysts often required a catalyst removal step in the manufacturing process to remove the residual inorganic material from the final polymer. These catalysts provided polyethylene with a narrower molecular weight distribution than the Cr-based Phillips catalyst and were more suitable for injection molding and film applications.

- Second generation Ti-based catalysts (ca. 1970–present) were based on significantly higher activity catalysts prepared with a titanium compound, a magnesium compound and aluminum alkyls as a cocatalyst. Because these second generation Ti-based catalysts were 10–100 times more active than first generation catalysts, the catalyst removal step was no longer required in the polymerization process. More importantly, the introduction of these high-activity second generation Ti-based ethylene polymerization catalysts led to the commercial introduction of linear low-density polyethylene (LLDPE) by chemical companies such as Dow (solution process) and Union Carbide (gas-phase process) on a significantly larger scale. In addition, these second generation Ziegler catalysts were also important in the growth of high-density polyethylene (HDPE) and LLDPE in injection molding and rotational molding applications and LLDPE in film applications.

The majority of the research summarized in this chapter will pertain to second-generation, magnesium-containing Ziegler catalysts due to their importance to the polyethylene industry today. However, a brief summary of first generation catalysts will be included due to the historical importance of these early ethylene polymerization catalysts from an academic and industrial perspective (see Appendix 2.1 for review references on first generation Ziegler catalysts).

2.3.1   First Generation Ziegler Catalysts for the Manufacture of Polyethylene

The first two commercial plants to manufacture low-pressure polyethylene were started up in 1956 using Ziegler's Mülheim process ($TiCl_4$ and $AlEt_3$) and in 1957 using the chromium-based Phillips catalyst.

In 1957, D. B. Ludlum and coworkers at the E. I. DuPont Company published an early paper on the characteristics of the Mülheim catalyst [11]. Ludlum investigated the polymerization of ethylene in a high-boiling inert solvent (decalin) using $TiCl_4$ and tri-isobutylaluminum (TIBA), Tri-n-hexyl aluminum (THA) or triethylaluminum (TEAL) at 100°C and 1–1.5 atm of ethylene pressure. The rate of ethylene uptake was found to decrease shortly after polymerization was initiated and Lunlum postulated that this was due to the encapsulation of polymerization sites by the polyethylene. The polymerization was also found to be first-order with regard to the ethylene pressure and the concentration of $TiCl_4$. The effect of the Al/Ti ratio on catalyst activity and the type of aluminum alkyl used (TIBA or TEAL) in the polymerization are shown in Figure 2.1.

Examination of Figure 2.1 shows that TIBA provided the more active catalyst with an activity of about 185 g PE/g Ti/hr/atm ethylene at an Al/Ti molar ratio near 3, while TEAl produced a catalyst with about 100 g PE/g Ti/hr/atm ethylene with an Al/Ti molar ratio of about 5.

Figures 2.2 and 2.3 provide additional information on the oxidation state of titanium at various Al/Ti ratios and the ethylene polymerization rate vs the average valence state of titanium, respectively.

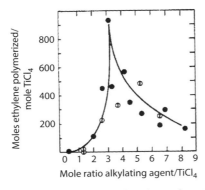

Figure 2.1 Moles of ethylene consumed/mole of $TiCl_4$ as a function of Al/Ti molar ratio for TEAL (o) and TIBA (●). Polymerization time 2 hr at 100°C and 1.5 atm of ethylene. Reprinted from [11] with permission from American Chemical Society.

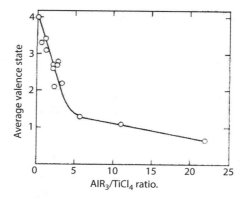

Figure 2.2 Dependence of average valence state of titanium on catalyst composition; catalysts prepared from $TiCl_4$ and triethylaluminum at 80oC in cyclohexane. Reprinted from [11] with permission from American Chemical Society.

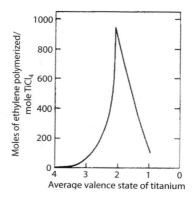

Figure 2.3 Dependence of catalytic activity on average valence state of titanium. Catalysts prepared from $TiCl_4$ and triethylaluminum or triisobutylaluminum. Reprinted with permission from [11]; American Chemical Society.

Hence, the research by Ludlum and coworkers suggested that the active center of the Ti-based catalyst used in the Mülheim commercial process is based on a Ti (+2) species.

2.3.2 Types of Metal Alkyls Investigated

As shown above, the first commercial Ziegler catalyst for the manufacture of polyethylene utilized triethyl aluminum (TEAL). However, early research involved extensive examination of various types of other metal alkyls that were used with titanium tetrachloride to promote ethylene polymerization. Some of these catalyst systems are summarized in Table 2.2.

Table 2.2 Summary of metal alkyls investigated for ethylene polymerization with $TiCl_4$.

Metal Alkyl	Reference
(n-Butyl)K	[12]
(Amyl)Na	[13]
$Li(CH_2)_5Li$	[14]
R_2Cd or RCdI	[15]
GaR_3	[16]

2.3.3   Soluble Titanium-Based Complexes for Ethylene Polymerization (1955–1960)

A wide variety of titanium-containing compounds as well as other transition metal compounds were examined in the period immediately following Ziegler's initial catalyst discovery based on $TiCl_4$ and $Al(C_2H_5)_3$, which is a heterogeneous catalyst system.

David Breslow of the Hercules Research Center recognized in the mid-1950s that efforts to identify a homogeneous ethylene polymerization catalyst system would be an important step in understanding the nature of the active site, as homogeneous systems are more easily characterized in terms of structure and better systems to investigate polymerization kinetics. Consequently, catalyst systems based on bis(cyclopentadienyl) titanium dichloride (Cp_2TiCl_2), which is soluble in toluene, were investigated between 1955–1960. The interaction of (Cp_2TiCl_2) with alkylaluminum compounds was reported by Natta et al. [17], Breslow et al. [18–20] and Chien [21]. Although Natta carried out his research in heptane (a nonsolvent for Cp_2TiCl_2), Natta found that the addition of triethylaluminum to the heptane slurry containing Cp_2TiCl_2 produces a soluble reaction product. Breslow and Chien carried out their work in toluene.

Natta et al. [17] reported that the interaction of Cp_2TiCl_2 with $Al(C_2H_5)_3$ in heptane at 70°C produces a blue solution with the simultaneous evolution of a gas as the suspended titanium compound slowly dissolves. Cooling this solution to −50°C provided a blue crystalline complex which was isolated. Elemental analyses indicated that the blue crystal had a composition of $Cp_2TiCl_2 Al(C_2H_5)_2$ with the proposed structure in Figure 2.4.

The presence of two ethyl groups in the proposed structure was verified by the elimination of two moles of ethane gas by addition of an alcohol to the blue solution containing this complex.

This heptane soluble complex (0.6 g) polymerized ethylene at 95°C and 40 atm of ethylene pressure for 20 hrs. Natta reported that only 8.3 g of a

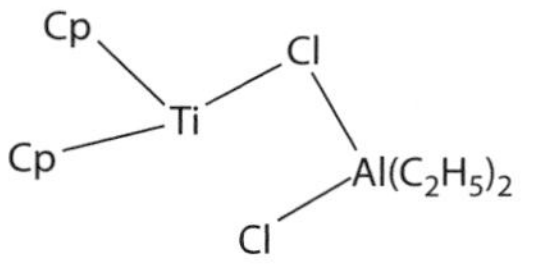

Figure 2.4 Proposed structure of Natta's blue crystalline compound.

relatively high molecular weight polyethylene was obtained, thus this particular catalyst exhibited very low activity ca. 0.02 gPE/g cat./hr/atm.

Breslow followed Natta's publication with a summary of research carried out at the Hercules Research Center a few years earlier than Natta's work in which very highly active soluble catalysts were found from the interaction of Cp_2TiCl_2 and $(C_2H_5)_2AlCl$ in toluene. Breslow and Newburg's catalyst [18] was formed at room temperature by adding 5 mmol of Cp_2TiCl_2 to toluene to form an orange solution. Next, the addition of 10 mmol of diethylaluminum chloride, $(C_2H_5)_2AlCl$ (DEAC), immediately turned the solution red, followed by a gradual change over a period of about an hour to a green color and finally a blue solution. If triethylaluminum (TEAL) was used in place of diethylaluminum chloride, the blue solution formed instantaneously, showing the stronger reducing behavior of TEAL compared to DEAC.

Breslow and Newburg noted that the red solution is a highly active ethylene polymerization catalyst, while the blue solution that formed after about one hour was a much poorer polymerization catalyst in the absence of oxygen. The visible spectrum of this solution was recorded and the details reported [19]. They concluded that the high activity catalyst involved a Ti(+4) complex, while the low activity catalysts were Ti(+3) complexes. The Ti(+3) complexes are blue. Therefore, Breslow and Newburg's blue solution also showed low ethylene polymerization activity, which agreed with Natta's data. In addition, they also isolated a blue crystalline Ti(III) complex, which also had poor ethylene polymerization activity. However, this blue solution exhibited high ethylene polymerization activity, as shown by the formation of 174 g of a very linear high molecular weight polymer in one hour, only if the ethylene feed contained 0.025% oxygen. Breslow and Newburg's oxygen-treated solution was brown.

Breslow and Newburg also examined Natta's blue crystalline complex and found that this complex also showed very high polymerization activity in the presence of a trace of oxygen. They rationalized his data by suggesting that the oxygen was required to oxidize the titanium +3 center in these complexes to a +4 oxidation state to achieve high activity. Later reports by Breslow and Long [19,20] confirmed that a titanium +4 complex was indeed the active species, as the complex $Cp_2(C_2H_5)TiCl{:}C_2H_5AlCl_2$ containing a

Ti-ethyl bond exhibited high ethylene polymerization activity comparable to the commercial Mülheim catalyst based on $TiCl_4$ and $Al(C_2H_5)_3$, which is a solid insoluble catalyst. At the time, this research demonstrated that a soluble catalyst possessed comparable activity to the insoluble Mülheim catalyst, thus showing that a solid surface was not required for a Ziegler-type polymerization.

One interesting feature of Breslow's research with these homogeneous catalyst systems was the type of polyethylene which was isolated. Polyethylene prepared with these catalysts was more linear, as indicated by a lower methyl content and a higher melting point of 137°C., and also possessed a narrower molecular weight distribution (MWD) than the polymer provided by the Mülheim catalyst. It is possible that some of these homogeneous systems may have been "single-site" catalysts, in which case this type of polyethylene would have had a MWD as indicated by the Mw/Mn value of 2 or one Flory component in the polymer.

Chien [21] reported on the kinetics of ethylene polymerization on a soluble catalyst based on Cp_2TiCl_2 and dimethylaluminum chloride, $(CH_3)_2AlCl$. Chien's work confirmed that the active species was a complex containing both Al and Ti, with the titanium in the tetravalent state. The chain initiation process was dependent on the Al/Ti ratio, initiation being slower at higher ratios, suggesting that the active complex may be in equilibrium with one or more inactive species.

Consequently, the research of Breslow and Chien was best explained by showing that the active species was formed by an alkylation reaction between the Ti complex and the aluminum alkyl, as shown in Equations 2.3 and 2.4.

$$Cp_2TiCl_2 + (C_2H_5)_2AlCl \rightarrow Cp_2TiCl_2:(C_2H_5)_2AlCl \text{ (complex)} \quad (2.3)$$

$$Cp_2TiCl_2:(C_2H_5)_2AlCl \text{ (complex)} \rightarrow Cp_2TiCl (C_2H_5):C_2H_5AlCl_2$$
(Alkyl exchange) $\quad (2.4)$

The initial complex in Equation 2.3 forms extremely rapidly, while the alkyl exchange reaction in Equation 2.4 to form the catalyst initiation complex takes place more slowly, but provides the titanium-carbon bond necessary for the ethylene propagation reaction to take place, leading to high molecular weight polyethylene. Therefore, the examination of these soluble titanium complexes provided the important data to explain the ethylene polymerization mechanism, which was detailed in a series of papers by Cossee [22–24].

From an historical perspective, it is important to point out how the research of Natta and Breslow in the mid-1950s that involved the interaction of Cp_2TiCl_2 with triethylaluminum and diethylaluminum chloride was followed up some 20 years later in a classic reaction by Tebbe and others. In their research, Cp_2TiCl_2 was reacted with trimethylaluminum, as shown below, to form the Tebbe reagent, which is the Ti/Al bridged compound shown in Equation 2.5.

$$Cp_2TiCl_2 + [AlMe_3]_2 \longrightarrow Cp_2Ti \underset{Cl}{\overset{CH_2}{<}} AlMe_2 + \tfrac{1}{2}[AlMe_2Cl]_2 \qquad 2.5$$

2.3.4   Mechanism of Polymerization

Active Center: Immediately after Ziegler's original discovery of the $TiCl_4/$ $Al(C_2H_5)_3$ catalyst system for ethylene polymerization, there was considerable debate over the metal responsible for the polymer growth. Because Ziegler had previously shown that polyethylene was produced by the aluminum center in the Aufbau reaction, which involved the interaction of ethylene with aluminum alkyls, it was not at all obvious which metal in the new catalyst system (Ti or Al) was the active center. It was very plausible at the time to suggest that the $TiCl_4$ modified the Aufbau reaction in such a way that the Al atom was the active center and that the $TiCl_4$ somehow provided much higher polyethylene molecular weight. Natta *et al.* originally believed that the polymeric chain grew from the Al-C bond and not from the Ti-C bond and offered several reasons for this conclusion [25,26].

However, research by Carrick *et al.* at Union Carbide Corporation research laboratory in Bound Brook, New Jersey [27], believed that chain growth occurred at the transition metal-carbon bond, and this belief soon became the position which is accepted today. Carrick's investigation examined the copolymerization behavior of ethylene and propylene with a series of transition metal compounds and found that the relative reactivity of propylene increased in the series $ZrCl_4 < TiCl_4 < VOCl_3 < VCl_4$ and were independent of the aluminum alkyl structure used in the copolymerization experiment. In addition, Carrick studied the copolymerization behavior of VCl_4 with ethylene and propylene and found no significant changes in the relative reactivity of the two olefins when the VCL_4 was activated with very different metal alkyls such as triisobutyl aluminum, diphenylzinc, dibutylzinc or methyl titaniumtrichloride, offering strong evidence that the metal alkyl is not the active center.

Beerman and Bestian [28] offered early evidence into the existence of a Ti-Carbon bond by reacting $TiCl_4$ with trimethylaluminum at 20°C to give the two alkylated products shown in Equation 2.6.

$$TiCl_4 + Al(CH_3)_3 \rightarrow CH_3TiCl_3 + CH_3TiCl_2 + Al(CH_3)_2Cl \qquad (2.6)$$

The amount of each Ti-containing species was controlled by the Al/Ti ratio employed.

The research by Beerman also demonstrated the relative stability of the Ti-methyl bond as compared to the relatively less stable Ti-ethyl bond that contains a hydrogen on the beta carbon and can, therefore, undergo beta-hydride transfer to the titanium metal and eliminate an ethylene molecule. This early research eventually lead in the 1970s to the identification of transition metal carbene complexes (M=CH2), which when reacted with olefins provide metallacyclobutanes [29].

By the mid-1960s, Natta and coworkers modified their view and recognized that the transition metal was the active center for polymer growth [30]. **Chain Propagation:** In 1960, one of the first detailed papers concerned with the mechanism for the polymerization of ethylene was published by Cossee [22–24]. Figure 2.5 illustrates the insertion of ethylene into a Ti-Alkyl bond, which is the chain propagation step required for the formation of a high molecular weight polymer molecule.

The active center in Figure 2.5 is shown as a titanium atom with an octahedral configuration with three chloride atoms in the equatorial position and one axial chloride atom and one axial alkyl group (R). Importantly, this octahedral configuration is: (a) coordinately unsaturated (exhibits a vacant coordination site in an equatorial site), which is needed in order for the ethylene molecule to coordinate into the first coordination sphere around the titanium atom, and (b) possesses a Ti-carbon bond formed from the exchange of an alkyl group from the aluminum alkyl and a chloride ion bonded to the titanium.

Cossee proposed the active site on the surface of the $TiCl_3$ lattice as illustrated in Figure 2.6

Figure 2.7 illustrates the relevant molecular orbital interactions between the transition metal center and ethylene as the ethylene coordinates to the active site. The transition state in Figure 2.7 is drawn as a dashed line, and is similar to what is presently designated as a metallocycle intermediate.

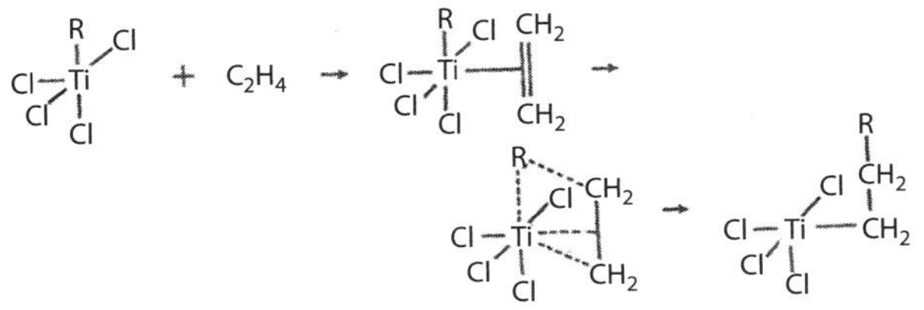

Figure 2.5 Mechanism for ethylene insertion into a Ti-Alkyl Bond. Reprinted from [22] with permission from Royal Society of Chemistry.

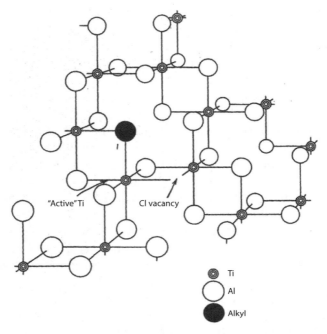

Figure 2.6 One layer of the crystal structure of alpha-TiCl₃ showing the Ti-alkyl bond and the chlorine vacancy forming the active center in the surface [1].

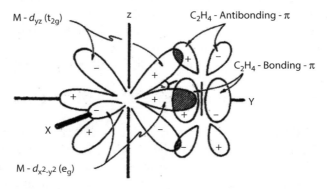

Figure 2.7 The molecular orbital interactions required for ethylene coordination to the transition metal [1].

Figure 2.7 shows that the ethylene is coordinated to the titanium center at the vacant octahedral position through sigma bonding involving the ethylene π-orbitals and the titanium $d_{x^2-y^2}$ orbital. The π–bond involves titanium d_{yz} orbital overlapping with empty π-antibonding orbitals on ethylene. The coordination of ethylene to the titanium active site is known as the Dewar-Chatt-Duncanson model of the metal-olefin bond [31]. The

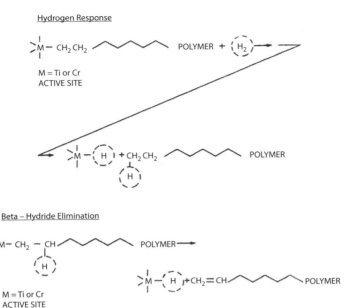

Figure 2.8 Termination of the polyethylene chain by molecular hydrogen (top) and by Beta-hydride elimination to the titanium center (bottom) [32].

coordination of ethylene to the titanium active site destabilizes the titanium-carbon bond, facilitating the insertion of ethylene into the titanium-carbon bond, promoting the growth of the polyethylene molecule.

Cossee also showed the molecular orbital energy diagram for the octahedral complex responsible for the ethylene propagation step. When ethylene is coordinated to the octahedral complex, new energy levels are shown by mixing metal d-orbitals and ethylene π-antibonding orbitals. Cossee's model predicts that only certain transition metals will be highly active as catalysts. According to this molecular orbital energy diagram, the catalyst activity does not depend on the presence of electrons in the d_{yz}, d_{xz} and d_{zy} orbitals [22–24].

Chain Termination: The polymerization chain termination step was the most readily characterized step in the polymerization sequence. Based on polymer end-group analyses, two basic steps have been written to explain the chain termination chemistry. One involves the interaction of molecular hydrogen with the active center, which creates a saturated end group on the polymer molecule, and the other a beta-hydride elimination involving the beta-carbon in the growing polymer chain attached to the active site, which produces an unsaturated end group. These termination steps are illustrated in Figure 2.8 [32].

2.4   Second Generation Ziegler Catalyst for the Manufacture of Polyethylene

Much of the discussion on the development of high-activity, second generation Ziegler catalysts will describe information provided from the patent literature. In the private sector, important technical developments are usually first disclosed during the process of obtaining intellectual property rights to protect the technology from being practiced by other companies.

Significant improvements in the activity of Ziegler-Natta catalysts occurred in the early 1970s when it was found that the preparation of catalyst precursors containing a magnesium compound, such as magnesium chloride or magnesium alkoxide, in conjunction with either $TiCl_4$ or $TiCl_3$, as well as the usual trialkylaluminum cocatalysts, improved catalyst efficiency by at least one or two orders of magnitude [33–42]. Catalyst efficiencies of 100–1000 kg of polymer/gram of titanium were reported. These magnesium/titanium-based catalysts are designated as second generation Ziegler-Natta catalysts. These new systems will be referred to as high-activity Ziegler catalysts. Because of the very high activity of these second generation catalysts, the residual catalyst does not need to be removed from the polymers and, consequently, catalyst removal steps are no longer necessary as a part of the manufacturing process. These developments provided the necessary technology to introduce a new type of polyethylene, referred to as linear low-density polyethylene (LLDPE) resins, which are manufactured utilizing low-pressure processes. The LLDPE resins are copolymers of ethylene and an α-olefin (1-butene, 1-hexene or 1-octene), which contain a sufficient amount of an α-olefin to reduce the polyethylene crystallinity and to provide polymers with densities of 0.91–0.94 g/cc. These second generation catalysts are also used to manufacture HDPE resins (density range of 0.94–0.97 g/cc) with a relatively narrow molecular weight distribution.

2.4.1   Early History of Ti/Mg-Based Catalysts – Solvay & Cie Catalyst

Starting in 1959, Solvay was one of the first companies to manufacture HDPE using the Ziegler catalyst and a slurry polymerization process and had a research laboratory active in the development of improved Ziegler catalyst systems.

2.4.1.1   Solvay & Cie Catalyst Details

U.S. Patent 3,624,059, "Process and Catalyst for the Polymerization and Copolymerization of Olefins," was issued on November 30, 1971, with a priority date of February 7, 1964, and assigned to the company Solvay & Cie. The inventors were Felix Bloyaert Boitsfort, André Delbouille, Jacques Stevens and Braine L'Alleund.

Catalyst Preparation: 20 grams of $MgCl_2 \cdot 6H_2O$ was heated to 200°C to form a dehydrated molten mass which was ground into a powder and then placed into a quartz tube and heated to 285°C under a dry nitrogen purge to form $Mg(OH)Cl$, which is then transferred to another reaction vessel where the $Mg(OH)Cl$ is reacted with excess $TiCl_4$ under refluxing conditions. A solid catalyst precursor was isolated that contained 13.7 mg Ti/g of support.

Polymerization: 1,500 mg of the catalyst precursor was added to a 1.5 liter autoclave using hexane as diluent and a mixture of tetrabutyl tin and aluminum trichloride as the catalyst activator. 30 g of polyethylene with a density of 0.962 g/cc was collected after a two hour polymerization at 80°C and 213 psig of ethylene. The Solvay & Cie catalyst exhibited an activity of about 1,500 g PE/g Ti.

Most likely, these early magnesium-containing Ziegler catalysts provided the necessary data that led to the discovery of additional very highly-active Ziegler catalysts, which were reported from about 1965–1980 from many different research laboratories around the world, with the common feature that all of these catalysts were based on magnesium and titanium compounds.

These developments provided several features that led to the introduction of linear low-density polyethylene (LLDPE) as a new type of commercial polyethylene on a large scale. In North America, Dow Chemical and Union Carbide were two of the first petrochemical companies that developed the commercial manufacture of LLDPE, with Dow employing a solution process and Union Carbide using gas-phase technology.

The important properties of these new catalysts were:

1. Sufficient activity so that catalyst residue remained in the finished polyethylene;
2. Good reactivity with other 1-olefins such as 1-butene, 1-hexene and 1-octene;
3. Easy control of the polyethylene molecular weight using hydrogen in the polymerization process as a chain transfer reagent;
4. The polyethylene produced possessed a relatively narrow molecular weight distribution with M_w/M_n values of about 3-4;
5. Methods were available to produce polyethylene particles with acceptable particle size and particle size distribution required for manufacture in the gas-phase process.

In order to better understand the importance of these new magnesium/titanium-based Ziegler catalysts it is necessary to discuss several United States patents issued to Farbwerke Hoechst, Mitsubishi Chemical Industries, Montedison Industries, and Union Carbide Corporation. These patents were issued in the United States between 1972–1981 and provided the catalyst technology that led to the introduction of LLDPE and HDPE as important commercial products that created significant growth in the polyethylene industry.

Based on the Solvay & Cie catalyst described above, other research groups across the globe reported the use of additional magnesium and titanium compounds that demonstrated the incorporation of a wide variety of other magnesium compounds and titanium/chlorine compounds into the catalyst preparation to provide high-activity ethylene polymerization catalysts.

The Hoechst patent described below, with a filing date in Germany of August 21, 1968, also represents another early development that may have initiated additional research efforts in other laboratories around the world. After about 1968, the identification of a large number of extremely active ethylene polymerization catalysts containing Mg/Ti/Chlorine was reported by many other research groups. Between about 1968 and the mid-1990s, the patent literature contained an enormous amount of new catalysts based on titanium and magnesium compounds from essentially all worldwide petrochemical companies involved in the polyethylene industry.

Some of the important catalysts that were reported in the patents described below will demonstrate the rapid pace of catalyst development that took place during this period.

(I)   "Process for the Polymerization of Olefins," United States Patent 3,644,318 issued on February 22, 1972 and assigned to Farbwerke Hoechst, of Frankfurt, Germany. The inventors of this patent were Bernard Diedrich and Karl Diether Keil.

This patent describes the reaction product obtained from the treatment of magnesium alcoholates such as $Mg(OEt)_2$ with tetravalent titanium compounds such as $TiCl_2(OR)_2$ or $TiCl_4$. Ethylene homopolymerizations and ethylene/1-butene copolymerizations were investigated and the polyethylene produced possessed a narrow molecular weight distribution as indicated from M_w/M_n values of 2–4 measured on the polyethylene samples. The polyethylene was described as especially suitable for the production of injection-molded articles.

Catalyst Preparation: 11 g of $Mg(OEt)_2$ was suspended in 50 ml of Diesel oil (boiling range of 130–160°C) and 200 ml of a 1 molar solution of $TiCl_2(O-isoC_3H_7)_2$ in cyclohexane was added. The suspension was refluxed for 7 hours, and then the suspension was washed six times with 150 ml portions of cyclohexane using a decantation procedure. The reaction product remained suspended in 250 ml of cyclohexane and the cyclohexane did not contain any titanium-containing species. A 10 ml portion of the suspended solid in cyclohexane contained 2.9 mmol (0.139 g) of titanium in the solid reaction product.

Ethylene Polymerization Results: 100 liters of diesel oil was added to a 150-liter reactor which was purged with dry nitrogen, and then the reactor was heated to 85°C; 400 mmol of triethylaluminum in 500 ml of diesel oil was added to the reactor, followed by 70 ml of the catalyst suspension described above, which was added to the reactor (20.3 mmol Ti, 0.97 g Ti), thus providing an Al/Ti molar ratio of 20. The polymerization was carried out for 7 hours at a total pressure of 7 atm with the gas phase containing 20% hydrogen; 41 kg of polyethylene was isolated, providing 11.7 kg of polyethylene/g of $Mg(OEt)_2$. This corresponds to about 42,300 g PE/g Ti or about 6.04 Kg PE/g Ti/hr. The polyethylene exhibited an M_w/M_n value of 2.9, which shows that the polyethylene had a very narrow MWD.

(II) "Catalyst and Process for Polymerization of Olefin," United States Patent 3,989,881 issued on November 2, 1976 and assigned to Mitsubishi Chemical Industries Ltd of Tokyo, Japan. The inventors of this patent were Kazuo Yamaguchi, Natsuki Kanoh, Toru Tanaka, Nobuo Enokido, Atsushi Murakami and Seiji Yoshida.

Catalyst Preparation: The catalyst was prepared using a 1-liter, four-necked flask equipped with a stirrer and under an inert atmosphere; 4.1 g of $TiCl_3(THF)_3$ dissolved into 130 ml of dry THF was added to the flask and the contents were heated to 60°C, and then 7.3 g of $MgCl_2$ complexed to THF was added to the flask and the contents were stirred for two hours to form a homogeneous solution, after which the solution was cooled to room temperature. Next, 500 ml of dry hexane was added dropwise to the solution to cause the precipitation of a powder. The powder was isolated by decanting the supernatant liquid and washing the powder three times with additional hexane followed by a decantation step. The powder was dried and the elemental analysis showed a composition of $Mg_4TiCl_{11}(THF)_6$ with a Ti content of 3.9 wt%.

Ethylene Polymerization Results: 500 ml of dry hexane was added to a 1-liter autoclave under a nitrogen purge, followed by the addition of 0.5 mmol of $Al(iso-C_4H_9)_3$ and 0.025 g of the catalyst powder described above. The autoclave was closed, heated to 90°C and the reactor was pressurized with 5 Kg/cm^2 of ethylene and 5 Kg/cm^2 of hydrogen. The polymerization was carried out for one hour and 232 grams of polyethylene powder were obtained. The activity of the catalyst powder under these polymerization conditions was 9,280 g PE/g catalyst. The titanium activity was 2.38 x 10^5 grams of polyethylene/g of titanium. The polyethylene possessed a relatively narrow molecular weight distribution as indicated by an M_w/M_n value of 2.9.

For comparison, a similar polymerization experiment was carried out in which the titanium compound was $TiCl_3(THF)_3$ and only 5.8 grams of polyethylene were produced, which corresponds to a titanium activity of 1.9 x 10^3. The catalyst containing titanium and magnesium was 125 times more active than the catalyst based on $TiCl_3(THF)_3$.

(III) "Catalysts for Polymerizing Olefins," United States Patent 4,124,532 issued on November 7, 1978 to Montedison S.p.A., Milan, Italy. The inventors of this patent were Umberto Giannini, Enrico Albizzati, and Franco Pirinoli.

This patent describes a series of complexes based on $TiCl_3$, $TiCl_4$ and anhydrous $MgCl_2$ with a wide variety of organic electron donors such as ethers, esters, alcohols, amines, and nitriles. The X-ray diffraction patterns and infrared spectra indicated that the isolated Mg/Ti/electron donor complexes were pure compounds.

Catalyst Preparation: (From Example 2); The catalyst was prepared by dissolving 6.92 g of anhydrous $MgCl_2$ into 100 ml of dry ethyl acetate and 7.48 g of $TiCl_3$ (hydrogen reduced grade) was dissolved into 72 ml of dry ethyl acetate. The two solutions were combined into one flask and the resulting solution was stirred for 4 hours at 60°C. A light green powder was isolated by evaporation of the solvent. The elemental analysis indicated a solid with the composition of $Mg_3Ti_2Cl_{12}$(ethyl acetate)$_7$ with Ti = 7.9 wt%.

Ethylene Polymerization Results: 1000 ml of dry heptane was added to a 3-liter stainless steel autoclave under a nitrogen purge followed by the addition of 2 ml of Al(iso-C_4H_9)$_3$ and 0.018 g (0.0014 g Ti) of the catalyst powder described above. The autoclave was closed, heated to 85°C and the reactor was pressurized with 9 atm of ethylene and 4 atm of hydrogen. The polymerization was carried out for four hours and 290 grams of polyethylene powder were obtained. The activity of the catalyst powder under these polymerization conditions was 4,030 g PE/g catalyst/hour or 51.8 Kg PE/g Ti/hr.

(IV) "Impregnated Polymerization Catalyst, Process for Preparing, and Use for Ethylene Copolymerization," United States Patent 4,302,565 issued on November 24, 1981 to Union Carbide Corporation, New York, NY (USA). The Inventors were George L. Goeke, Burkhard Wagner and Frederick J. Karol. U.S. Patent 4,302,566 issued on the same date to the same inventors describes additional gas-phase polymerization data. This patent describes the impregnation of a tetrahydrofuran solution containing $TiCl_4$ and anhydrous $MgCl_2$ into the internal pore structure of an amorphous silica support in order to prepare a suitable catalyst supported on silica that could be utilized in the Union Carbide fluidized-bed, gas-phase process (trademarked as the UNIPOL Process) that was being used commercially at the time to prepare HDPE with chromium-based catalysts. It is important to note that the role of the silica for the $MgCl_2/TiCl_4/THF$ complexes impregnated into silica was for the silica to act as a template to produce polyethylene particles with acceptable particle size and particle size distribution suitable for the gas-phase process.

The silica supported catalyst was prepared in a multi-step synthesis.

(a) An amorphous silica, such as Davidson Grade 952 silica, is dried in a nitrogen-purged fluid-bed glass column at about 800°C for several hours. This dry silica is then added to a flask fitted with a stirrer and under an inert atmosphere, isopentane is added to the flask to provide a slurry, and then the silica is treated with 4 to 8 wt% of triethylaluminum. The isopentane is removed under a nitrogen purge to produce a free-flowing powder.

(b) In a 12-liter flask equipped with a mechanical stirrer are placed 41.8 g (0.439 mol) of anhydrous MgCl2 and 2.5 liters of dry tetrahydrofuran (THF), and the contents of the flask are heated to 60°C. Next, 27.7 g (0.146 mol) of $TiCl_4$ is added to the flask dropwise over ½ hour in order to completely dissolve the solids. Finally, a solid material is isolated by crystallization or precipitation with the elemental composition of $Mg_3TiCl_{10}(THF)_{6.7}$ and is described as a catalyst precursor.

(c) In a similar flask as described above are placed 146 g of the catalyst precursor and 2.5 liters of dry THF, and the contents of the flask are heated with stirring to 60°C to form a solution. Next 500 g of dry silica described in step (a) are added to the flask and stirring is continued for ¼ hour. Finally, the THF is removed with a dry nitrogen purge to form a free-flowing powder with the same particle size and particle size distribution as the starting silica.

(d) Finally, the silica-impregnated catalyst intermediate described in step (c) is added to a flask under an inert atmosphere, and isopentane is added to the flask to form a slurry, and then an alkyl aluminum compound (AlR_3) is added to the slurry in sufficient quantity to provide an Al/Ti molar ratio of 4.5 to 7.5 depending on the type of aluminum alkyl used. Examples with the alkyl group ethyl, isobutyl, n-hexyl and n-octyl are provided. This finished silica-supported catalyst material is designated as an activated catalyst precursor that contains about 20 wt% of the catalyst powder described in (b).

2.4.2   Gas-Phase Fluidized-Bed Polymerization

The activated catalyst precursor was evaluated in a continuous, gas-phase fluid-bed reactor to produce a linear low-density ethylene/1-butene copolymer (LLDPE) with Melt Index ($I_{2.16 Kg}$) of 2.9 and a density of 0.921 g/cc, utilizing a 1-butene/ethylene and hydrogen/ethylene gas-phase molar ratio of 0.375 and 0.266, respectively, and a production rate (space time yield) of 5.3 lbs. PE/hr/ft^3 of reactor volume. The granular polyethylene had a residual titanium level of 3–5 ppm and high bulk density of 26.2 lbs./ft^3 when the catalyst precursor was preactivated with Tri(n-hexyl) aluminum at an Al/Ti molar ratio of 6.6.

A summary of the activity of the Mg/Ti-based catalysts identified in the four patents summarized above is shown in Table 2.3.

The activity of these second generation Ziegler catalysts is 10–100 times more active on a titanium basis than the first-generation Ziegler catalysts, which created significant growth opportunities in the polyethylene industry.

Table 2.3 Titanium-based activity of various second generation Ziegler catalysts.

U.S. Patent Example (year issued)	Titanium-based Activity Kg PE/g Ti/hr[a]	Company/Location
I (1972)	6.0	Hoechst; Frankfurt, Germany
II (1976)	240	Mitsubishi; Tokyo, Japan
III (1978)	52	Montedison; Milan, Italy
IV (1981)	250	Union Carbide; New York, USA

[a] Ethylene pressure in these experiments ranged from about 70–130 psi of ethylene with various levels of hydrogen.

2.4.3   Impact of High-Activity Mg/Ti Ziegler Catalysts on the Polyethylene Industry

In early 1980, Union Carbide expanded its manufacturing facilities in Seadrift, Texas, to manufacture the new grades of polyethylene summarized in Table 2.1. In addition, Exxon Chemical Company and Mobil Chemical Company were early licensees of the Union Carbide UNIPOL process in order to construct polyethylene manufacturing complexes in Saudi Arabia (Jubail, Saudi Arabia, was an Exxon site and Yanbu, Saudi Arabia, was a Mobil site), which came on stream in the early 1980s. In addition, Exxon also expanded its polyethylene manufacturing facilities in Baytown, Texas, while Mobil expanded its polyethylene manufacturing complex in Beaumont, Texas. Most of the polyethylene manufactured in Saudi Arabia was used to service European markets.

2.4.4   Overview of Particle-Form Technology

A subject that applies to polyethylene manufactured by either the slurry or gas-phase process, where the polyethylene is produced as a solid particle containing the catalyst residue, is the morphology of the solid particle. This discussion applies to any catalyst type regardless of the transition metal (Ti, Cr, Zr) used to prepare the catalyst.

2.4.4.1   Historical Introduction

The development of transition-metal-based, low-pressure ethylene polymerization technology resulted in the introduction of new process methods used for the commercial manufacture of polyethylene. Two of these low-pressure processes were the slurry polymerization process and the gas-phase process, both of which produce granular polyethylene particles in which the polyethylene is removed directly from the process as solid particles, which is referred to as particle-form technology.

The term "particle-form technology" is the process of forming granular polyethylene particles that replicate the shape of the original catalyst particles. Hogan and Banks at Phillips Petroleum Company initially developed particle-form technology in the mid-1950s for chromium-based catalysts. However, the process did not become commercial until about 1961, as the first commercial-scale plant for the manufacture of HDPE started up on December 31, 1956 used a solution process in which the polyethylene is manufactured as a solution at high temperature (ca. above 135°C). The first commercial plant operating at lower temperature in which the polyethylene particle remains as a solid particle (slurry process) started up in 1961 [43].

Hogan and Banks prepared chromium-based catalysts that were supported on silica by reacting a chromium compound with a porous silica at elevated temperatures. A strong chemical interaction between the chromium compound and the silica surface resulted [43]. The silica utilized for the chromium-based catalyst is an amorphous material of high porosity. Such silicas have an average particle size of 20–90 μm, a pore volume of ~1.5 cc/g and a surface area of ~300 m²/g, which produce polyethylene particles with a suitable size. During the polymerization process, the catalyst particle is suspended in an organic liquid (for example, isobutane), which does not dissolve the polyethylene. Hence, one silica-supported catalyst particle generates one polymer particle in which the original silica particle has been sufficiently fragmented to form residual catalyst particles less than 0.1 μm in size within the polymer particle. This catalyst-fragmentation process eliminates visible catalyst residue in the finished polyethylene. These small fragments of catalyst particles are encapsulated into the granular polymer particle. Because of the high activity of the Phillips catalyst, catalyst removal steps are not required as part of the polyethylene manufacturing process and the polyethylene is removed from the organic liquid by filtration. Polymer particles are approximately 500–2,000 microns in size.

In the gas-phase process, the second-generation, high-activity Ziegler-Natta catalysts also employ particle-form technology. Therefore, the particle size, the particle size distribution and the porosity of supported second generation Ziegler-Natta catalysts need to be controlled over a certain range to produce granular polyethylene particles of a suitable size in the gas-phase process. Such requirements improve the reliability of these catalysts in the gas-phase polymerization process. For example, very small catalyst particles produce very small polymer particles which may cause reactor fouling, while polymerization on very large catalyst particles generates a high level of heat within the growing polymer particle that can melt polymer particles, and thus foul the polymerization process.

Therefore, suitable supports for second generation Ziegler-Natta catalyst systems must meet similar requirements in order to perform satisfactorily.

The particle size of the finished catalyst precursor should range from about 1–250 μm (a preferred range of 10–100 μm). Spherical catalyst precursor particles are preferred but not necessary. Catalyst particles should provide granular polymer particles of 500–2000 μm in diameter. Polymer particles of this size are optimal for the operation of the slurry and gas-phase polymerization processes.

It is important that the supported catalyst particles possess sufficient porosity so that the active titanium species are uniformly distributed throughout the entire volume of the support. This is necessary so that the catalyst particles fragment uniformly during the polymerization process and produce catalyst residue particles less than 0.1 μm in size. Insufficient support fragmentation can produce visible inorganic gels in the finished polymer products and must be avoided. Catalysts with a surface area of 10–300 m^2/g have sufficient porosity so that the original catalyst particle will fragment in an acceptable manner.

2.4.5   Growth of the Polymer Particle

Early research carried out with first generation Ziegler catalysts also aroused a great deal of interest in understanding the growth of the polymer particle. It was found very early on in Ziegler research that the polymer particle possessed the same shape as the original catalyst particle, which in the Ziegler research area was referred to as particle replication. Consequently, industrial laboratories explored methods that could be used to control the particle morphology (morphology referring to the size, shape, and porosity) of the Ziegler catalyst particle such as solid $TiCl_3$. Early research findings in particle growth and replication were summarized by Boor [44], Wunderlich [45] and Chanzy et al. [46].

Depending on the preparation method, $TiCl_3$ exists in four crystalline modifications designated alpha, beta, gamma and delta. Various forms of $TiCl_3$, which are prepared by reduction with aluminum alkyls, have unique architectures. These forms are not large single crystals, but are large secondary particles which are actually made up of much smaller primary particles that are 20-40 microns in diameter and possess a nearly spherical shape but have porosity due to empty octahedral holes in the lattice [47]. This type of $TiCl_3$ is shown in Figure 2.9.

Figure 2.9 illustrates the titanium atoms in the porous lattice as the solid dots with the lines connecting the solid dots outlining one Ti_8 unit. The Cl ions are not shown. The lattice defects are shown in Figure 2.9 as small squares that represent octahedral holes which provide the lattice with sufficient porosity to allow ethylene to completely penetrate the lattice. The growth of polyethylene in this interior part of the lattice provides sufficient shear forces to shatter the lattice into small fragments, which leads to the growth of the polyethylene particle. Figure 2.10 illustrates the process where the small catalyst fragments are distributed into a large polymer particle and held together by a "web" type structure of polyethylene [48].

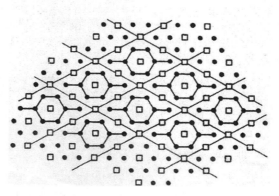

Figure 2.9 Illustration of a porous $TiCl_3$ particle in which the porous particles are made up of Ti_8Cl_{24} units containing lattice octahedral holes providing porosity on the molecular level. Reprinted from [44] with permission from Elsevier Publishing.

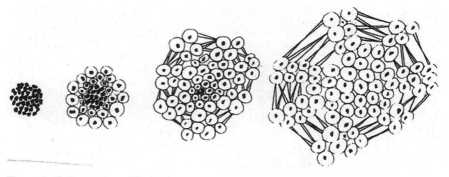

Figure 2.10 Formation of "cobweb" morphology in a polyethylene particle. Reprinted from [44] with permission from Elsevier Publishing.

Figure 2.10 shows that the original catalyst particle fragments into polyethylene subparticles based on the original size of the Ti_8 units in the solid lattice, thus causing the complete disintegration of the catalyst particle.

Burkhard E. Wagner and coworkers at Union Carbide characterized the fragmentation of silica-supported catalysts [49]. They proposed that catalysts supported on silica undergo a similar fragmentation pattern in which the silica pore volume fills with polyethylene and the shear forces of the particle growth fragments the silica particle into microspheroidal aggregates of 0.05–0.1 microns in diameter.

2.4.6   Catalyst Polymerization Kinetics and Polyethylene Particle Morphology

The polyethylene particle morphology (particle shape, size and density) is an important property for high-activity Ziegler catalysts used in the gas-phase, fluidized-bed process. As discussed previously in patent description IV, U.S. Patent 4,302,565 issued to Union Carbide Corporation on November 24, 1981, magnesium/titanium complexes discovered in the 1970s (patents described in II and III above) were impregnated into silica in order to provide a catalyst system that could be injected into the gas-phase process as a solid material.

The silica provided a suitable particle size for the catalyst so that the catalyst could be easily fed into the reactor. In this case, the silica is used primarily as a host material to deposit the Mg/Ti/Cl/THF compound into the silica as a template to grow polyethylene particles. The polymerization characteristics of the solid Mg/Ti/Cl/THF complex are similar to the silica-supported material, suggesting that the role of silica after the complex has been deposited into the silica pore structure is to provide finished polyethylene catalyst particles that produce polyethylene particles with acceptable morphology.

The removal of the heat of polymerization from the polymer particle to the fluidization gas is an important factor in forming polymer particles with good morphology. Therefore, the initial polymerization rate of the catalyst particle during the first few minutes after the catalyst particle is introduced into the reactor determines how well the silica particle is fragmented during the early stages of the polymerization process.

If the initial polymerization rate of the catalyst is too high, then the efficient transfer of the heat of polymerization within the catalyst particle to the fluidization gas is prevented and the silica fragmentation process produces a large polymer particle (perhaps partially melted) with poor morphology that interferes with the fluidization process within the fluidized-bed reactor. If the polymer particle becomes overheated, particle agglomeration will foul the reactor and destroy the fluidization process. In a severe case, extremely large polymer particles will form polyethylene logs, shutting down the reactor. Many high-activity Mg/Ti-containing catalysts exhibit a very high initial rate of polymerization that decays very rapidly with time in the presence of sufficient levels of 1-butene, 1-hexene and 1-octene to produce LLDPE. This type of kinetic profile makes it more difficult to produce granular polyethylene particles with acceptable morphology in the gas-phase reactor.

Therefore, the overall catalyst productivity of commercial catalysts in the gas-phase process is limited by constraints to only produce granular polyethylene particles with good particle morphology. Both the catalyst composition and the polymerization conditions within the gas-phase reactor need to be controlled to achieve the acceptable particle morphology in order that the commercial process may operate without process problems (i.e., development of fused aggregates of polyethylene inside the polymerization volume or fouling of the reactor walls, the reactor bottom gas distribution system or the product discharge system).

Ethylene partial pressure within the gas-phase process is an important variable used to control particle morphology. For the manufacture of LLDPE, ethylene partial pressure is limited to a value that maintains acceptable particle morphology. This constraint lowers catalyst productivity. A commercial catalyst with a productivity of at least 1,500 to 2,000 g PE/g of catalyst is necessary for a commercial reactor. One important goal in polyethylene research is to provide a better catalyst system where catalyst productivity may be increased, yet maintaining acceptable particle morphology. Catalysts with productivities of > 3,000 g PE/g catalyst are preferred.

2.4.7 Magnesium-Containing Compounds that Provide High-Activity Ziegler Catalysts

A wide variety of different magnesium compounds have been investigated to produce highly-active ethylene polymerization catalysts. The characteristics of the polymers produced with highly-active Mg/Ti-based catalysts are determined by the type of magnesium-containing compound used in the catalyst preparation. But the important characteristics of the finished solid catalyst material include polymerization kinetic profile, particle size, particle size distribution, and particle shape (spherical or irregular).

The finished catalyst material can be prepared using a variety of methods such as precipitation reactions, ball-milling techniques or spray-drying methods. Preparation temperatures, addition rates, holding times and stirring rates all affect the nature of the finished catalyst. If the finished catalyst is not prepared on silica, then the magnesium compound serves the dual role of a support and an activity enhancer. Examples of magnesium compounds (R = alkyl; X = halide) that have either been used as supports or as precursors to form the supports include: (1) magnesium chloride, $MgCl_2$ (anhydrous, hydrated, alcohol adducts); (2) magnesium alkoxides, $Mg(OR)_2$; (3) magnesium alkyls, MgR_2; (4) mixtures of magnesium and aluminum alkyls, MgR_2/AlR_3; and (5) Grignard reagents, RMgX. Typically, magnesium alkyls will react with the titanium chloride compound ($TiCl_4$ or $TiCl_3$) during the preparation of the catalyst precursor to form the high-activity catalyst. In some cases a specific Mg/Ti/Cl/electron donor crystalline compound may be formed during the catalyst preparation that can be characterized with a crystalline structure, or a more complex, less defined solid material is isolated, as is the case with ball-milling methods.

2.4.8   Additional Preparation Methods for Catalyst Precursors

Patent examples I–IV above involved the preparation of high-activity Mg/Ti Ziegler catalysts using either $MgCl_2$ or $Mg(OEt)_2$ as the magnesium source. However, a wide variety of additional catalysts can be prepared using magnesium alkyl compounds. Some of these methods are discussed below.

The preparations involve a magnesium alkyl (MgR_2, MgR_2/AlR_3 or RMgX) and a titanium-chloride compound as the source of Ti, but mixtures of a titanium chloride compound and a titanium alkoxide compound have also been used.

2.4.8.1   Reduction of $TiCl_4$ with Organomagnesium Compounds

Two early examples of providing a high-activity ethylene polymerization catalyst by reducing $TiCl_4$ with a magnesium-alkyl-containing compound are illustrated below.

Shell scientists R. N. Haward, A. N. Roper and K. L. Fletcher [41] published a paper in 1973 that demonstrated the importance of reducing $TiCl_4$ with a wide variety of Grignard reagents such as butyl magnesium halides, phenyl magnesium halides and alkyl magnesium alkoxides to prepare highly-active Ziegler type catalysts. The data clearly showed that the reduction of $TiCl_4$ with these organomagnesium compounds produced ethylene polymerization catalysts that were more active, by a factor of ten, than the first generation Ziegler catalysts. Characterization of the catalyst precursors suggested that titanium was present as Ti(III) and Ti(II), with catalysts containing 41% Ti(II) retaining very high ethylene polymerization activity. In addition, X-ray diffraction data on the catalyst precursors and elemental analyses of the catalyst prepared with butyl magnesium chloride had the composition $TiCl_{2.4}$:1.5 $MgCl_2$: 0.6 (dibutylether).

The reduction of $TiCl_4$ with diphenylmagnesium to produce highly-active ethylene polymerization catalysts was reported by scientists L. Petkov, R. Kyrtcheva, Ph. Radenkov and D. Dobreva [38,39] at SEC Neftochim Research Centre in Bourgas, Bulgaria. This catalyst exhibited an activity of about 30,000 g PE/gTi for a 0.25 hr polymerization at 75°C and ethylene pressure of 5 kg/cm^2.

2.5   Catalysts Prepared on Silica

Amorphous silica may be obtained commercially over a broad range of product specifications from a number of chemical companies across the globe. A typical silica suitable for ethylene polymerization catalysts has a particle size of between about 5–200 microns, a pore volume of about 1.4–3.0 cc/g and a surface area of 100–500 meter^2/g.

Amorphous silica is an excellent support for ethylene polymerization catalysts for the slurry polymerization process and the gas-phase process where the particle size of the finished polyethylene granular particle is an important consideration for both maintaining excellent reactor continuity, thus avoiding costly reactor shutdowns, and providing a granular polyethylene solid that can be conveyed easily within the polyethylene manufacturing facility before the polyethylene is converted into the common large pellets for shipping to customers. There are two techniques that are used to prepare a high-activity, magnesium- and titanium-containing catalyst into silica: a physical impregnation technique and a chemical impregnation technique.

2.5.1   Physical Impregnation of a Soluble Mg/Ti Precursor into the Silica Pores

As discussed previously, U.S. Patents 3,989,881 and 4,125,532, contain a large number of magnesium/titanium/chlorine/electron donor complexes that are suitable candidates for the preparation of a wide variety of high-activity catalysts for ethylene polymerization catalysts supported on silica, using the physical impregnation method. In this method, the Mg/Ti/Cl/electron donor complexes are prepared in the particular electron donor solvent and then this solution is added to silica, after which the electron donor solvent is evaporated to cause the solid complex to be deposited into the silica pore structure. The silica only acts as a carrier of the Mg/Ti/electron donor complex, which becomes the active species once the complex is reacted with an aluminum alkyl.

An example of the physical impregnation method is provided by U.S. Patent 4,302,565, which was discussed above. The dissolution of a magnesium compound and a titanium compound in a suitable solvent, usually an electron donor solvent such as tetrahydrofuran (THF), can lead to the formation of a new species (catalyst precursor) that provides a highly-active Ziegler system when activated. Removing the solvent at elevated temperatures in the presence of a support material such as silica will precipitate the precursor into the pores of the silica. In this method it is necessary that the precursor remains dissolved until the amount of remaining solution is equal to or less than the pore volume of the porous silica support. If this is achieved, then the particle size distribution of the silica material is the same before and after the precursor has been impregnated. If compounds precipitate outside of the silica pores, then the precursor material will contain a variety of solid catalyst particles of various sizes that will most likely lead to the fouling of the reactor and loss of reactor continuity, causing a reactor shut-down. Very small catalyst particles are particularly damaging, as such particles will produce small polymer particles (polymer fines) during the polymerization process. Such polymer particles tend to adhere to the reactor walls and will foul the polymerization process. This is particularly important in the gas-phase polymerization process.

2.5.2   Chemical Impregnation of Silica

In this method, the silica is used as a reagent in which a magnesium and/or aluminum alkyl compound is reacted with the silica surface using a non-coordinating solvent such as isopentane or hexane, providing a silica-supported magnesium alkyl intermediate that may be treated with an electron donor compound, followed by $TiCl_4$, which is a liquid source of titanium that is soluble in non-coordinating solvents. In this method, care must be taken to use only the amount of magnesium alkyl that will be fixed to the silica surface without any residual magnesium alkyl remaining in the non-coordinating solvent. The patents discussed in V–IX below demonstrate the versatility of supporting high-activity Mg/Ti catalysts onto silica using the chemical impregnation method.

An early patent that describes the chemical impregnation of silica was filed on August 1, 1979 and is summarized below in (V).

(V)   "Polymerization Catalyst," United States Patent 4,263,171 issued on April 21, 1981 to Mitsuzo Shida, Thomas J. Pullukat and Raymond E. Hoff and assigned to Chemplex Company, Rolling Meadows, IL (USA).

This patent describes the preparation of silica-supported catalysts in which a dried silica was reacted with a mixture of magnesium and aluminum alkyls (heptane solution of dibutylmagnesium/triethyl aluminum) followed by treatment with $TiCl_4$ or a catalyst in which the dried silica was treated with only a magnesium alkyl (butyl, ethyl magnesium – BEM) followed by treatment with $TiCl_4$.

In Example 1, Davison Grade 952 silica was dried at 600°C for five hours in a glass column in which the silica was fluidized with a nitrogen flow. Then, 2.2 g of dry silica contained in a small flask equipped with a magnetic stirring bar was prepared. The contents of the flask were cooled to 0°C, followed by the addition of 7.0 mmol of dibutylmagnesium and 1.08 mmol of triethyl aluminum dissolved in 13.8 ml of heptane, and the contents of the flask were stirred for 30 minutes. Next, 0.75 ml (6.7 mmol) of TiCl$_4$ was added to the flask and stirring continued for another 30 minutes. Finally, the flask was placed into an oil bath heated to 90°C and the contents of the flask were dried using a nitrogen purge to yield a dark brown free-flowing powder.

The ethylene polymerization of this catalyst was carried out in an autoclave reactor at 221°F in isopentane as the slurry solvent in the presence of triisobutylaluminum as cocatalyst and 50 psig of hydrogen and sufficient ethylene to achieve a total reactor pressure of 550 psig. The catalyst activity was 10,540 g of PE/g of catalyst/hr, which corresponded to an activity of 146,000 g PE/g Ti/hr. The granular polyethylene product obtained was considered suitable for a particle-form slurry process such as the Phillips slurry process. The polyethylene sample displayed a Melt Index (I$_{2.16\,kg}$) value of 0.70 and a High Load Melt Index (I$_{21.6\,kg}$) value (HLMI) of 31 with a HLMI/MI ratio of 45, which indicates that the polyethylene molecular weight distribution was of an intermediate value.

This patent also provided an example where an excess of butyl, ethyl magnesium was added to a Davison Grade 952 silica dried at 850°C which was slurried in heptane with the excess magnesium compound extracted using a decantation step after allowing the magnesium-supported silica to precipitate. The sample was washed several times with dry heptane using the same decantation procedure. After this step, the magnesium-supported intermediate was reacted with an excess of TiCl$_4$ and the excess TiCl$_4$ was removed with a similar heptane washing procedure. The decantation procedure was needed to insure that the catalyst was completely contained within the silica pores.

(VI) "Catalyst and Method," United States Patent 4,335,016 issued on June 15, 1982 to Robert A. Dombro and assigned to Chemplex Company, Rolling Meadows, IL (USA).

This patent teaches a chemical impregnation of silica method where a dried silica is slurried into a non-coordinating solvent such as heptane, reacted with a heptane solution of butyl, ethyl magnesium (BEM) to provide a first intermediate material that is then reacted with a silicon compound with the general formula (EtO)$_n$SiMe$_{4-n}$, where n is 1 to 4 to provide a second intermediate material. This secondary intermediate material is then reacted with $TiCl_4$ to provide a silica-supported catalyst. The silicon-containing compound reacts with the silica/magnesium alkyl and does not produce any solid material that is not impregnated into the silica.

In Example 12 of this patent, 10 g of silica (pore volume of 1.7 cc/g and surface area of 300 meter^2/g) previously dried at 1100°F was added to a reaction vessel containing 60 cc of pure n-heptane. Next, 22.5 mmol of n-butyl, ethyl magnesium (BEM) was added to the reaction flask followed by 19.8 mmol of $(EtO)_2SiMe_2$ and then 59 mmol of $TiCl_4$. The contents of the flask were heated to 90–100°C. The finished silica-supported catalyst was isolated by decanting the supernatant liquid, and then repeating the decantation step with several washings of the silica with additional heptane, and isolating a dry free-flowing powder after removing the residual heptane with a nitrogen purge. The elemental analyses of the finished catalyst was 4.0 wt% Ti; 18.4 wt% Cl providing a Cl/Ti molar ratio of 6.

The catalyst was evaluated in ethylene homopolymerization and copolymerization (using 1-butene as comonomer) experiments using isobutane as diluent, a polymerization temperature of 220°F, 550 psig total reactor pressure, and 20 psig of hydrogen with tri-isobutyl aluminum as cocatalyst. The polymerization experiments produced a final granular polyethylene powder. In an ethylene homo-polymerization test, the catalyst activity was 1576 g PE/g catalyst/hr that corresponds to 39,400 g PE/g catalyst under these polymerization conditions. In other polymerization tests, polyethylene molecular weight was varied over a Melt Index range of 0.4 to 10.4, where the higher MI ethylene/1-butene products were suitable for injection molding grades of polyethylene. These polymer samples were evaluated in a continuous particle-form slurry loop reactor.

(VII) "Polymerization Catalyst, Production and Use," United States Patent 4,558,024 issued on December 10, 1985 to Steven A. Best and assigned to Exxon Research & Engineering Co., Florham Park, NJ (USA).

This patent teaches a chemical impregnation of silica method where a dried silica is slurried into a non-coordinating solvent such as hexane, and then the silica is treated in a variety of methods to alter the chemical impregnation technique. A few examples are summarized in Table 2.4

Polymerization data showed that ethylene/1-butene copolymers were prepared over a density range from 0.919–0.943 and a Melt

Table 2.4 Examples of the chemical impregnation of silica method.

Example	Reaction Sequence[a]
1	SiO_2/EADC/BEM/$TiCl_4$
2	SiO_2/BEM/EADC/$TiCl_4$
3	SiO_2/BEM/$TiCl_4$/EADC
4	SiO_2/(BEM + 1-butanol)/$TiCl_4$/EADC

[a] EADC = ethyl aluminum dichloride; BEM = butyl ethyl magnesium.

Index range from 1.25–6.12, which would be suitable for polyethylene film and injection molding applications. The polyethylene molecular weight distribution was relatively narrow as indicated by Melt Flow Ratio values (HLMI/MI) reported between 26.0 and 33.5. The narrower MWD exhibited by the polymer samples would offer improved film and injection molding properties.

(VIII) "High-Activity Polyethylene Catalysts Prepared with Alkoxysilane Reagents," United States Patent 5,470,812 issued on November 28, 1995 to Robert I. Mink and Thomas E. Nowlin and assigned to Mobil Oil Corporation, Fairfax, VA (USA).

This patent describes an ethylene polymerization catalyst supported on silica using a chemical impregnation method where a dried silica is slurried in hexane and heated to 50–55°C; and then the silica/hexane slurry is treated, sequentially, with dibutylmagnesium (DBM), a tetraalkylorthosilicate, $(RO)_4Si$; [R is ethyl (TEOS) or n-butyl (TBOS)] and then treated with $TiCl_4$. The catalyst properties depend on the type and amount of tetraorthosilicate used in the catalyst preparation. The catalyst preparation took place in one vessel and did not require a filtration step.

Catalyst Preparation: 7.0 g of Davison Grade 955 silica, previously dried at 600°C for 16 hours, were placed into a 200 ml glass flask containing a magnetic stirring bar, and the flask was under a slow purge with dry nitrogen. Hexane (90 ml) was added to the flask and the flask was placed into an oil bath set at 50–55°C. Next, 7.0 mmols of dibutylmagnesium (DBM) in heptane were added to the flask and the contents of the flask were stirred for one hour; then 4.6 mmol of tetraethylorothsilicate (TEOS) were added to the flask and stirring was continued for one hour. Next, $TiCl_4$ (7.0 mmol) were added to the flask and stirring was continued for an additional hour. After this time, the solvents were removed by distillation to obtain 10.0 grams of a free-flowing catalyst powder that contained 3.26 wt% Ti.

Table 2.5 Polymerization data of various catalyst formulations.

Catalyst	Silane Reagent (mmol/g silica)	Catalyst Productivity (g PE/g cat/hr/100 psi ethylene)	HLMI (21.6 kg)	1-hexene mol%	MFR (HLMI/MI)
1	none	1830	23.5	2.7	41.5
2	TEOS (0.66)	4200	18.5	3.4	25.8
3	TBOS (0.66)	6850	21.0	3.3	26.0

Note: TEOS is tetraethylorthosilicate; TBOS is tetrabutylorthosilicate.

Polymerization: All polymerizations were carried out for 60 minutes in a 1.6-liter stainless steel autoclave containing 500 ml of dry heptane, 250 ml of 1-hexene and 3.0 mmol of cocatalyst (trimethylaluminum – TMA) at 85°C under a stirring rate of 900 rpm. The reactor contained 12–20 psig of hydrogen and ethylene was supplied to the reactor to increase the reactor pressure to 120 psig. The polymerization data of various catalyst formulations are summarized in Table 2.5.

Catalysts 2 and 3 show that both TBOS and TEOS provide high-activity catalysts with a titanium-based activity of 210,000 g PE/g Ti and 130,000 g PE/g Ti, for TBOS- and TEOS-based catalysts, respectively. The important attribute of these catalysts is the relatively narrow polyethylene molecular weight distribution, as indicated by MFR values of about 26.

(IX) "High Activity Polyethylene Catalysts Prepared with Alkoxysilanes," United States Patent 6,291,384 issued on September 18, 2001 to Robert I. Mink and Thomas E. Nowlin and assigned to Mobil Oil Corporation, Fairfax, VA (USA).

This patent teaches the importance of preparing the same catalysts discussed in patent (VIII) above on silica calcined at a temperature at or above 800°C in order to manufacture ethylene/1-hexene copolymers with a very narrow molecular weight distribution. The catalysts shown in Table 2.6 were evaluated in a continuous gas-phase fluidized-bed reactor with triethylaluminum as cocatalyst to produce LLDPE polyethylene samples with a density of 0.918 g/cc.

The polyethylene samples obtained from Catalysts 3 and 4 were compounded on a Brabury mixer with a suitable additive package and then the samples were fabricated into 1.0 Mil LLDPE film using a Gloucester extruder (430°F) operating with a 100 mil die-gap; 2:1 blow-up ratio and a 25-inch frost line height and operating at a fabrication rate of 250 lbs PE/hr. The film mechanical properties; dart

Table 2.6 High-activity polyethylene catalysts prepared with alkoxysilanes.

Catalyst	Silica Calcination Temperature °C	TEOS Loading (mmol/g silica)	MFR
1	600	0.66	26–27
2	865	0.66	23–24
3	600	0.44	30–31
4	800	0.44	26–27

Note: TEOS is tetraethylorthosilicate; MFR is HLMI/MI and directly correlates with the MWD of the sample.

Table 2.7 Properties of 1 mil LLDPE film.

Sample from	Catalyst 3	Catalyst 4
MI, dg/min	0.9	0.9
Density (g/cc)	0.917	0.917
MFR	30–31	26–27
FDA Ext wt%	3.7–4.0	3.4–3.7
DDI, F50, g	180–250	400–500
MDT, g/mil	300–360	400–475

drop impact (DDI, g) and machine direction tear strength (MDT, g/mil) found for the LLDPE sample produced with Catalyst 4, relative to Catalyst 3, are shown in the Table 2.7.

This data summarized the improved film toughness properties obtained from the LLDPE sample prepared with Catalyst 4.

2.6   Characterization of Catalysts Prepared with Calcined Silica, Dibutylmagnesium or Triethylaluminum and TiCl$_4$

As outlined in the patents discussed above, highly-active Ziegler catalysts may be prepared on silica via a chemical impregnation method from the interaction of magnesium alkyls or aluminum alkyls with porous silica that has been calcined over a range of temperatures typically from about 300°C to 800°C [50].

Supporting high-activity Ziegler catalysts on silica using a chemical impregnation method is more complex than the physical impregnation method. The metal alkyls react with the surface hydroxyl groups (Si-OH) to yield the corresponding alkane and a silica/metal intermediate. For example, dibutylmagnesium (DBM) reacts with silica to liberate butane and to form an intermediate in which a Mg-R group is chemically attached to the silica. The amount of DBM or triethylaluminum (TEAL) that can react with the silica depends upon the silica calcination temperature, as shown in Table 2.8. This data was obtained by reacting an excess of either DBM or TEAL with the corresponding silica which was slurried in hexane and heated to 55°C. After one hour, the solution was cooled to room temperature and the silica reaction product was isolated by filtering to collect the intermediate reaction product, which was washed several times with excess hexane and then dried to obtain a finished material, which was then analyzed for Mg or Al.

The results of these experiments are summarized in Figure 2.11 and Table 2.8.

The infrared spectra of the silicas that were calcined at 300, 600 and 800°C before and after treatment with DBM are shown in Figure 2.12. The silicas calcined at 600 and 800°C exhibit a single peak at 3746 cm^{-1}, which has been assigned to isolated hydroxyl groups on these silicas. The infrared spectrum of the silica calcined at 300°C contains both a singlet peak due

Table 2.8 Fixation of DBM or TEAL onto silica surface.

Silica calcination Temperature	Hydroxyl content mmol OH/g silica	DBM fixed mmol Mg/g silica	TEAL fixed Mmol Al/g silica
800°C	0.45	1.22	1.14
600°C	0.72	1.74	1.30
300°C	1.39	2.21	1.74

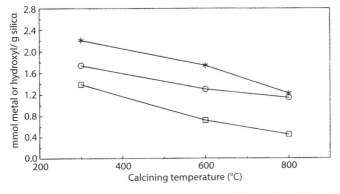

Figure 2.11 Fixation of Mg from Dibutylmagnesium or Al from triethylaluminum to silica calcined at 300, 600 or 800°C [50].

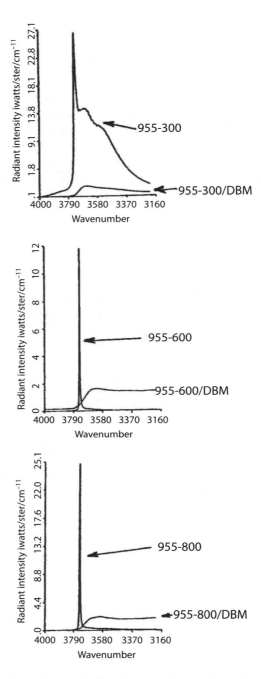

Figure 2.12 Infrared spectra in the hydroxyl region of silica calcined at 300, 600 or 800°C before and after treatment with Dibutylmagnesium [50].

to relatively isolated hydroxyl groups and a broad absorption from 3200 to 3840 cm^{-1} due to the interaction of neighboring hydroxyl groups on the silica surface (hydroxyl bonding interactions) leading to a broad peak. After the silicas were treated with DBM, the infrared data show that all isolated hydroxyl groups have reacted and that a few neighboring hydroxyl groups in the 300°C silica may remain.

The data in Table 2.8 show, as expected, the higher the silica calcination temperature, the lower the surface hydroxyl concentration and the Mg or Al content of the silica/metal alkyl reaction product. But the data also indicate that the amount of DBM or TEAL that can react with silica greatly exceeds the surface hydroxyl concentration on each of the silicas. Hence, additional sites on the silica surface are available for reaction with metal alkyls other than surface hydroxyl sites. Most likely the metal alkyl compounds coordinate to the silica surface through an electron pair on the oxygen atom of a siloxane (Si-O-Si) linkage. In addition, DBM reacts with each silica to a greater extent than does the TEAL.

Further evidence for a second site comes from the fact that DBM will also fix to silica in which the hydroxyl groups have been completely converted to trimethylsilyl groups as shown below.

$$Si\text{-}OH + Me_3SiNMe_2 \rightarrow Si\text{-}O\text{-}SiMe_3 + Me_2NH \qquad (2.7)$$

The Mg content from the interaction of silylated treated silicas, calcined at 300, 600 and 800°C, reacted with an excess of DBM as described above is shown in Table 2.9.

The data shown in Table 2.9 may be rationalized by suggesting that surface siloxane linkages on the silica surface can fix DBM to the surface through a lone pair of electrons on the available oxygen atom. Consequently, a silica containing hydroxyl groups and treated with excess DBM contains two different types of magnesium groups bonded to the silica through both the silica-hydroxyl group and a siloxane group.

The silica-supported metal alkyl intermediates described above were converted into ethylene polymerization catalysts by reacting each intermediate with an excess of TiCl$_4$, filtered to remove the silica-containing

Table 2.9 Amount of Mg from DBM fixed to silylated silica.

Silica Calcination Temperature (°C)	Fixed Mg mmol Mg/g silica
800	0.85
600	0.69
300	0.65

reaction product, washed with excess hexane and dried to produce finished catalyst precursors. Some polymerization data is shown in Table 2.10

The data in Table 2.10 indicate that Catalysts 1 and 3 that were treated with the silylation reagent to convert Si-OH groups to Si-O-SiMe$_3$ groups provided much more active catalysts after been treated with DBM and TiCl$_4$ than the non-silylated Catalysts 2 and 4. Catalyst 5 in which the silica was treated with the silylation reagent did not react with TiCl$_4$ and, therefore, had no polymerization activity, while Catalyst 6, which did contain Si-OH groups reacted with TiCl$_4$ and was not treated with DBM, provided a catalyst with very low activity. Consequently, Catalysts 1–4 treated with DBM were about 7 to 60 times more active than Catalyst 6, which was not treated with DBM.

The silicas calcined at 300, 600 and 800°C and then reacted with Me$_3$SiNMe$_2$ also exhibited no residual hydroxyl groups, as indicated by their IR spectrum [50]. However, when these silylated silicas were reacted with DBM, the silylated silicas had a Mg concentration of 0.65, 0.69 and 0.85 mmol/g of silylated silica, respectively, demonstrating the presence of a second site of attack for the fixation of the DBM molecule to the silica surface. The interaction of the silylated silica/magnesium intermediates (Catalysts 1 and 3 in Table 2.10) with TiCl$_4$ provides catalyst precursors with the higher activity.

In summary, the addition of a magnesium alkyl such as DBM, that is soluble in a non-coordinating solvent such as heptane, to a slurry of silica in a non-coordinating solvent results in the fixation of two different magnesium moieties onto the silica surface. These silica/magnesium intermediates may then be reacted with TiCl$_4$ or reacted sequentially with

Table 2.10 Polymerization data of catalysts prepared with various silica intermediates.

Data Point	Silica Calcination Temp. (°C)	Silylation step	Metal Alkyl	Ti wt%	Productivity kg PE/g Ti/hr/100 psi
1	300	yes	DBM	2.65	151.0
2	300	no	DBM	6.73	12.3
3	600	yes	DBM	0.78	191.0
4	600	no	DBM	5.93	22.8
5	600	yes	None	nil	nil
6	600	no	None	2.83	3.2

Polymerization conditions: 1.6 liter stainless steel autoclave, 400 ml hexane, 200 ml 1-hexene 85°C; 600 rpm; 3.0 mmol TEAL; one hour; 20 psig hydrogen, 120 psig total pressure; 20–50 mg catalyst.

an electron donor and TiCl$_4$ [51] to produce a Ziegler catalyst precursor impregnated onto the silica surface. Catalyst activities for these systems may be quite high. However, in this method care must be taken not to add more metal alkyl than can be fixed onto the silica surface, unless the excess metal alkyl in the supernatant layer is removed via filtration or decantation. Otherwise, upon addition of TiCl$_4$ not only will the silica/magnesium intermediate react with the TiCl$_4$, but so will the excess magnesium alkyl in the supernatant layer, leading to the precipitation of a non-silica-containing solid. In this case, the catalyst precursor would actually be a mixture of this precipitated non-silica-containing solid and the treated silica-containing solid, and must be avoided in order to produce a granular polyethylene particle with an acceptable particle size distribution.

2.6.1 Spray-Drying Techniques

Another technique used to prepare high-activity Mg/Ti-based solid catalyst precursors is the spray-drying method. In this method, the reagents are either dissolved into solution or melted at an elevated temperature and then sprayed through specially designed nozzles to form liquid droplets that are dried rapidly to create solid particles. Particle size and particle size distribution are controlled by the spray-drying conditions.

A method of producing spherically-shaped catalysts for either a slurry or gas-phase polymerization process is a spray-drying technique. In this technique a solution containing the catalyst components is spray-dried through a spray nozzle that creates individual liquid droplets in which the solvent used to form the solution containing the catalyst components is rapidly removed to form individual solid particles. These solid particles need to exhibit sufficient mechanical strength (i.e., resist both crumbling and compacting) so that they are not easily fragmented before they are used in a polymerization process, and an acceptable particle size and particle-size distribution that provides a polyethylene particle of a suitable size. For example, spray-dried catalyst particles in the 10–200 micron range possess a suitable particle diameter that provides finished polymer particles which are acceptable in a commercial process.

U.S. Patent 3,989,881 issued to Mitsubishi Chemical Industries Ltd. and U.S. Patent 4,124,532 issued to Montedison S.p.A., which were previously discussed, provide good examples of solutions that contain anhydrous MgCl$_2$ and TiCl$_3$ or TiCl$_4$ in an electron donor solvent that could be spray-dried into solid particles of suitable size for a slurry or gas-phase process. Scientists at Union Carbide found that a solution of anhydrous MgCl$_2$ and TiCl$_4$ in tetrahydrofuran (THF), that also contains a fumed silica sold commercially as CAB-O-SIL, could be spray-dried into a spherical solid particle for operation in a gas-phase reactor. In Example 1 of U.S. Patent 4,293,673 [66], a first solution containing 1.0 liter of THF, 71 g anhydrous magnesium chloride and 90 g of fused silica (CAB-O-SIL) with a particle size of 0.007 to 0.05 microns, and a second solution containing 13.4 g of anhydrous magnesium chloride and 8.9 ml of $TiCl_4$ dissolved into 0.8 liter of THF, were combined to provide a slurry where all components were soluble in THF, except the fused silica. This slurry was spray-dried into a drying chamber under a nitrogen atmosphere maintained at 112°C with an apparatus that had two spray nozzles, an annular ring diameter of 0.10 inches, and an atomization pressure of 10 psi. The solid particles collected had an average particle diameter of 25 microns. These solid particles were then slurried into isopentane, and an aluminum alkyl compound was added to the isopentane to achieve an Al/Ti ratio of 4 to 8 in order to preactivate the catalyst precursor particles; and then the preactivated particles were dried to a free-flowing powder by removing the solvents. The preactivated particles were utilized in a gas-phase, fluid-bed reactor to produce spherical granular polyethylene particles with a diameter of about 400–1,000 microns. Polyethylene particles with a diameter < 74 microns were usually less than 0.6 wt% and granular resin bulk density was about 21–28 lbs/ft^3 in most samples.

This type of spray-dried, preactivated catalyst particle can manufacture HDPE in a slurry polymerization process or both HDPE and LLDPE in a gas-phase process.

Another early example of utilizing a spray-drying technique for the production of spherical catalyst particles containing $MgCl_2$ and $TiCl_4$ with high catalyst activity is discussed in U.S. Patent 3,953,414 issued to Paolo Galli, Giovanni Di Drusco and Saverio De Bartolo on April 27, 1976 and assigned to Montecatini Edison S.p.A., Milan, Italy. These scientists spray-dried molten $MgCl_2$:$6H_2O$ through a 0.34 mm spray nozzle and then rapidly cooled the liquid droplets with a stream of nitrogen. After the spray-drying process, the particles were collected into a particle size of 53–105 microns that were then heated to 130°C to remove 2/3 of the hydrated water to form $MgCl_2$:$2H_2O$. Next, the $MgCl_2$:$2H_2O$ particles were reacted with an excess of liquid $TiCl_4$ at about 138°C, filtered, washed with additional $TiCl_4$, and then washed five times with heptane and dried to yield a solid catalyst particle. The finished catalyst particles possessed a good resistance to crumbling, a porous structure with a surface area of 33.7 meter^2/g, a pore size with a mean radius of 59 angstroms and an elemental composition of 2.95 wt% Ti, 60 wt% Cl, 20.5 wt% Mg and 2.85 wt% H_2O. The catalyst showed a very high activity of 41,000 g PE/g catalyst/hr (1.4 x

10^6 g PE/g Ti). The polymerization was carried out in a 4.5-liter autoclave containing 2.0 liters of heptane, 4 g tri-isobutyl aluminum as cocatalyst at 85°C, 5.5 kg/cm² of ethylene, 7.5 kg/cm² hydrogen with 0.00452 g catalyst powder. The 4 hr polymerization yielded 740 g of HDPE. The polymer particles were spherical with a 1–2 mm diameter.

2.6.2 Ball-Milling Techniques

Solids such as magnesium chloride are mechanically agitated with metal spheres (usually stainless steel balls) contained in a closed metal container. The magnesium solid may also be ball-milled in the presence of an electron donor. Usually the ball-milled solid is reacted with $TiCl_4$ to form the catalyst precursor.

Ball-milling is a technique in which solid crystalline materials are physically degraded over a prolonged period (5–300 h) by rotating or vibrating a closed container of stainless steel balls with a solid material such as $MgCl_2$. The crystalline defects produced by the ball-milling of the solid material are exhibited by significant changes in the X-ray diffraction pattern of the solid material. Galli *et al.* published a detailed paper on the mechanical grinding of anhydrous $MgCl_2$ containing $TiCl_4$ [52].

Figure 2.13 shows the X-ray diffraction pattern after 20 hours of milling $MgCl_2$ in the presence of $TiCl_4$. The lattice disorder of the milled solid is apparent from the shape of the X-ray pattern and is due to the disorder in the packing of the chloride anions. Ball-milling time is important in determining the amount of crystalline disorder obtained in the solid. Catalyst activity increases rapidly with ball-milling time up to about 20 hours (160 kg PE/g Ti) and reaches a maximum activity after about 80 hours (250 kg

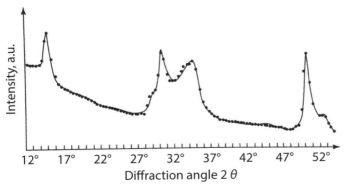

Figure 2.13 X-ray diffraction pattern of $MgCl_2$ (+$TiCl_4$) 20 hr activated (solid line) fitted by the proposed structural model. Reprinted from [52] with permission Elsevier Publishing.

PE/g Ti). Additional milling up to 200 hours has little effect on catalyst activity. Crystallite size also decreases with ball-milling time from ~100 Å after 20 hours to ~50 Å after 80–200 hours.

2.6.3   Characterization of High-Activity Ti/Mg-based Ziegler Catalyst Precursors

This section provides some specific examples of various catalyst precursors outlined above that have been characterized more completely to determine catalyst structure and other properties.

The interaction of titanium tetrachloride complexes and magnesium chloride complexes at various Mg/Ti molar ratios in THF produced crystalline $MgCl_2/TiCl_4/THF$ complexes, which were highly active ethylene polymerization catalysts when combined with an aluminum alkyl. The U.S. Patent 3,989,881 issued on November 2, 1976 to Mitsubishi Chemical Industries Ltd. [53] and U.S. Patent 4,124,532 issued on November 7, 1978 to Montedison S.p.A. [54] provide catalyst compositions for producing highly-active ethylene polymerization catalysts, but do not provide details of the structure of these complexes used for the catalysts. Mitsubishi showed that if $TiCl_4(THF)_2$ in THF was added to $MgCl_2(THF)_{1.5}$ in THF at a Mg/Ti molar ratio of 1.3 a solid material is formed, which when combined with tri-isobutylaluminum, produced an ethylene polymerization catalyst with an activity of 4.8 kg PE/g cat. The molecular weight distribution of the resin was relatively narrow, a M_w/M_n value of 2.9 [53].

In 1984, P. Sobota disclosed the structure of a crystalline material that was isolated from saturated solutions of $MgCl_2(THF)_2$ and $TiCl_4(THF)_2$ in THF at a Mg/Ti molar ratio of 2 that were mixed to form yellow crystals which were isolated [55]. The stoichiometry of these yellow crystals was shown to be $Mg_2TiCl_8(THF)_7$ formed via the reaction below.

$$TiCl_4(THF)_2 + 2\ MgCl_2(THF)_2 + THF\ (solvent) \rightarrow Mg_2TiCl_8(THF)_7 \qquad (2.8)$$

The structure of this complex contains the ionic species $[Mg_2Cl_3(THF)_6]^+$ $[TiCl_5(THF)]^-$, as shown by a X-ray crystal structure determination shown in Figure 2.14.

In this structure there is a di-octahedral magnesium cation in which the two Mg^{2+} atoms are bridged by three Cl atoms. The oxygen atoms of three THF molecules are coordinated to each Mg^{2+} atom. The Ti atom in the anion is pseudo-octahedrally coordinated with five Ti-Cl bonds and a Ti-O bond from one ligand of THF.

Karol and coworkers also discussed the structure and properties of the crystalline species $[Mg_2Cl_3(THF)_6]^+$ $[TiCl_5(THF)]^-$ [56].

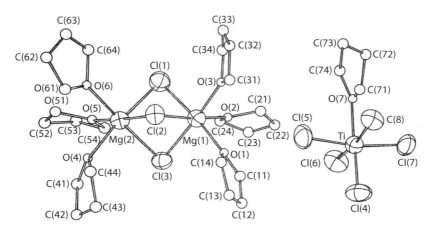

Figure 2.14 Crystal structure of $Mg_2TiCl_8(THF)_7$. Reprinted from [55] with permission from Royal Society of Chemistry.

Table 2.11 Results for complexes based on ethanol and benzonitrile.

Composition	Activity (gPE/gTi)
$MgTiCl_5(Ethanol)_6$	133,000
$Mg_3TiCl_{12}(benzonitrile)_7$	74,200

2.6.4   Additional Electron Donor Complexes

In U.S. Patent 4,174,429 [54], Montedison characterized a large number of Mg/Ti/Cl/electron donor complexes based on a wide variety of ligands such as ethyl acetate, ethanol, ethyl benzoate, benzonitrile, ethylene carbonate, n-butyl acetate and diethyl ether, using $TiCl_3$ prepared from hydrogen reduction of $TiCl_4$.

In Example 7 of the '429 patent a complex with the composition of $Mg_3Ti_2Cl_{12}(ethyl\ acetate)_7$ exhibited an ethylene polymerization activity of 167,000 gPE/gTi at 10 atm ethylene, 3 atm hydrogen at 85°C and a 4 hr polymerization time, which was 50 times more active than the titanium-only complex $TiCl_3$:ethyl acetate, demonstrating the importance of the magnesium compound in improving polymerization activity. Results for complexes based on ethanol and benzonitrile are shown in Table 2.11.

The structures of a wide variety of additional crystalline complexes based on titanium chlorides were discussed in a detailed publication by Greco and coworkers [57]. Complexes containing anions such as $(Ti_2Cl_{10})^{2-}$; $TiCl_6^{2-}$; and $TiCl_5[OP(C_6H_5)Cl_2]^{1-}$ were isolated and evaluated as ethylene polymerization catalysts.

2.6.5   Catalysts Based on Magnesium Diethoxide and TiCl$_4$

In the 1970s, L. L. Böhm of Hoechst AG in Frankfurt, Germany, made many significant contributions in preparing high-activity Mg/Ti catalyst [58]. Böhm investigated a heterogeneous catalyst prepared by adding solid Mg(OC$_2$H$_5$)$_2$ to liquid TiCl$_4$ and heating at temperatures > 100°C, followed by filtration and washing steps, which resulted in the formation of a solid powder containing 8.5 wt% Ti. A redistribution reaction was proposed as shown in Equation 2.9.

$$Mg(OC_2H_5)_2 + TiCl_4 \rightarrow MgCl_2 + Ti(OC_2H_5)_2Cl_2 \qquad (2.9)$$

Specific reaction conditions determined the extent of reaction that had taken place. The particle size distribution of this powder is shown in Figure 2.15 and was determined to be 7–30 μm with a median of 18.5 μm. The surface area of the powder was 60 m^2/g, which indicated that the powder had a high porosity with a pore volume of 1.16 cm^3/g.

Polymerization results showed that the high activity of this catalyst was due to an exceptionally high number of active sites (~70% of the titanium atoms), which was at least a factor of 10 higher than the early Ziegler catalysts. The particle size distribution of the polymer was identical to that of the catalyst precursor, with the median value of the polymer shifting to much higher values depending on the polymerization time. This data, illustrated in Figure 2.15, suggested that every catalyst particle formed one polymer particle.

Böhm's data obtained from electron microscopy also showed that the original catalyst particles were completely disintegrated during the polymerization process, as no catalyst residues were detected in the polymer. Hence, catalyst residue must be smaller than ~50 Å.

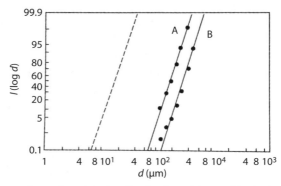

Figure 2.15 Comparison of particle size distribution of the catalyst (---) and the granular polyethylene for two polymerization times shown as lines A and B. Reprinted from [58] with permission from Elsevier Publishing.

Additional data reported by Böhm showed that the interaction of the triethylaluminum cocatalyst with the solid catalyst precursor took place in less than 5 minutes and that catalyst activity was a function of the Al/Ti molar ratio, with higher activity at an Al/Ti ratio of about 8–30. Longer reduction times of up to four hours did not affect catalyst activity.

Catalysts based on Grignard reagents and bulky magnesium alkyls and $TiCl_4$ were developed by Shell research scientists R. N. Haward, A. N. Roper and K. L. Fletcher at Carrington Plastics Laboratory in the UK [60]. The reaction of Grignard reagents such as butylmagnesium chloride in diethyl ether or didodecyl magnesium in iso-octane with $TiCl_4$ generated highly-active catalysts in which the titanium center had been reduced to Ti(II) and Ti(III) by the magnesium compound. Haward prepared a variety of catalysts by slowly adding $TiCl_4$ to a solution of the magnesium alkyl compound at 0–60°C. Their method involved magnesium alkyl compounds that were soluble in hydrocarbons by using magnesium alkyls with large alkyl groups or by solvating a magnesium alkyl with a suitable ether or amine at a 1:1 or 2:1 molar ratio. Hence, the addition of $TiCl_4$ to such solutions produced a solid catalyst component. For these systems, aluminum alkyl cocatalysts with large alkyl groups such as tri-n-octyl aluminum were three times more active than catalysts activated with triethylaluminum.

Two important conclusions from this investigation were: (1) catalyst systems prepared by reducing $TiCl_4$ with organomagnesium compounds are much more active (factor of 3–10) than catalysts prepared by reducing $TiCl_4$ with aluminum alkyls, and (2) organomagnesium compounds reduced the titanium center to up to 41% Ti(II) and the catalyst systems still maintained a very high activity.

2.6.6   Spherical Magnesium-Supported Catalyst Particles

Scientists at Montedison prepared a spherical support of $MgCl_2 \cdot 3C_2H_5OH$ by combining anhydrous $MgCl_2$, anhydrous ethanol, Vaseline oil, and silicone oil and then heating the mixture at high stirring speeds to 120°C until the $MgCl_2$ was completely dissolved [61,62]. Eventually the mixture was discharged into a vessel containing anhydrous heptane and cooled. After filtering the mixture and drying the solid, the spherical support was then subjected to a thermal treatment until a partial dealcoholation was obtained. A residual ethanol content of ~35 wt% was obtained, corresponding to an ethanol:Mg molar ratio of 1.1:1. This spherical support was treated with an electron donor (ethyl benzoate or succinates) and $TiCl_4$. This procedure produced a catalyst precursor with spherical particles. The morphology of the polymer particles using this catalyst precursor was also spherical [63].

2.6.7   Catalysts Prepared with Grignard Reagent/TiCl$_4$ with and without Silica

Catalysts based on ethylmagnesium chloride in THF and TiCl$_4$ were prepared in the presence and absence of silica [64]. The silica was a Davison (W. R. Grace) Commercial Grade 952 silica with an average particle size of ~45 μm and was dried at 200°C. The silica was used as a template to control the particle size distribution of the catalyst precursor.

Catalyst Prepared with Silica: This catalyst was prepared in two steps. In the first step, ethylmagnesium chloride in THF (2.1 M) was added dropwise to the silica, which was slurried in hexane and stirred rapidly while under reflux conditions. Next, the solvents were removed by distillation to provide a white free-flowing powder. In the second step, the white powder was slurried in heptane containing TiCl$_4$ and the slurry was refluxed for 2 hr and then allowed to cool to room temperature. The silica-supported catalyst was isolated by filtration, washed with excess hexane and dried again to obtain a free-flowing powder.

Catalyst Prepared without Silica: This catalyst was prepared by adding ethylmagnesium chloride in THF (2.1 M) to hexane under refluxing conditions, and then solvents were removed by distillation to produce a solid residue. Next, dry heptane containing TiCl$_4$ was added to the solid residue and the slurry was heated to refluxing conditions, then the slurry was cooled and the reaction product isolated by filtering.

2.6.8   Polyethylene Structure

One common feature of ethylene/1-olefin copolymers prepared with high-activity Mg/Ti-based Ziegler catalysts is the nature of the intermolecular branching distribution obtained with these catalysts. The polymer composition consists of a heterogeneous distribution of the 1-olefin (usually 1-butene, 1-hexene or 1-octene) in the molecular structure. Hence, the 1-olefin content of individual polyethylene molecules varies over a wide range from very low levels to very high levels of 1-olefin content (i.e., ethyl branches obtained from the incorporation of 1-butene, butyl branches from 1-hexene or hexyl branches from 1-octene). This is due to variable ethylene/1-olefin reactivity ratios found from different types of active sites (referred to as multi-site catalysts) in these Mg/Ti-based catalysts. The high-activity Ziegler catalysts identified over the years differ from each other in the degree of heterogeneous branching distribution over a relatively small range. One example of a method used to determine the value of the various reactivity ratios in a high-activity Ziegler-type catalyst is discussed below.

2.6.9   Characterization of Reactivity Ratios in Multi-site Mg/Ti Catalysts

Solvent extraction methods were used by Y. V. Kissin to fractionate ethylene/1-hexene copolymers prepared with the Mg/Ti catalyst described above [64] in which ethylmagnesium chloride and $TiCl_4$ were used to prepare high-activity catalysts with and without silica.

Each catalyst was evaluated in ethylene/1-hexene copolymerization reactions at 80°C in the presence of the cocatalyst TEAL at various 1-hexene/ethylene ratios. The ethylene reactivity ratio r_1 was determined for each catalyst from the simplified copolymerization equation [65].

$$m_e/m_h = r_1 M_e/M_h \qquad (2.10)$$

where	m_e = mole fraction of ethylene in the copolymer
	m_h = mole fraction of 1-hexene in the copolymer
	M_e = molarity of ethylene in the slurry solvent
	M_h = molarity of 1-hexene in the slurry solvent
where	$r_1 = k_{ee}/k_{eh}$
	k_{ee} is the rate ethylene adds to an ethylene bound polymer chain
	k_{eh} is the rate 1-hexene adds to an ethylene bound polymer chain

The catalyst prepared on silica exhibited very high reactivity with 1-hexene as indicated by a r_1 value of 26, whereas the catalyst prepared without silica exhibited a r_1 value of 125. Hence, the silica-supported catalyst reacts approximately five times better with 1-hexene than the catalyst prepared without the silica.

In addition, the copolymer produced from each of these catalysts was fractioned into five different fractions using hexane, heptane and octane and the comonomer content was determined for each fraction. The results are shown in Table 2.12.

These data explain the relatively heterogeneous branching distribution of ethylene/1-olefin copolymers prepared with Ti/Mg Ziegler catalysts. For example, the catalyst prepared without silica contained at least five active sites that exhibited a range of r_1 values from ~19 to ~180, which varies by a

Table 2.12 Fractionation of ethylene/1-hexene copolymers.

	Unfractioned polymer	Octane insoluble	Octane soluble	Heptane soluble	Hexane soluble	Cold hexane soluble
Catalyst without silica						
Content	100	21.2	60.8	14.0	2.9	1.1
C_H, mol%[a]	3.0	2.2	2.5	7.1	9.7	17.7
r_1	125 (average)	180	160	53	38	19
Catalyst on silica						
Content	100	71.5	12.2	4.6	4.0	7.6
C_H, mol%[a]	3.6	1.2	5.4	9.1	9.6	~25
r_1	26 (average)	90	19	11	10	4

[a] 1- Hexene content in copolymer, mol%.

factor of nine with the catalyst displaying an average r_1 of 125. The octane insoluble fraction was produced by an active center that reacted very poorly with 1-hexene (r_1 = 180), while the cold hexane soluble copolymer fraction was produced by an active center that reacted relatively well with 1-hexene, as indicated by an r_1 value of 19. The catalyst prepared on silica exhibited active sites with a range of r_1 values from 4 to 90, a factor of about 25, with the average r_1 value for this catalyst of 26. This data clearly demonstrates the hetergenerous branching distribution of these copolymers.

2.7   Kinetic Mechanism in the Multi-site Mg/Ti High-Activity Catalysts

2.7.1   Introduction

Methods used to more completely characterize the kinetic behavior of high-activity Ziegler catalysts in terms of the number and behavior of each of these multi-site catalysts will be summarized below [67,68].

The standard kinetic scheme for ethylene/α-olefin copolymerization reactions usually includes several chain initiation reactions, chain propagation reactions, and chain termination reactions as shown below. In this scheme, C^* represents an active center and m is the number of ethylene and/or comonomer units in a polymer chain.

Chain propagation reactions:

$$C^*\text{-E-Polymer} + E \rightarrow C^*\text{-E-E-Polymer}$$
$$C^*\text{-E-Polymer} + M \rightarrow C^*\text{-M-E-Polymer}$$
$$C^*\text{-M-Polymer} + E \rightarrow C^*\text{-E-M-Polymer}$$
$$C^*\text{-M-Polymer} + M \rightarrow C^*\text{-M-M-Polymer}$$

Chain transfer reactions:
Chain transfer reactions with hydrogen:

$$C^*\text{-E-Polymer} + H_2 \rightarrow C^*\text{-H} + \text{Polymer}$$
$$C^*\text{-M-Polymer} + H_2 \rightarrow C^*\text{-H} + \text{Polymer}$$

Chain transfer reaction with ethylene:

$$C^*\text{-E-Polymer} + E \rightarrow C^*\text{-E} + \text{Polymer}$$

Chain initiation reactions:
Chain initiation reaction after chain transfer to ethylene (E):

$$C^*\text{-E } (m = 1) + E \rightarrow C^*\text{-E-E } (m = 2)$$

Chain initiation reaction with ethylene after chain transfer to hydrogen (H):

$$C^*\text{-H} + E \rightarrow C^*\text{-E } (m = 1)$$

Chain initiation reaction after chain transfer to α-olefin (M):

$$C^*\text{-M } (m = 1) + E \rightarrow C^*\text{-E-M } (m = 2)$$

Chain initiation reaction with α-olefin after chain transfer to hydrogen:

$$C^*\text{-H} + M \rightarrow C^*\text{-M } (m = 1)$$

However, ethylene polymerization reactions in the presence of supported Ti-based Ziegler catalysts often exhibit behavior that is unusual compared to the polymerization reactions of α-olefins such as propylene, 1-butene, etc. This behavior cannot be explained by the standard kinetic scheme shown above.

Examples of unusual behavior found with ethylene polymerization reactions involving a supported Ti-based Ziegler catalyst include the following [68]:

1. Ethylene is the most reactive olefin as found from the values of its reactivity ratios in copolymerization reactions. Although ethylene is much more reactive than propylene (3–5 times) and hexene (>50 times), direct comparisons of polymer yields and activities show that ethylene never obtains the level of reactivity expected from its relative reactivity in copolymerization reactions.

2. Ethylene copolymerization reactions always proceed at substantively higher rates than ethylene homopolymerization reactions. Since ethylene is the most reactive olefin, addition of an α-olefin such as 1-hexene would be expected to decrease the polymerization reaction, not increase the rate [73].

3. The reaction order n with respect to the partial pressure of ethylene (P_E) or to the concentration of ethylene in solution (C_E) in the rate of an ethylene homopolymerization reaction (R_{pol}) greatly exceeds first order:

$$R_{pol} = k \bullet (P_E)^n \text{ or } R_{pol} = k \bullet (C_E)^n$$

For example, ethylene homopolymerization reactions were carried out in slurry at different ethylene concentrations. Plotting the total yield as a function of ethylene concentration gave n as high as 1.7–2.0. Also, by varying the P_E in gas-phase reactions and plotting the relative rate vs pressure, a reaction order of 1.8 was determined.

4. Addition of hydrogen significantly decreases the ethylene polymerization rate in contrast to the hydrogen effect observed in propylene polymerization reactions [69].

Any proposed kinetic mechanism should be able to explain the above unusual features.

2.7.2   Multi-center Sites

A new approach was employed by Y. V. Kissin and coworkers for understanding the ethylene polymerization kinetics and polymerization mechanism for supported Ti-based Ziegler catalysts based on examining the molecular weight distribution data of the polymers produced with these catalysts [70,68,71,72]. These catalysts contain several types of active centers that produce different types of polymer molecules at different rates. Each active center has its own propagation rate and decay rate, produces polymer molecules of different molecular weights, and has different incorporation rates of comonomer into the polymer chains. Thus, each center produces a group of polymer molecules with a narrow molecular weight distribution of M_w/M_n of 2.0, according to the Flory theory. The fraction of the polymer molecules produced by a single type of active center is called a Flory component.

Details of the method developed by Kissin to determine the number of different active sites and the polymerization kinetics of each type of site were published in the *Handbook of Transition Metal Polymerization Catalysts* [68]. Some important conclusions from this publication are summarized below.

(a) Ethylene homopolymerization in the absence of hydrogen exhibits a very low initial rate of polymerization that slowly increases in polymerization rate over the first 40 minutes, after which the polymerization rate is remarkably constant over a two hour period. A similar polymerization in the presence of hydrogen exhibits a similar polymerization profile except that the rate of polymerization is about 50% lower. Figure 2.16 illustrates ethylene homopolymerization with and without hydrogen.

(b) An ethylene/1-hexene copolymerization in the presence of relatively high levels of 1-hexene (2.21 Molar) exhibits a very high initial rate of polymerization and reaches a maximum rate of polymerization in 2–4 minutes and then shows a relatively fast decay over a polymerization time of 60 minutes. Polymerization rate increases with increasing 1-hexene concentration. The addition of 1-hexene greatly increases the catalyst activity [73]. This effect is shown in Figure 2.17.

(c) The Mg/Ti-based catalyst has four or five different active sites as determined by the number of Flory components

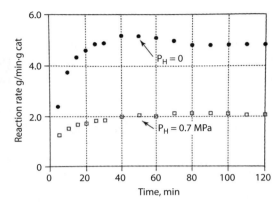

Figure 2.16 Effect of hydrogen on kinetics of ethylene homopolymerization at 80°C with an ethylene concentration of 0.26 M [66].

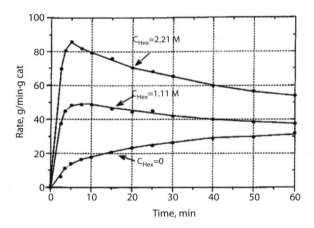

Figure 2.17 Comparison of polymerization kinetics at 85°C, Ethylene concentration 0.55 M without 1-hexene and with 1-hexene at 2.21M and 1.11 M. Copolymer compositions were 1.7 mol%; 0.6 mol% and 0.0 mol% from top to bottom [66].

identified in the molecular weight distribution profile recorded using gel permeation chromatography (GPC). Ethylene homopolymers resolve into four Flory components, i.e., four different active sites, while ethylene/1-hexene copolymers provide GPC profiles that resolve into five Flory components. Figure 2.18 shows the resolution of the

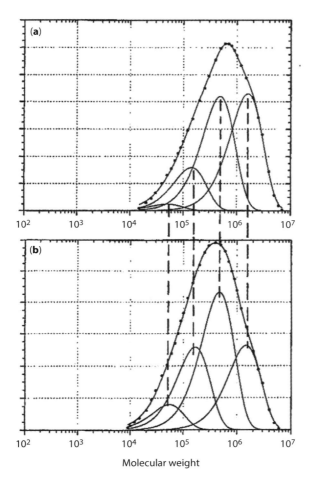

Figure 2.18 GPC curves of two ethylene homopolymers produced at two ethylene concentrations; (A) 0.86 Molar and (B) 0.25 Molar in a 4 hr polymerization. Each curve was resolved into four Flory components [66].

GPC curves obtained from ethylene homopolymers at two ethylene concentrations.

(d) The five active sites identified from the GPC curves obtained from ethylene/1-hexene copolymers each exhibit a very different kinetic profile. The active site that reacts relatively much better with 1-hexene exhibits a very high initial rate of polymerization and decays almost completely in 20 minutes. Active sites that react relatively very poorly with 1-hexene exhibit a very slow initial rate of polymerization that increases steadily over 60 minutes. The kinetic

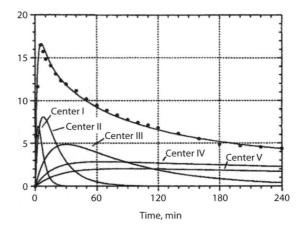

Figure 2.19 Kinetics of ethylene/1-hexene copolymerization reaction at 85oC in absence of hydrogen. The points are experimental data and the lines are calculated rates for each of the five active centers [66].

Table 2.13 1-Hexene content of the polyethylene produced by each type of active site.

1-Hexene Content (mol%)	Active Center
8.5–10	I
4.5–5.3	II
0.6	III
0.3	IV
0.2	V

profiles of each of the five types of active sites found are shown in Figure 2.19. These active centers are identified by the numerals I–V in the order of increasing molecular weight of the Flory components they produce.

A summary of the 1-hexene content of the polyethylene produced by each type of active site is shown in Table 2.13.

Kissin proposed a kinetic mechanism to explain the many unexpected features of ethylene polymerization with and without hydrogen and/or comonomer. This mechanism is shown in Figure 2.20 and experimental details supporting it are explained [68].

The multicenter site analysis revealed that some of the active centers behave differently in the presence of an α-olefin such as 1-hexene. For example:

1. The reaction order n for each center with respect to C_E in ethylene homopolymerization reactions was significantly greater than first order.
2. In ethylene/α-olefin copolymerization reactions, those centers that incorporate an α-olefin into the polymer chain poorly (Centers IV and V) retain this high reaction order (n > 1), while the reaction order changes to first order for those centers that incorporate an α-olefin into the polymer chain well (Centers I, II, and III).
3. In homopolymerization reactions hydrogen suppresses the activity of all the centers, whereas in ethylene/α-olefin copolymerization reactions, hydrogen does not suppress the activity of those centers that incorporate an α-olefin into the polymer chain well (Centers I, II, and III), but still suppresses the activity of Centers IV and V.
4. The introduction of an α-olefin into a polymerization reaction increases the activity of those centers that incorporate an α-olefin into the polymer chain well (Centers I, II, and III), but does not affect Centers IV and V.

These observations are specific to ethylene polymerization reactions. Other α-olefin polymerization reactions do not exhibit this type of behavior. In some cases, opposite behavior is observed. For example, propylene polymerization activity with the same catalysts increases in the presence of hydrogen but decreases when higher α-olefins are introduced into the reaction medium.

Kissin *et al.* have proposed a kinetic scheme that accounts for the unusual kinetic features of ethylene polymerization reactions. The scheme is shown in Figure 2.20.

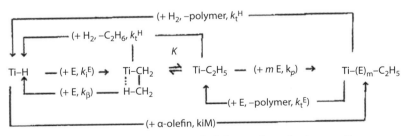

Figure 2.20 Proposed kinetic mechanism of ethylene polymerization reactions proposed by Y.V. Kissin [68].

The key assumption in the scheme is that the Ti-CH$_2$CH$_3$ species is unique in that it is relatively stable. This stability may arise from a strong β-agostic interaction between the hydrogen atom of the methyl group and the Ti atom. Other Ti-alkyl species proposed in the scheme do not possess this stability. In this scheme, formation of the Ti-CH$_2$CH$_3$ species occurs as a result of an ethylene insertion reaction into the T-H bond or after a chain transfer reaction with ethylene.

References

1. L. Reich, and A. Schindler, *Polymerization by Organometallic Compounds*, Interscience Publishers, John Wiley and Sons, New York, Chap. 1, pp. 1-6, 1966.
2. J. Boor, Jr., *Ziegler-Natta Catalysts and Polymerizations*, Academic Press, pp. 19-25, 1979.
3. K. Ziegler, E. Holzkamp, H. Breil, and H. Martin, *Angew. Chem.*, Vol. 67, p. 541, 1955. Note: U.S. patents issued were 3,546,133 issued to K. Ziegler, E. Holzkamp, H. Breil and H. Martin on Dec. 8, 1970 and U.S. 3,903,017 issued Sept. 2, 1975.
4. J. Boor, Jr., *Ziegler-Natta Catalysts and Polymerizations*, Academic Press, p. 23, 1979.
5. R.A.V. Raff, and J.B. Allison, *Polyethylene*, Interscience Publishers, Inc., New York, pp. 1-18, 1956.
6. J. Boor, Jr., *Ziegler-Natta Catalysts and Polymerizations*, Academic Press, pp. 23-25, 1979.
7. G. Natta, *Science*, Vol. 147, p. 261, 1965.
8. G.N. Gaylord, and H.F. Mark, *Linear and Stereoregular Addition Polymers*, Interscience Publishers, John Wiley and Sons, New York, 1959.
9. S.E. Horne, *Ind. Eng. Chem.*, Vol. 48, p. 784, 1956.
10. H.R. Sailors, and J.P. Hogan, History of polyolefins, *J. Macromol. Sci. Chem.*, Vol. A15, Iss. 7, pp. 1377-1402, 1981.
11. D.B. Ludlum, A.W. Anderson and C.E. Ashby, *J. Am. Chem. Soc.*, Vol. 80, p. 1380, 1957.
12. M. Frankel, et al., *J. Poly. Sci.*, Vol 39, p. 347, 1959.
13. M. Frankel, et al., *J. Poly. Sci.*, Vol 33, p. 141. 1958.
14. A.L.J. Raum, UK Patent 841,527, July 20, 1960.
15. K. Kocheshkov, et al., *Poly. Sci. USSR (English)*, Vol. 4, p. 1564, 1963.
16. T. Tanaka, et al., *Kogyo Kagaku Zasshi*, Vol 73, p. 1061, 1970.
17. G. Natta, P. Pino, G. Mazzanti, and U. Giannini, *J. Am. Chem. Soc.*, Vol. 79, p. 2975, 1957.
18. D. Breslow, and N.R. Newburg, *J. Am. Chem. Soc.*, Vol. 79, p. 5072, 1957.
19. D. Breslow, and N.R. Newburg, *J. Am. Chem. Soc.*, Vol. 81, p. 81, 1959.

20. W.P. Long, and D. Breslow, *J. Am. Chem. Soc.*, Vol. 82, p. 1953, 1960.

21. J.C.W. Chien, *J. Am. Chem. Soc.*, Vol. 81, p. 86, 1959.

22. P. Cossee, *Tetrahedron Letter*, Vol. 17, p. 12, 1960.

23. P. Cossee, *J. Catalysis*, Vol. 3, p. 80, 1964.

24. P. Cossee, *Trans. Faraday Society*, Vol. 58, p. 1226, 1962.

25. G. Natta, et al., *Chim. Ind. (Milan)*, Vol. 39, p. 19, 1957.

26. G. Natta, and G. Mazzanti, *Tetrahedran*, Vol. 8, p. 86, 1960.

27. W.L. Carrick, F.J. Karol, G.L. Karapinka, and J.J. Smith, *J. Am. Chem. Soc.*, Vol. 82, p. 1502, 1960.

28. C. Beerman, and H. Bestian, *Angew. Chem.*, Vol. 71, p. 618, 1959.

29. J.P. Collman, et al., *Principles and Applications of Organotransition Metal Chemistry*, University Science Books, Chap. 9, 1987.

30. J. Boor, Jr., *Ziegler-Natta Catalysts and Polymerizations*, Academic Press, pp. 342-349, 1979.

31. G. Henrici-Olive, and S. Olive, *Angew. Chem. Int.*, Vol. 10, No. 2, 1971.

32. T.E. Nowlin, "Low pressure manufacture of polyethylene," in: *Progress in Polymer Science, Vol. 11*, Pergamon Press, pp. 29-55, 1985.

33. K. Yamaguchi, N. Kanoh, T. Tanaka, N. Enokido, A. Murakami, S. Yoshida, (Mitsubishi Chemical Industries Ltd.) US Patent 3,989,881, 1976.

34. U. Giannini, E. Albizzati, S. Parodi, F. Pirinoli, (Montedison S.p.A.) US Patent 4,124,532, 1978.

35. T.E. Nowlin, R.I. Mink, F.Y. Lo, T. Kumar, *J. Polymer Science: Part A: Polymer Chemistry*, Vol. 29, p. 1167, 1991.

36. G.L. Goeke, B.E. Wagner, F.J. Karol, (Union Carbide Corporation) US Patent 4,302,565, 1981.

37. L.L. Bohm, *Polymer*, Vol. 19, p. 562, 1978.

38. L. Petkov, R. Kyrtcheva, P. Radenkov, D. Dobreva, *Polymer*, Vol. 19, p. 567, 1978.

39. L. Petkov, R. Kyrtcheva, P. Radenkov, D. Dobreva, *Polymer*, Vol. 14, p. 365, 1973.

40. D.G. Boucher, I.W. Parsons, R.N. Haward, *Makromol.*, Vol. 175, p. 3461, 1974.

41. R.N. Haward, A.N. Roper, K.L. Fletcher, *Polymer*, Vol. 14, p. 365, 1973.

42. T.E. Nowlin, Y.V. Kissin, K.P. Wagner, *J. Polymer Science: Part A: Polymer Chemistry*, Vol. 26, p. 755, 1988.

43. J.P. Hogan, and H.R. Sailors, *J. Macromol. Sci. Chem: Part A*, Vol. 15, p. 1377, 1981.

44. J. Boor, Jr., *Ziegler-Natta Catalysts and Polymerizations*, Academic Press, pp. 180-212, 1979.

45. B. Wunderlich, *Adv. Polymer Sci.*, Vol. 5, p. 568, 1968.

46. H.D. Chanzy, et al., *Critical Review Macromol. Sci.*, Vol. 1, Iss. 3, p. 315, 1973.

47. V.W. Buls, and T.L. Higgins, *J. Polymer Sci., Part A-1*, Vol. 8, pp. 1025 and 1037, 1970.

48. R.J.L. Graff, G. Kortleve, and C.G. Vonk, *J. Polymer Sci., Part B*, Vol. 8, p. 735, 1970.

49. B.E. Wagner, et al., *Macromolecules*, Vol. 25, pp. 3910-3916, 1992.

50. T.E. Nowlin, R.I. Mink, F.Y. Lo, T. Kumar, *J. Polymer Science: Part A: Polymer Chemistry*, Vol. 29, p. 1167, 1991.

51. R.I. Mink, T.E. Nowlin, (Mobil Oil Corporation) US Patent 5,260,245, 1993.

52. P. Galli, P. Barbe, G. Guidetti, R. Zannetti, A. Martorana, A. Marigo, M. Bergozza, A. Fichera, *Eur Polym. J.*, Vol. 19, pp. 19-24, 1983.

53. K. Yamaguchi, N. Kanoh, T. Tanaka, N. Enokido, A. Murakami, S. Yoshida, (Mitsubishi Chemical Industries Ltd.) US Patent 3,989,881, 1976.

54. U. Giannini, E. Albizzati, S. Parodi, F. Pirinoli, (Montedison S.p.A.) US Patent 4,124,532, 1978 and US Patent 4,174,429, 1979.

55. P. Sobota, J. Utko, T. Lis, *J. Chem. Soc. Dalton Tran.*, Iss. 9, p. 2077, 1984.

56. F.J. Karol, K.J. Cann, and B.E. Wagner, "Developments with high-activity titanium, vanadium and chromium catalysts in ethylene polymerization," in: W. Kaminsky and H. Sinn, Eds., *Transition Metals and Organometallics as Catalysts for Olefin Polymerization*, Springer-Verlag Berlin, p. 149, 1988.

57. A. Greco, G. Bertolini, and S. Assoreni, *Journal of Applied Polymer Science*, Vol. 25, pp. 2045-2061, 1980.

58. L.L. Bohm, *Polymer*, Vol. 19, pp. 553, 545 and 562, 1978.

59. A.V. Kryzhanovskii, I.I. Gapon, S.S. Ivanchev, *Kinetics Catalysis*, Vol. 31, p. 90, 1990.

60. R.N. Haward, A.N. Roper, K.L. Fletcher, *Polymer*, Vol. 14, p. 365, 1973.

61. M. Ferrais, F. Rosati, S. Parodi, E. Giannetti, G. Motroni, E. Albizzati, (Montedison S.p.A.) US Patent 4,399,054, 1983.

62. M. Sacchetti, I. Cuffiani, G. Pennini, (Montell Technology Company bv) US Patent 5,578,541, 1996.

63. G. Morini, G. Balbontin, Y.V. Gulevich, R.T. Kelder, H.P. Duijghuisen, P.A. Klusener, F.M. Korndorffer, (Montell Technology Company bv) WO 0063261, 2000.

64. T.E. Nowlin, Y.V. Kissin, K.P. Wagner, *J. Polymer Science: Part A: Polymer Chemistry*, Vol. 26, p. 755, 1988.

65. G.E. Ham, *High Polymers, Copolymerization, Vol XVIII*; Interscience Publishers, p. 119, 1964.

66. A.D. Hamer, F.J. Karol. U.S. Patent 4,293,673 issued October 6, 1981 and assigned to Union Carbide Corporation.

67. Y.V. Kissin, A.J. Brandolini, *J. Polym. Sci. Part A: Polym. Chem.*, Vol. 37, p. 4273, 1999.

68. Y.V. Kissin, R.I. Mink, and T.E. Nowlin, *Handbook of Transition Metal Polymerization Catalysts*, R. Hoff and R.T. Mathers, Eds., John Wiley & Sons, Inc., Chap. 6, pp. 131-155, 2010.

69. Y.V. Kissin, *J. Mol. Catal.*, Vol. 46, p. 220, 1989.

70. Y.V. Kissin, R.I. Mink, T.E. Nowlin, *Topics in Catalysis*, Vol. 7, p. 69, 1999.

71. Y.V. Kissin, *Makromol. Chem. Macromol. Symp.*, Vol. 66, p. 83, 1993.

72. Y.V. Kissin, *Makromol. Chem. Macromol. Symp.,* Vol. 89, p. 113, 1995.

73. D.C. Calabro, F.Y. Lo, *In Transition Metal Catalyzed Polmerizations*, R.P. Quick, Ed., Cambridge Univ. Press: New York, p. 729, 1988.

Appendix 2.1

Review References for First Generation Ziegler Catalysts

N.G. Gaylord and H.F. Mark, *Linear and Stereoregular Polymers*, Interscience Publishers, New York, 1959.

F.W. Breuer, L.E. Geipel, and A.B. Loebel, "Catalyst systems," in: *Crystalline Olefin Polymers, Part I*, R.A.V. Raff and K.W. Doak, Eds., Interscience Publishers; John Wiley and Sons Inc., Chap. 3, 1965.

L. Reich and A. Schindler, *Polymerization by Organometallic Compounds*, Interscience Publishers; John Wiley and Sons Inc., 1966.

J. Boor, Jr., *Ziegler-Natta Catalysts and Polymerizations*, Academic Press, 1979.

3

Chromium-Based Catalysts

3.1 Part I – The Phillips Catalyst

Ethylene polymerization catalysts based on chromium as the active center are primarily used in the low-pressure manufacture of high density polyethylene (HDPE) with a relatively broad molecular weight distribution. This type of polyethylene is primarily used for blow-molding applications for bottles and large containers and for extrusion applications for pipe and HDPE film for merchandise bags.

Table 3.1 summarizes a few examples of end-use applications for chromium-based polyethylene in terms of the density, molecular weight distribution and molecular weight of the HDPE for various product applications.

3.1.1 Early History of the Phillips Catalyst

The first chromium-based ethylene polymerization catalyst [1] was discovered by John P. Hogan and Robert L. Banks of Phillips Petroleum Company. On March 4, 1958, an historic U.S.Patent 2,825,721was issued to them, and assigned to the Phillips Petroleum Company, in which an olefin

109

Table 3.1 Examples of product applications for HDPE.

Product Type	Melt Index (MI) $(I_{2.16})^a$	Flow Index (FI) $(I_{21.6})^b$	MFR $(FI/MI)^c$	Density (g/cc)
Blow molding	0.4	35	80–110	0.954
Milk bottle	0.9	80	80–90	0.962
Pipe	0.1	10	>100	0.955
HMW-HDPE Film	0.05	5-9	>100	0.953

[a]Melt Index is an inverse relative measure of polymer MW.
[b]Flow Index is a measure of the polymer processability.
[c]Melt Flow Ratio of the FI and MI values is a relative measure of the polymer MWD where MFR values between 80–120 indicate that the MWD is relatively broad. Melt Index, Flow Index, MFR and density are only approximate values for each application. The milk bottle resin is an ethylene homopolymer, while the other products contain a small amount of comonomer (1-butene or 1-hexene) which is used to reduce polymer crystallinity.

polymerization catalyst based on the interaction of chromium compounds with high surface area inorganic supports was reported[1].

The catalyst, designated since the mid-1950s as simply the "Phillips" catalyst, was prepared in a two-step process in which the chromium compound was impregnated into silica and/or alumina to provide an intermediate material, which was activated by heating this intermediate material with dry air to 750–1500°F. This patent was remarkable in the broad terms in which the catalyst was evaluated and the wide variety of polymeric materials that were described. For example, homopolymerizations involving each 1-olefin from ethylene through 1-octene were performed, as well as copolymers in which ethylene was copolymerized with at least one other olefin such as propylene and 1-butene. These types of polymers, but involving other types of catalysts, were later developed into commercially important materials such as isotactic polypropylene, linear low density polyethylene (designated as LLDPE) and ethylene/propylene (EP) rubbers. The LLDPE is primarily ethylene/1-butene, ethylene/1-hexene and ethylene/1-octene copolymers that contain sufficient (approximately 2–4 mol%) comonomer to provide a semicrystalline material with a density of 0.91–0.93 g/cc, while EP rubbers contain high levels of propylene

---

[1] This patent was a continuation-in-part of two other applications filed on January 27, 1953, as Serial Number 333,576 and December 20, 1954, as Serial Number 476,306, which were abandoned. This detail is necessary to understand that the research was carried out prior to the development of other olefin polymerization catalysts and was needed to eventually assign the composition of matter patent on isotactic polypropylene to Phillips Petroleum Company.

to provide an elastomeric material. However it is important to understand that titanium-based catalysts are used commercially to manufacture isotactic polypropylene and most LLDPE because of product advantages over similar products produced with Cr-based catalysts. Another important aspect of this patent was the wide variety of test methods developed in order to better understand the mechanical properties of these new polymers. Tests were described that probed the brittleness, impact strength, flexibility and adhesive properties of these new materials.

In order to put the discovery of the Phillips catalyst into perspective, it is necessary to discuss the type of ethylene polymerization research that was taking place in the early 1950s. Table 3.2 summarizes the three different ethylene polymerization catalyst systems that were discovered in the early 1950s by three independent research laboratories [2].

A paper by Sailor and Hogan [2] provides a very detailed summary on the history of polyolefins which includes research reported from 1897–1940 in which linear polyethylene was prepared by various scientists without directly polymerizing ethylene. This early research clearly demonstrated that a linear polymethylene was prepared with a relatively high molecular weight as indicated by a crystalline melting point of 132–134°C. For example, linear polymethylene was prepared by diazomethane decomposition and by hydrogenation of carbon monoxide over ruthenium and cobalt catalysts.

This paper also provides the chronological details of the three catalyst systems outlined in Table 3.2 in which the Standard Oil (Indiana) catalyst was the first ethylene polymerization catalyst identified [3]. However, as pointed out by Sailor and Hogan, the Standard Oil catalyst was commercialized more slowly than the other two catalyst types, and because of higher operating costs, the Standard Oil process was shut down in 1973. The other two catalyst systems listed in Table 3.2 account for most of the polyethylene manufactured today.

Table 3.2 Important dates for the development of low-pressure polyethylene.

Process Type	Active Center (metal)	Approximate Date of Laboratory Development	Filing Date of Patent Application
Standard of Indiana	Molybdenum or Cobalt Molybdate on alumina	Last half of 1950	April 28, 1951
Phillips	Chromium	Last half of 1951	January 27, 1953
Ziegler	Titanium	November, 1953	November 27, 1953

One additional aspect of the Sailor and Hogan paper is important to highlight as it pertains to the litigation process that took place in the Federal District Court (Delaware) in order to resolve the dispute in awarding the composition of matter patent for isotactic polypropylene to Phillips Petroleum on January 11, 1980.

Although the United States Patent and Trademark Office awarded the composition of matter patent to Montecatini on February 6, 1973 (U.S. Patent 3,715,344), the Federal Court later ruled that Phillips Petroleum scientists were the first to prepare isotactic polypropylene between October 9, 1951 and April 16, 1952, and were, therefore, awarded the composition of matter patent for isotactic polypropylene.

However, the enormously important research carried out later by Natta to modify the original Ti-based Ziegler catalyst to provide isotactic polypropylene in much higher yield than the Cr-based catalyst, remains one of the important achievements in polymer chemistry.

The Nobel Prize in Chemistry was awarded to Karl Ziegler and Giulio Natta in 1963 for their research in developing olefin polymerization catalysts primarily based on titanium compounds and aluminum alkyl compounds required for the catalyst initiation process. However, by 1963 the details on the discovery of the Cr-based catalyst system in which the chromium compound was supported on amorphous silica were widely published and commercially important for the manufacture of HDPE. In the view of this author, the 1963 Nobel Prize awarded in chemistry should also have included two additional scientists, John P. Hogan and Robert L. Banks from Phillips Petroleum.

3.1.2   Preparation of the Phillips Catalyst

The Phillips catalyst is prepared by impregnating a chromium compound into a high surface area silica, such as Davison Grade 952 silica, with a pore volume of about 1.6 cc/g and a surface area of about 300 m^2/g. The type of chromium compound used as the chromium source does not affect the behavior of the finished catalyst after the activation (oxidation) step [4]. Chromium(III) acetate, ammonium chromate or dichromate and chromium oxide (CrO_3) are possible sources of chromium. Sufficient chromium is used to yield about 0.5 to1.0 wt% Cr in the final catalyst.

Hogan has postulated that a surface chromate species is formed by the interaction of surface hydroxyl groups (Si-OH) on the silica with CrO_3[5].

A dichromate structure may also be present to some extent, and these two species are illustrated in Figure 3.1, as they exist on a silica surface after the high temperature oxidation step. In order to chemically bond the

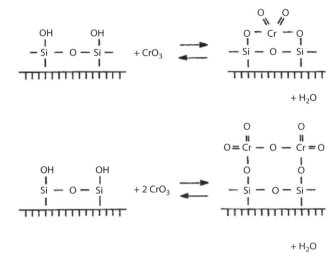

Figure 3.1 Chemisorption of CrO$_3$ onto silica after the high temperature calcination step with dry oxygen [6].

chromium oxide onto the silica support most efficiently, it is necessary to heat the chromium-impregnated intermediate in dry air and remove the water of condensation in order to drive the equilibrium to the right by a mass action effect. Most likely, the relative amount of the dichromate species correlates directly with the amount of chromium impregnated into a particular catalyst.

Figure 3.1 shows the chromium atom in a Cr(VI) oxidation state, but the chromium center needs to be reduced in order to produce an ethylene polymerization center. This reduction step can be carried out using hydrogen, carbon monoxide or ethylene. The most common method is to use the ethylene that is present in an ethylene polymerization reactor. The time required to carry out this reduction process within the polymerization reactor results in a delay in ethylene consumption once the oxidized chromium species is exposed to ethylene. This delay is usually several minutes, depending on the specific catalyst composition, and is referred to as an induction period.

This reduction process is illustrated in Figure 3.2 and is shown to produce a coordinately unsaturated chromium center as a Cr(II) atom [6]. Depending on the reducing agent, water, carbon dioxide or formaldehyde is formed as a byproduct of reduction.

Hydrogen is the least preferred reducing agent, as it forms water as the oxidation product which may remove the chromium species from the silica surface via a hydrolysis reaction. Carbon monoxide has been used in various studies to determine the final oxidation state(s) of the reduced chromium in order to gain an understanding of the oxidation state needed to initiate ethylene polymerization. In addition, carbon monoxide has been utilized during the catalyst preparation step, prior to air oxidation, to provide a modified Phillips catalyst that can produce relatively lower molecular weight polyethylene [7].

In an experiment where the chromium-impregnated silica was previously oxidized, Hogan compared the chromium reduction reaction in the presence of carbon monoxide and hydrogen. The oxidized chromium samples were heated at 1.5°C/min in the presence of CO or H_2. Carbon monoxide was found by Hogan to begin to rapidly reduce chromium at 150°C, while with hydrogen, the reduction process began at 205°C and proceeded at a slower rate. With CO the reduction reaction was complete in about 100 minutes, while with hydrogen, the reduction process appeared to be incomplete due to the formation of water as the oxidation product, which most likely began to remove chromium from the surface of the silica by a hydrolysis reaction [5].

Reduction of the Phillips catalyst with ethylene at 135°C was investigated by Carrick and Baker [8]. Their data are shown in Table 3.3. Oxidative cleavage of the ethylene double bond, with the formation of the

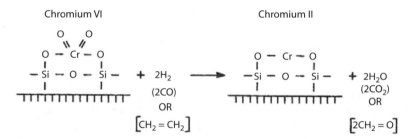

Figure 3.2 Reduction of chromium with either hydrogen, carbon monoxide or ethylene [6].

Table 3.3 Chromium reduction in presence of ethylene at 135°C[a].

Ethylene Contact Time (minutes)	%Cr(II)	%Cr recovered
3	85	89
6	96	87
10	86	52[b]

[a]Catalyst was 0.5 wt% Cr contained in 1 atm Ethylene
[b]Low Cr recovery in this experiment may be due to onset of polymerization and encapsulation of the chromium.

corresponding aldehyde as the primary oxidation product, was reported. The data indicate that most of the chromium is reduced to the +2 oxidation state and that most of the chromium was recovered in the reduced product after 3 or 6 minutes of contact time, which is the induction time before the polymerization process begins. The low recovery after 10 minutes of reduction was attributed to the onset of polymerization.

3.1.3   Unique Features of the Phillips Catalyst

There are primarily two unique features of the Phillips catalyst that have been the focus of much of the research regarding this catalyst since its development in late 1951.

3.1.3.1   Control of Polyethylene Molecular Weight

The Phillips catalyst illustrated by Figures 3.1 and 3.2 provides a catalyst that produces a relatively high molecular weight polyethylene under commercial polymerization conditions, which are approximately a polymerization temperature of 90–110°C and an ethylene partial pressure of 100–200 psi. Unlike Ziegler-type catalysts, the Phillips catalyst does not react with hydrogen as a chain transfer agent. The chain termination step for the Phillips catalyst involves only beta-hydride elimination from the growing polymer chain attached to the active site, to an ethylene molecule coordinated to the same active site. This process is thought to produce a $Cr\text{-}CH_2CH_3$ bond on the active site, which can initiate the growth of a new polymer molecule, and a finished polymer molecule with a terminal double bond. This process is illustrated in Figure 3.3.

Consequently, for the standard Phillips catalyst the polymerization temperature is the primary process variable used in a commercial reactor to control the molecular weight of the polyethylene produced in the polymerization reactor. However, the degree of molecular weight control is relatively small over the temperature range useful in the commercial process. Therefore, a great deal of research has been carried out since the mid-1950s to provide a modified catalyst that produces a broader range of polyethylene molecular weights.

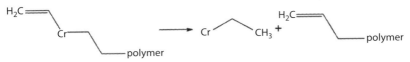

Figure 3.3 Beta-hydride elimination as a chain transfer mechanism for molecular weight control.

3.1.3.2   Initiation of Polymerization at the Active Center

Another unique feature of the Phillips catalyst is the mechanism responsible for the onset of the polymerization reaction.

Figure 3.2 shows the reduction of the chromium center that is necessary to begin the ethylene polymerization process. However, the mechanism responsible for a chromium center to initiate the polymerization process is not clearly understood.

Unlike Ziegler-type catalysts, based mostly on titanium, where an aluminum alkyl cocatalyst is responsible for the formation of a Ti-alkyl bond and initiation of the ethylene polymerization process, chromium-based catalysts do not require such a cocatalyst.

Figure 3.4 shows the alkylation process for a titanium-based ethylene polymerization catalyst that is widely accepted as the initiation process for Ziegler-type catalysts.

3.1.3.3   Possible Initiation Steps for Cr-Based Catalyst

The chromium center may be initiated by formation of a chromium-hydrogen bond (Cr + H-source $\rightarrow$ Cr-H) to provide a Cr-hydride, which then reacts with ethylene (Cr-H + CH_2=$CH_2 \rightarrow$ Cr-CH_2-CH_3) to form a Cr-alkyl to initiate the polymerization process. One possible source of the hydrogen atom is a surface hydroxyl group (Si-OH) known to be present on the silica surface.

As pointed out by Groppo and coworkers [9], if one assumes that a Cr-alkyl species is required in order to initiate the polymerization process, then ethylene plays three important roles as: (i) a reducing agent to create a Cr species in a lower oxidation state as a coordinatively unsaturated active chromium precursor, (ii) as an alkylation agent, and (iii) as a monomer for chain propagation.

If ethylene is responsible for the alkylation step to initiate the polymerization process then one possibility is that two ethylene molecules may add to the chromium center to form a metallocycle, which may initiate the ethylene polymerization through a ring-opening reaction after a beta-hydride transfer to the chromium. This initiation mechanism is illustrated in Figure 3.5.

Intermediate 1 results from the oxidative addition of two ethylene molecules to the coordinatively unsaturated Cr(II) center. Formation of Intermediate 2 involves a beta-hydride transfer to the chromium center from

$$Ti - Cl \ + \ AlR_3 \ \longrightarrow \ Ti - R \ + \ AlR_2Cl$$

Figure 3.4 Alkylation of Ziegler type catalyst.

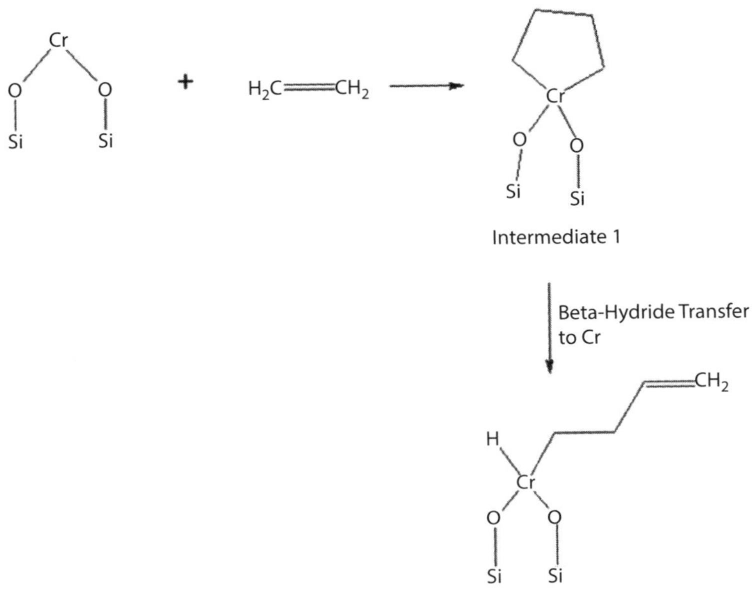

Figure 3.5 Possible mechanism to initiate ethylene polymerization in the Phillips catalyst.

the metallocycle species shown as intermediate 1. Ethylene coordination to Intermediate 2 followed by ethylene insertion into the Cr-carbon bond, would result in the growth of a polymer molecule similar to the mechanism proposed for Ziegler-type catalysts. However, this type of initiation reaction provides an initial polymer molecule with unsaturation at both ends of the polymer molecule, assuming that beta-hydride elimination is the primary termination reaction. But all other molecules provided by this type of active site would contain a methyl group at one end and an unsaturation group at the terminal end. As pointed out by Hogan [5], several thousand molecules produced at the same site would so dilute the initial, non-typical molecule with unsaturation at both ends, that the non-typical molecule would not be detected by analytical measurements.

3.1.4   Characterization of Polyethylene Produced with the Phillips Catalyst

As discussed previously, the contents of the original Phillips patent was remarkable in the broad range of polymer compositions that were described in the patent. The details of the various types of polyethylene that were made with the Phillips catalyst were described by Hogan in a 1964 publication [10]. Unlike low-density polyethylene (LDPE) that was available at the time utilizing a high-pressure, free-radical polymerization process, the Phillips catalyst provided a linear ethylene homopolymer with a much higher degree of crystallinity and, therefore, a much higher polymer density (ca. 95% crystallinity and a density of 0.96 g/cc), which was designated at the time as high-density polyethylene (HDPE). The LDPE is a highly branched polymer with a significantly lower degree of crystallinity and a lower polymer density. As stated by Hogan [10] in his recollection about the research that took place at the Phillips Research Center in the mid 1950s:

..... "Although the high crystallinity of the new polyethylenes made them useful in many new applications, it was apparent that an even broader use spectrum would be made possible by modifying the crystallinity to fit the use. The addition of short-chain branching by copolymerization of ethylene with alpha-olefins has provided the control over crystallinity that was needed."

The data of Hogan show the relationship between polymer density and crystallinity, and the polymer composition of ethylene/propylene and ethylene1-butene copolymers and density are summarized in Figures 3.6 and 3.7.

The data illustrated in Figures 3.6 and 3.7 led to the introduction of ethylene/1-butene copolymers as the predominate polyethylene for commercial applications. In addition, in order to produce a polyethylene with a density of 0.92 g/cc, about 4.5 mol% of 1-butene is required, which is approximately the composition of ethylene/1-butene copolymers manufactured today as linear low-density polyethylene with Ziegler-based catalysts.

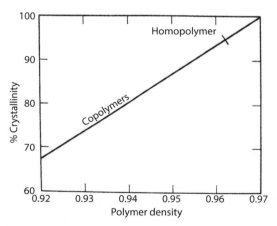

Figure 3.6 Relationship between density and crystallinity in linear polyethylene and copolymers involving ethylene and a second 1-olefin [10].

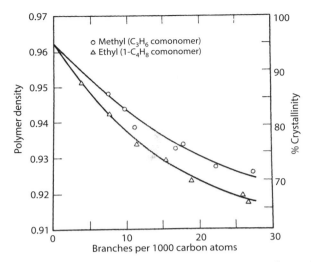

Figure 3.7 Effect of short-chain branches on polyethylene density and crystallinity. (o) Methyl branch from ethylene/propylene copolymers; (▼) Ethyl branch from ethylene/1-butene copolymers [10].

3.1.5   Improvements to the Phillips Catalyst

The first important modification to the standard Phillips catalyst was the addition of a titanium compound to the catalyst preparation prior to the high-temperature oxidation step. This research was carried out in the mid 1960s (reported J. P. Hogan and D. R. Witt in U.S. Patent 3,622,521) and provided a new catalyst formulation that produced polyethylene over a wider range of molecular weights, with the primary emphasis on providing lower molecular weight polymer. The modified catalyst was prepared by adding titanium tetraisopropoxide (Ti-$(OC_3H_7)_4$) dissolved in cyclohexane, to a silica-impregnated intermediate material that had been previously impregnated with CrO_3 and dried at 200-400°F. After the addition of titanium tetraisopropoxide, this titanium-containing intermediate was dried at 300°F. In a final step, the silica support containing both chromium and titanium compounds was oxidized in air at 1300°F for five hours.

The modified catalyst formulation was then tested in an ethylene polymerization experiment carried out at 225–235°F in isobutane at a total pressure of 500 psig. A control catalyst prepared under similar conditions without the titanium compound was also examined in a similar polymerization experiment. The results showed that the Melt Index of the polyethylene prepared with the titanium-containing catalyst was 3.5, while the control catalyst produced polyethylene with a Melt Index of only 0.48. Hence, the Melt Index data showed that the molecular weight of the polymer prepared with the titanium-containing catalyst was significantly lower than the molecular weight of the polymer prepared with the control catalyst[2]. Relatively higher Melt Index values indicate a lower molecular weight polymer sample.

Additional characterization of these titanium-containing Phillips catalysts was reported by Pullukat and coworkers [11] in which it was proposed that the chromium atom is not directly bonded to the silica, as is the case with the standard Phillips catalyst. Figure 3.8 illustrates the bonding of chromium to the silica in the standard Phillips catalyst and the titanium-modified Phillips catalyst.

The interaction of titanium tetraisopropoxide with CrO_3 was investigated in refluxing carbon tetrachloride, which gives the Cr(III) species $Cr[(RO)_3\text{-}Ti\text{-}O]_3$. The presence of such a compound indicates that the titanium tetraisopropoxide need not react with the surface hydroxyl group on silica in a separate step, but may react with the CrO_3 first, and this intermediate material would react with the surface hydroxyl groups to lead to the type of structure shown in Figure 3.8, Structure (II).

In addition, the titanium-containing Phillips catalyst provided an ethylene polymerization catalyst with higher activity and a shorter polymerization induction time than the standard Phillips catalyst. This data are shown in Figure 3.9.

As shown in Figure 3.9, the titanated catalyst exhibits some activity immediately after the addition of ethylene to the reactor and reaches approximately

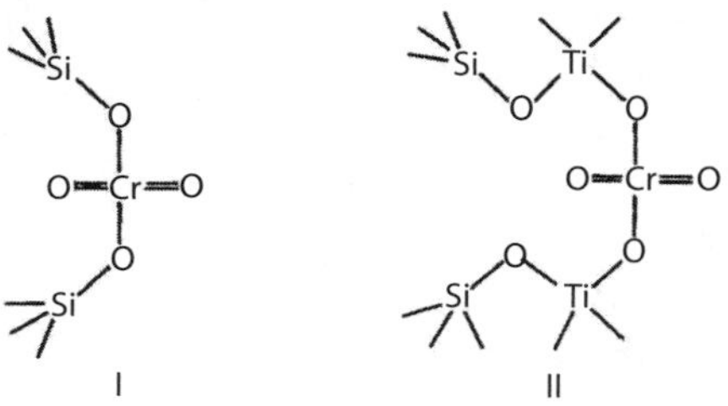

Figure 3.8 Structure (I) illustrates the manner in which Chromium is bonded to silica in the standard Phillips catalyst while Structure (II) illustrates the Chromium bonded to the silica through the O-Ti-O linkage for the Titanium-Modified Phillips Catalyst. Reprinted from [11] with permission from John Wiley and Sons.

---

[2] In the polyethylene industry, the molecular weight of polyethylene is measured by determining a Melt Index (MI) value. Melt Index is the amount of molten polyethylene that flows in ten minutes through a specific orifice maintained at 190°C with a weight of 2.16 Kg, positioned directly on the molten polyethylene sample. The test method is ASTM D-1238-condition E and the test value is referred to as the Melt Index, MI or $I_{2.16}$ or simply I_2.

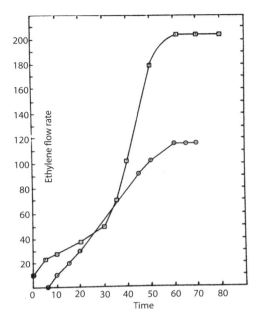

Figure 3.9 Ethylene polymerization rates for chromium on silica (o) and for chromium/titanium on silica (square symbol). Ethylene flow rate in arbitrary units per gram of catalyst. Reprinted from [11] with permission from John Wiley and Sons.

twice the polymerization rate than the standard Phillips catalyst after about 50 minutes of polymerization. The standard Phillips catalyst exhibits no activity until after about five minutes after the introduction of ethylene to the polymerization reactor. Note that the polymerization kinetics shown in Figure 3.9 are typical for the Phillips-type catalyst in that the polymerization rate exhibits a steady increase over the course of at least one hour, which is approximately the catalyst residence time in a commercial reactor. This type of reactivity profile also favors the formation of polyethylene particles with excellent morphology.

In another investigation of chromium and titanium-based catalysts, Hoff and coworkers showed that the Melt Index of polyethylene produced with titanium-containing chromium catalysts was very dependent on catalyst preparation variables [7].

A series of catalysts were prepared in which the CrO_3- (2 wt%) impregnated silica was fluidized in a quartz tube with nitrogen, heated to 170°C, and then liquid titanium tetraisopropoxide was added to the fluidized bed. After ten minutes, carbon monoxide was added to the fluidizing stream and the entire contents of the quartz tube was heated to 700°C and maintained for five hours. Finally, the carbon monoxide was purged out of the fluidized bed, and dry air was used to fluidize the silica-based material as the final oxidizing stream. This final step was carried out at various temperatures between 400 and 700°C.

A second series of catalysts were prepared in exactly the same manner, except these catalysts were reduced with carbon monoxide at 400°C after the air oxidation step. Polyethylene was prepared at 105°C, with each of these finished catalysts using isobutane as the slurry solvent and a total pressure of 550 psi. The Melt Index of each polyethylene was determined and the results are illustrated in Figure 3.10.

Examination of Figure 3.10 clearly shows that both types of titanium-chromium catalysts provide polyethylene with a significant increase in Melt Index (lower molecular weight) if the air oxidation temperature step is carried out at about 425–550°C. However, the catalysts prepared by reacting the titanium tetraisopropoxide with CrO_3 using carbon monoxide at 700°C, followed by an air oxidation step at about 500°C, provide polyethylene with a Melt Index value of about 45 under the polymerization conditions outlined. Consequently, this type of catalyst was able to produce significantly lower molecular weight polyethylene than the catalysts that were reduced with carbon monoxide at 400°C as the final catalyst preparation step.

Additional characterization of the Phillips catalyst was reported by McDaniel and Welch in 1983 in a series of three papers [12–14] that

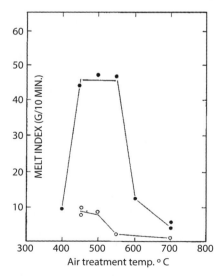

Figure 3.10 Effect of air oxidation temperature on polymer melt index: (●) catalysts without final CO reduction step; (o) catalysts with final CO reduction step at 400°C. Reprinted from [7] with permission from John Wiley and Sons.

provided a better understanding of the parameters that affected the characteristics of the Phillips catalyst.

In one set of experiments [12], a series of catalysts were prepared on a silica-titania support using a one-step thermal activation step in which the finished catalysts contained 1 wt% Cr. The effect of the temperature employed in the thermal activation step was investigated over a temperature range of 595–925°C and was shown to have a significant effect on the ethylene polymerization kinetics and on the weight average molecular weight of the polyethylene. The kinetic data are summarized in Figure 3.11.

Examination of Figure 3.11 clearly shows that although each catalyst required the same induction time before the onset of ethylene polymerization, the activity of each catalyst increased very significantly with increasing activation temperature over the range investigated. McDaniel and Welch attribute this increase in catalyst activity to changes in the support hydroxyl content which is known to take place at these elevated temperatures. Perhaps even more importantly, the weight average molecular weight of the polyethylene produced with each catalyst decreased with increasing activation temperature. For example, the polymer molecular weight (M_w) decreased from approximately 300,000 to 100,000 g/mol as

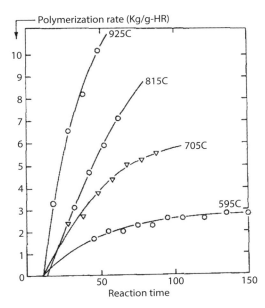

Figure 3.11 Polymerization kinetics of Cr-based catalysts supported on silica-titania and activated with one thermal activation step at the indicated temperatures. Polymerization rate from 1 to 10 Kg/g catalyst/hr vs polymerization time from 0–150 minutes. Reprinted from [12] with permission from Elsevier Publishing.

the activation temperature increased from 595°C to 925°C, respectively. This data was rationalized by suggesting that the Phillips catalyst contained at least two types of active sites, one type provided the relatively high molecular weight polymer and the second type of site provided a relatively low molecular weight polymer. The relative amount of each type of active site was controlled by the temperature of the activation step which primarily determined the surface hydroxyl content in the finished catalyst.

In a second series of catalysts involving a two-step, thermal activation procedure, McDaniel prepared finished catalysts in which the support was calcined first in air or carbon monoxide. Next, the chromium compound was added to the calcined support as an anhydrous solution and then this intermediate material underwent a second thermal activation step in air that was carried out to oxidize the supported chromium compound. These catalysts were evaluated under the same polymerization conditions as the one-step thermal activation catalysts discussed above. The relative molecular weight of the polyethylene produced with each type of catalyst was determined and the results are illustrated in Figure 3.12.

Examination of Figure 3.12 shows that both catalyst types prepared with the two-step thermal activation procedure (data sets A and B) provide polyethylene with a significantly lower molecular weight than the catalysts

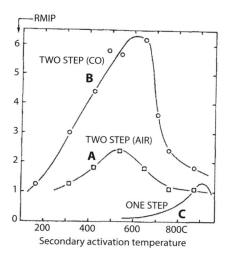

Figure 3.12 Relative Melt Index potential (RMIP) vs secondary catalyst activation temperature. RMIP is the melt index of the polyethylene sample normalized by the Melt Index of the standard Phillips catalyst containing 1 wt% Cr and activated with one thermal treatment in air at 870°C. Melt Index is inversely proportional to polymer MW. Reprinted from [12] with permission from Elsevier Publishing.

prepared with the one-step thermal activation step (data set C). Moreover, the catalysts in which the support was calcined at 870°C in carbon monoxide (data set B) provided polymer with the lowest molecular weight.

Finally, in a rather surprising finding, McDaniel showed that the type of compound(s) used in the first thermal activation step greatly affected the molecular weight of the polyethylene produced with a particular catalyst. Catalysts prepared with a support that was thermally treated at 870°C with carbonyl sulfide, then impregnated with dicumene chromium (0) to provide a material with 0.5 wt% Cr, and then activated in oxygen between 315 and 650°C, produced polyethylene with a very low molecular weight.

Additional methods of increasing the catalyst activity and the range of the molecular weight of the polyethylene produced by the Phillips catalyst were reported [13] that involved reducing the chromium at 700–900°C with carbon monoxide or carbon disulfide (CS_2), followed by reoxidation in dry air at a lower temperature. This reduction/reoxidation technique was effective on silica-only supports and silica-titania supports. Pullukat and Hoff investigated a similar reduction/reoxidation technique with chromic titanate catalysts and this data is illustrated in Figure 3.9. McDaniel postulated that the reduction step promotes a redistribution of the chromium atoms on the surface into larger aggregates, and that the reoxidation step redistributes the chromium atoms on the support surface by reacting with oxide linkages on the support surface.

McDaniel, Welch and Dreiling reported [14] a detailed investigation into the effect of adding titanium to the finished Phillips catalyst employing a wide variety of techniques that included:(a) co-precipitating the titania into the silica support, (b) adding a titanium compound such as a titanate ester $(Ti(OR)_4)$ to the silica support and then calcining this intermediate in an oxidizing environment, or (c) reacting the silica support with titanium tetrachloride, hydrolyzing this intermediate and then reacting this new intermediate with CrO_2Cl_2 to anchor the chromium to the titanium-containing support as a chromate (VI) compound. This third technique was carried out at 200°C and provides the same structure as shown in Figure 3.8, Structure (II), without using a high-temperature activation step. Evaluation of this catalyst provided relatively low molecular weight polyethylene as indicated by a relative melt index potential (RMIP) of 5.

3.1.6 Review Articles for the Phillips Catalyst

The discussion of the Phillips catalyst system outlined above can only provide a very limited summary of one of the most important industrial catalysts used in the petrochemical industry. Therefore, the four review articles listed below are recommended for any scientist interested in a more complete understanding of the enormous amount of research that has been reported over the past 60 years.

Over the past 50 years, there have been many review articles published that provide important technical summaries of the vast amount of research that has taken place around the world into this important catalyst. Some of these publications include:

1. J. P. Hogan discussed the early research in detail in a 1964 publication [10] and in a review article discussing the "History of Polyolefins" [2].

2. McDaniel updated the technology in a 1985 publication, "Supported Chromium Catalysts for Ethylene Polymerization" [15].

3. More recently, McDaniel published an extensive summary in 2010, "Review of Phillips Chromium Catalyst for Ethylene Polymerization," in the *Handbook of Transition Metal Polymerization Catalysts*; Chapter 10, pages 291-446, edited by Ray Hoff and Robert T. Mathers and published by John Wiley & Sons, Inc. This article has 464 references and is considered by this author as the most comprehensive discussion of the Phillips catalyst [16].

4. Finally, an excellent review article published in 2005 by E. Groppo, C. Lamberti, S. Bordiga, G. Spoto and A. Zecchina from the University of Torina, which includes 355 references, provides an excellent comprehensive summary of research into the Phillips catalyst in both industrial and academic laboratories that has taken place since the initial discovery of the catalyst in1951 [9].

Each of these publications is highly recommended for anyone interested in a more complete discussion of the Phillips catalyst.

3.2 Part II – Chromium-Based Catalysts Developed by Union Carbide

In the 1960s, scientists at the Union Carbide Corporation developed two additional silica-supported, chromium-based catalysts that are used in a gas-phase process for the manufacture of HDPE. One catalyst is based on

Bis(triphenylsilyl)chromate, $[(C_6H_5)_3SiO]_2CrO_2$ and an aluminum alkyl and the other catalyst is based on chromocene, $Cr(C_5H_5)_2$. These two Cr-based catalysts, in addition to a Phillips-type catalyst, were used in the 1960s to commercialize the gas-phase process for the manufacture of various grades of HDPE. This process was licensed by Union Carbide around the world under the trade name UNIPOL Process.

3.2.1   Bis(triphenylsilyl)chromate Catalyst

The catalyst based on Bis(triphenylsilyl)chromate, $[(C_6H_5)_3SiO]_2CrO_2$, was reported by Carrick and coworkers at Union Carbide [17] and is prepared on a dehydrated (110–800°C) support material of silica/alumina or only silica. Catalysts prepared on silica require the addition of an aluminum alkyl to the catalyst preparation step to achieve high activity, while catalysts prepared on silica/alumina [19] do not require an aluminum alkyl in the preparation procedure.

A key attribute of the Bis(triphenylsilyl)chromate catalyst deposited into silica is that this catalyst does not require a high temperature activation (oxidation) step in the catalyst preparation. Consequently, this catalyst is activated in the polymerization reactor by the ethylene, which results in an induction time similar to the Phillips catalyst before the onset of polymerization. However, because the Bis(triphenylsilyl)chromate catalyst on silica utilizes an aluminum alkyl such as diethylaluminum ethoxide (DEAEO), it is believed that the finished catalyst does not contain Cr(VI), but contains chromium in lower oxidation states; probably Cr(II), Cr(III) and/or Cr(IV). Tables 3.4 and 3.5 summarize some polymerization data for

Table 3.4 Ethylene polymerization with Bis(triphenylsilyl)chromate on silica/alumina[a].

Silyl Chromate	Pressure (Psig)	Silica/ Alumina[a](g)	Temp (°C)	Time (hrs)	Yield PE(g)	MI (12)	FI/ MI[c]	Activity[d]
1.00[b]	20,000	None	150	6	55	0.04	120	0.15
0.10	600	0.3	160	6	90	0.7	86	75
0.20	600	1.5	153	2	110	0.4	110	140
0.05	600	1.0	135	6	110	0.3	137	185

[a]Dehydrated at 500°C.

[b]Grams of Bis(triphenylsilyl)chromate.

[c]Flow Index (FI) is melt index with 21.6 kg wt; FI/MI is directly proportional to PE molecular weight distribution.

[d]g polyethylene/g of Bis(triphenylsilyl)chromate/hour/300 psig total pressure.

Table 3.5 Ethylene polymerization with Bis(triphenylsilyl)chromate on silica[a].

Silyl Chromate	Al Alkyl	Temp (°C)	Time (hrs)	Yield PE (g)	Flow Index (FI)	Activity[c]
0.010[b]	Et_2AlOCH_3	90	1.5	181	9.3	12,100
0.010	$Et_2AlOC_2H_5$	90	1.5	195	14.2	13,000
0.010	$Et_2AlOC_3H_7$	90	1.0	136	9.1	13,600
0.010	$Et_2AlOC_4H_9$	90	1.0	140	11.1	14,000

[a]0.4 g silica dehydrated at 200°C; 90°C; 0.18 mmol Al.
[b]Grams of Bis(triphenylsilyl)chromate.
[c]g polyethylene/g of Bis(triphenylsilyl)chromate/hour/300 psig total pressure.

the Bis(triphenylsilyl)chromate catalyst supported on silica/alumina and silica, respectively [17].

Examination of Table 3.4 shows that the deposition of Bis(triphenylsilyl)chromate into porous silica/alumina support increased catalyst activity by at least a factor of 500. The FI/MI ratio of 86–137 indicates that the molecular weight distribution of polyethylene prepared with this catalyst is relatively very broad compared to the Phillips catalyst, which may provide significant product advantages in certain applications over a similar grade of polyethylene prepared with the Phillips catalyst.

Examination of Table 3.5 shows that Bis(triphenylsilyl)chromate supported on silica and then treated with various aluminum alkyls is about 100 times more active than similar catalysts prepared on silica/alumina supports without an aluminum alkyl. On a chromium basis, the activity of the silica-supported catalyst is approximately 170,000 g PE/g Cr under the slurry polymerization conditions reported in Table 3.5. Under commercial polymerization conditions in which the effect of impurities is relatively less, the catalyst activity is approximately 1×10^6 g PE/g Cr, assuming a catalyst productivity of approximately 3 Kg PE/g silica-supported catalyst with the catalyst containing 0.3 wt% Cr.

The silica-based catalysts in Table 3.5 also require a much lower polymerization temperature than the catalysts on silica/alumina to provide HDPE products with a commercially useful molecular weight.

Consequently, the commercially important Bis(triphenylsilyl)chromate catalyst is prepared on silica in a two-step process. First, a solution of Bis(triphenylsilyl)chromate in isopentane is added to the dried silica to produce a slurry in which the Bis(triphenylsilyl)chromate is completely fixed to the silica surface. After the deposition process, an aluminum alkyl such as diethylaluminum ethoxide is added to the first intermediate and the catalyst is dried to produce a free-flowing powder. Unlike the Phillips catalyst, the molecular weight of the polymer produced with the Bis(triphenylsilyl) chromate-based catalyst can be controlled by using mixtures of aluminum alkyls in the catalyst preparation procedure (Figure 3.13) or, more importantly, by using hydrogen as a chain transfer agent in the polymerization process (Figure 3.14) [17].

Another important feature of the Bis(triphenylsilyl)chromate catalyst is the broader molecular weight distribution (MWD) of the HDPE produced with this catalyst. A comparison of the MWD of the polyethylene produced with the Bis(triphenylsilyl)chromate catalyst and the Phillips catalyst is shown in Figure 3.15 [17].

Examination of the molecular weight distribution of the HDPE produced with the Bis(triphenylsilyl) chromate catalyst indicates that this polyethylene contains a relatively large polymer component with a lower molecular weight, and a relatively smaller amount of polymer with a higher molecular weight, than the HDPE produced with the Phillips catalyst. For certain applications such as a HMW-HDPE film product, this broader MWD provides easier processability of the polyethylene during the film-forming process and improved mechanical properties such as a higher dart impact strength of the film. In other applications such as the milk bottle resin, this broader MWD is a disadvantage, as smoke and odor problems during the blow-molding process used to fabricate the milk bottle would be unacceptable.

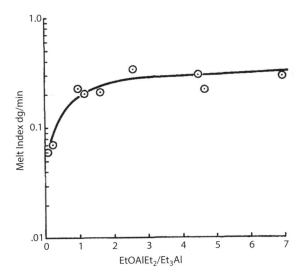

Figure 3.13 Effect of diethylaluminum ethoxide to triethylaluminum ratio on polymer Melt Index (I2.16) for the Bis(triphenylsilyl) chromate catalyst supported on silica. The aluminum alkyl mixture is added after the deposition of the chromate compound onto the silica. Reprinted from [17] with permission from John Wiley and Sons.

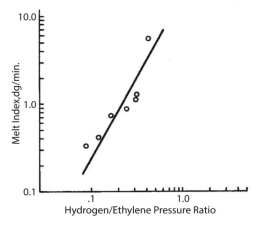

Figure 3.14 Effect of hydrogen/ethylene gas phase ratio on Melt Index (MI) for the Bis(triphenylsilyl) chromate supported on silica with diethylaluminum ethoxide as cocatalyst. Reprinted from [17] with permission from John Wiley and Sons.

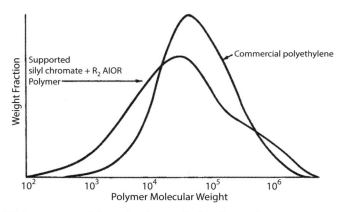

Figure 3.15 Comparison of the molecular weight distribution for HDPE prepared from the Bis(triphenylsilyl) chromate catalyst and the Phillips catalyst labeled as the commercial polyethylene curve. Reprinted from [17] with permission from John Wiley and Sons.

In 2010, Cann *et al.* reported a modified Cr-based catalysts in which diethyl aluminum ethoxide was added to the catalyst preparation step to provide catalysts that produced relatively high-density polyethylene (ca. 0.94–0.96 g/cc) with a very broad molecular weight distribution (Mw/Mn values between about 20–35) with improved stress-crack resistance, as measured by a slow crack growth test, which is an important property for polyethylene pipe applications [18].

A paper by Lesnikova *et al.* [20] investigated, using elemental chemical analysis and IR spectroscopy, the deposition of Bis (triphenylsilyl) chromate into dehydrated silica and the subsequent addition of an aluminum alkyl to the Bis(triphenylsilyl)chromate/silica intermediate. Characterization of the Bis(triphenylsilyl)chromate/silica intermediate indicated that the Bis(triphenylsilyl)chromate was fixed to the silica surface through hydrogen bonds between the silica-hydroxyl group (Si-OH) and the phenyl rings in Bis(triphenylsilyl)chromate, as shown in Figure 3.16, Step (A). The reaction shown in Figure 3.16, Step (B), that attaches the chromium compound to the silica through a covalent bond by eliminating triphenyl hydroxyl silane was shown not to occur.

Lesnikova also investigated the interaction of triethylaluminum with the hydrogen-bonded adduct, which involves Bis(triphenylsilyl)chromate bonded to the dehydrated silica intermediate material illustrated in Figure 3.16, Step (A). The liberation of ethane, the detection of phenyl-containing silicon compounds in solution, elemental analysis and IR

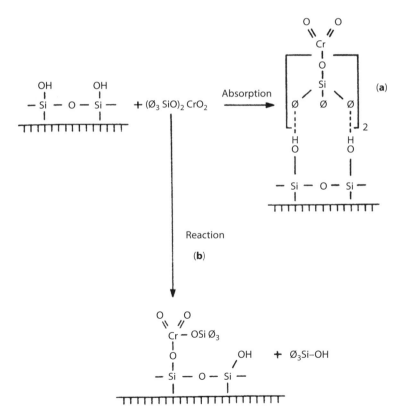

Figure 3.16 Possible modes of deposition of Bis(triphenylsilyl) chromate onto silica. (A) shows the adduct as proposed by Lesnikova [20]. Pathway (B) does not occur [6].

spectroscopy results strongly suggested that the chromium was bonded to the silica through Si-O-Al-O-Cr linkages in a reduced oxidation state involving Cr(II), Cr(III) and Cr(IV) complexes [20,6].

3.2.2   Chromocene-Based Catalyst

A catalyst based on chromocene, Bis(cyclopentadienyl)chromium, was developed in the 1960s at Union Carbide by G. L. Karapinka and coworkers[3] and was the first commercial chromium-based catalyst that was prepared with an organochromium compound containing Cr-carbon bonds in the starting material. In addition, the starting material based on Cr(II), did not need to be oxidized to a Cr(VI) species to obtain a high activity ethylene polymerization catalyst.

This Cr-based catalyst has several characteristics that make this particular catalyst unique in comparison to the Phillips catalyst and the Bis(triphenylsilyl)chromate catalyst. These unique characteristics relative to these two other Cr-based catalysts include:

(a) Polyethylene prepared with the chromocene catalyst has a relatively narrow molecular weight distribution. Consequently, HDPE prepared with the chromocene-based catalyst is not used in product applications shown in Table 3.1 that require polyethylene with a relatively broad MWD, but can be used in product applications, such as injection and rotational molding, that require polyethylene with a relatively narrow MWD. The chromocene-based HDPE underwent a loss of market share in these product applications in the late 1970s when Union Carbide scientists developed a Ziegler-type catalyst supported on silica for the UNIPOL gas-phase process [21]. These new catalyst developments provided Ziegler-based HDPE and LLDPE products with a narrower MWD than the HDPE produced with the chromocene-based catalyst and, consequently, these Ziegler-based products exhibited improved product properties, thus eliminating the chromocene-based HDPE from most applications.

---

[3] Belgian Patent 723775 (1968) issued to Union Carbide Corporation; G. L. Karapinka, U.S. Patent 3,709,853 issued Jan. 7, 1973 to Union Carbide Corporation, which was a continuation in part of applications filed in Nov. 1967, Feb. 1968 and Oct. 1968 which were abandoned.

(b) The chromocene catalyst exhibits very high reactivity with hydrogen as a chain transfer reagent. Hence a very wide range of polyethylene molecular weights can be prepared with this catalyst. This catalyst characteristic was responsible for making high Melt Index ($I_{2.16}$) HDPE products (MI values of 5–50) that are required for injection molding and rotational molding applications. In addition, the polymer produced has a very high degree of saturation with a low degree of terminal unsaturation.

(c) The chromocene catalyst reacts very poorly with other 1-olefins so that only ethylene homopolymers are manufactured with this catalyst in a commercial process. This catalyst property greatly limits the potential applications for polyethylene prepared with the chromocene-based catalyst as polymer, with densities lower than approximately 0.94 g/cc, prepared with either 1-butene,1-hexene or 1-octene as a comonomer, not being possible under commercial operating conditions. Injection molding grades of LLDPE were easily manufactured with Ziegler-type catalysts and, therefore, Ziegler-type catalysts could be more easily transitioned within a commercial reactor to manufacture the entire range of injection molding products (HDPE and LLDPE products) without resorting to a more complex transition to a chromocene catalyst.

Karol et al. [22] reported that the deposition of chromocene onto a previously dehydrated support material takes place under an inert atmosphere at room temperature. A dark red solution of the chromocene compound in hexane is prepared and then this solution is added to the flask containing the dried support material. The contents of the flask are stirred vigorously under an inert atmosphere for approximately 30 minutes. During this time, the silica develops a black color and the hexane solution becomes clear, indicating that the chromocene compound has been completely deposited into the silica support in what is best described as a chemical impregnation technique involving a condensation reaction in which cyclopentadiene is found in the inert solvent (hexane) used to prepare the catalyst. The catalyst formation step proceeds by the interaction of chromocene with a surface hydroxyl group (Si-OH), as shown in Figure 3.17.

The elimination of approximately one cyclopentadiene per chromocene is reported [22] at a deposition temperature of 25°C after one hour using 0.010 g chromocene in 30 ml decane and 0.5 g of Davison Grade 952 silica previously dehydrated at 800°C.

Another possible deposition reaction [24] is shown in Figure 3.18, where the Cr(II) species undergoes a reaction with two surface hydroxyl groups to form a Cr(IV) species containing a Cr–H bond. A deposition reaction involving two adjacent surface hydroxyl groups reacting with chromocene could also result in the elimination of both cyclopentadienyl groups to give a chromium (II) species similar to the structure shown in Figure 3.2. However, if this were the case, this catalyst should be similar to the Phillips catalyst, which is not observed [25].

The reaction shown in Figure 3.18 also requires two adjacent surface Si–OH groups in order to proceed. This species does contain a Cr–H bond that can initiate the ethylene polymerization reaction by coordination of ethylene to the chromium center, followed by migratory insertion into the Cr–H bond, but such a species has not been identified.

However, as pointed out by K. H. Theopold [26], it is not at all obvious how the supported chromium species illustrated in Figure 3.17 might support ethylene polymerization, because this chromium species does not contain a Cr alkyl or Cr–H moiety to undergo migratory ethylene insertion to produce high molecular weight polymer.

Nonetheless, Karol et al. [25] and Lunsford et al. [27]postulate that the chromium species in Figure 3.17 is the predominant structure on silica calcined at 800°C, because silica calcined at 800°C is believed to contain only isolated silica-hydroxyl groups, therefore the deposition reaction shown in Figure 3.18 is not possible.

The deposition of chromocene into Grade 56 silica that was previously dehydrated at 100°C–800°C was investigated [25] using either decane or

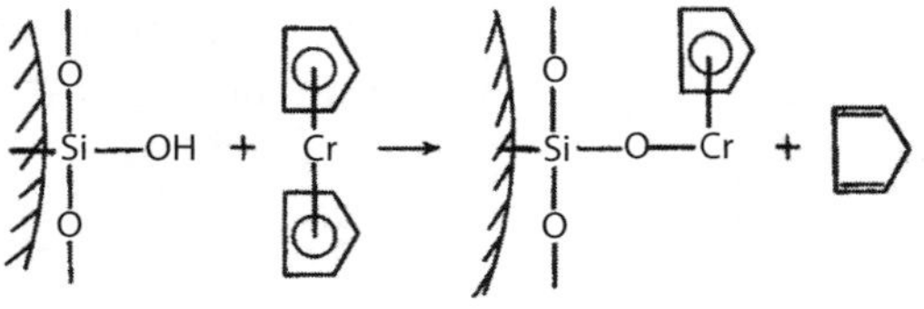

Figure 3.17 The deposition of chromocene [Cp2Cr(II)] onto a dehydrated silica support containing Si-OH groups [22,23]. Reprinted from [23] with permission from Prof. K. H. Theopold.

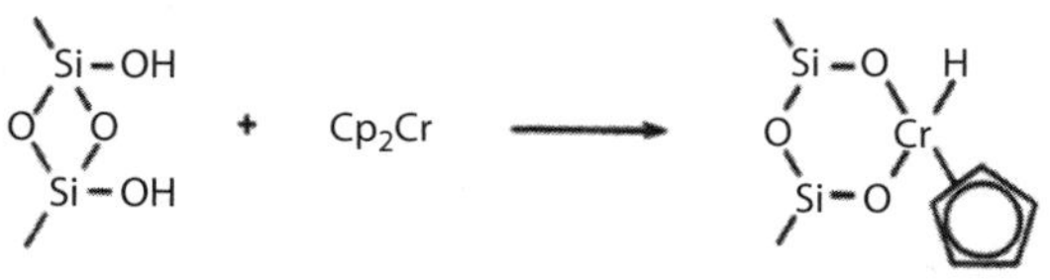

Figure 3.18 The deposition of chromocene onto a dehydrated silica support in which two Si-OH groups are utilized to bond the chromium to the silica support.

toluene as the solvent. The results of this investigation are summarized in Table 3.6 for silica dehydrated at 800°C, and the results illustrated in Figure 3.20 are for Grade 56 silica dehydrated at various temperatures.

Examination of Table 3.6 shows that significantly higher amounts of chromocene can be fixed to the silica surface, relative to the number of surface hydroxyl groups, using decane as solvent in place of toluene (data points 1 and 3). With toluene as solvent, data points 1 and 2 show that increasing deposition temperature from 25°C to 55°C increases the amount of chromocene fixed to the silica surface by approximately 55%. A deposition temperature of 25°C in toluene only utilizes approximately 75% of the available SiOH groups, suggesting that only the chromium species shown in Figure 3.17 is present. However, under all other deposition conditions summarized in Table 3.6, the amount of chromium fixed to the silica surface is in excess of the Si-OH content. Consequently, Karol postulates, based on the amount of cyclopentadiene liberated during the deposition process and the amount of chromocene fixed to the silica surface, that the chromocene in excess of the amount of Si-OH groups on the silica is coordinated to the silica surface through the oxygen atom in the siloxane linkage (Si-O-Si) shown in Figure 3.19.

This siloxane/chromocene adduct shown in Figure 3.19 probably forms a relatively weak chemical bond. Hence, Data Point 3 in Table 3.6 shows that the amount of chromocene that reacts with a surface hydroxyl group (0.45 mmol/g) and the amount of chromocene that coordinates to the siloxane linkage (0.43 mmol/g) are approximately the same. The addition of toluene to the decane solution in Data Point 4 of Table 3.6 suggests that the aromatic ring in toluene can displace most of the chromocene weakly bonded to the siloxane linkage.

Table 3.6 The deposition of chromocene into Grade 56 silica dehydrated at 800°C.

Data Point	Solvent (slurry)	Temperature (°C)	Chromocene (mmol/g silica)	Excess Chromocene (mmol/g silica)[a]
1	Toluene	25	0.35	None
2	Toluene	55	0.54	0.09
3	Decane	25	0.88	0.43
4	Decane then Toluene[b]	25	0.60	0.15

[a] Grade 56, 800°C silica has ca. 0.45 mmol Si-OH/g.
[b] Toluene added to decane solution.

Examination of Figure 3.20 shows two interesting features from the chromocene deposition experiments on silica that were dehydrated at 100°C and 600°C. For silica dehydrated at 100°C, the solvent effect is not observed. Most likely, this is due to the removal of both of the cyclopentadienyl rings in chromocene, by reacting with the high concentration of silica hydroxyl groups on this type of silica. The structure shown in Figure 3.19 is not present after the deposition process. The chemically attached, disubstituted chromium species prepared on silica dried at 100°C

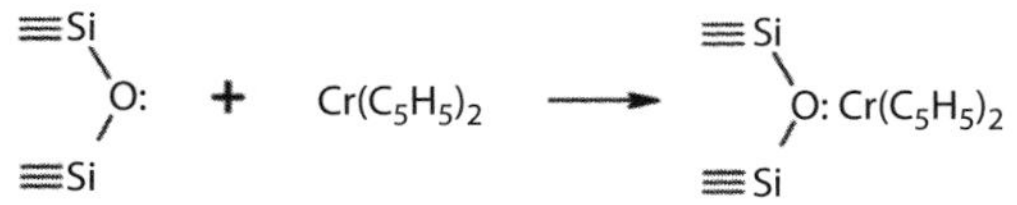

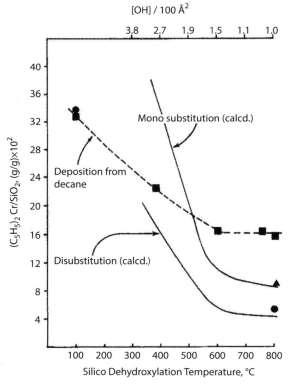

Figure 3.19 The proposed structure of chromocene coordinated to a silica siloxane linkage for silica dehydrated at 800°C [25].

Figure 3.20 The effect of the silica dehydration temperature on the amount of chromocene fixed to the silica surface: (●) toluene at 25°C; (▲) toluene at 55°C; (■) Decane at 25°C. Reprinted from [25] with permission from Elsevier Publishing.

is inactive or only weakly active in ethylene polymerization. For silica dehydrated at 600°C through 800°C, the amount of chromocene fixed to the silica surface using decane as solvent is independent of the dehydration temperature with all three data points in Figure 3.20 showing that approximately 0.88 mmol chromocene/g silica is fixed to the silica surface. This most likely indicates that the silica surface is saturated with chromocene in these three particular cases with chromium present on the surface, as shown in Figures 3.17 and 3.19. Catalysts prepared on silicas dried above approximately 600°C show relatively high activity in ethylene polymerization experiments [22]. Finally, the three data points in Figure 3.20 involving silica calcined at 800°C are from Table 3.6 discussed above.

3.2.3   Hydrogen Response of the Chromocene-Based Catalyst

Figure 3.21 shows the two chain termination steps for a polymer molecule with the chromocene-based catalyst. The chromocene-based catalyst favors a chain transfer reaction with hydrogen (top reaction) over the beta-hydride elimination (bottom reaction), which accounts for the wide range of molecular weights that can be prepared with the catalyst and the high degree of methyl end-groups found in the polymer. An ethylene homopolymer with a M_n of 15,900 contains 1.68 CH_3/1000C with a total unsaturation level of only 0.13 C=C/1000 carbons. At a polymerization temperature below 90°C, the chromocene-based catalyst provides a highly-linear, highly-saturated ethylene homopolymer. However, at polymerization temperatures above 100°C, Lunsford et al. [27] report that unsaturated, branched polymers or oligomers are produced, suggesting a significant change in the catalyst active site and/or polymerization mechanism.

Examination of Figure 3.22 shows that the chromocene-based catalyst has extremely high hydrogen response resulting in a highly-saturated polymer chain with almost exclusively methyl end groups, while the catalyst

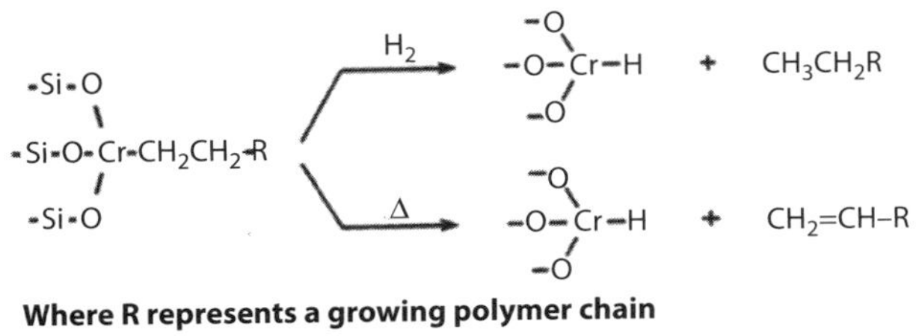

Where R represents a growing polymer chain

Figure 3.21 Chain transfer reactions that are used to control polyethylene molecular weight. The chromocene-based catalyst exhibits very high hydrogen response. Cp ligand on Cr not shown.

based on Bis(triphenylsilyl)chromate terminates almost exclusively by beta-hydride transfer. Both termination reactions are shown in Figure 3.21. For comparison purposes, with a Ziegler-type catalyst based on a Mg/Ti/Cl complex and an aluminum alkyl, both chain transfer reactions take place to a significant extent depending on the polymerization conditions.

3.2.4   Effect of Silica Dehydration Temperature on the Chromocene-Based Catalyst

The silica dehydration temperature has a significant effect on catalyst activity with a catalyst prepared on a silica dehydrated at 670°C being approximately 12 times more active than a catalyst prepared on a silica dehydrated at 200°C. This data [22] is summarized in Table 3.7.

The activity of the chromocene catalyst prepared on 670°C silica is approximately 35,000 g PE/g Cr. A significant increase in catalyst activity is found in a commercial reactor where very high purity raw materials are employed [28] and operate at a higher ethylene pressure. Under these conditions the chromocene catalyst can achieve an activity of 0.5–1.0 x 10^6 g PE/g Cr.

Thermal aging of the chromocene-based catalyst in an inert atmosphere at 100–600°C resulted in the complete loss of the cyclopentadienyl ligands that provides a modified catalyst with only 10–30% of the original activity and introduces a high level of terminal unsaturation into the polyethylene [29].

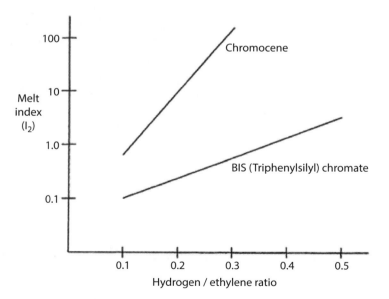

Figure 3.22 Melt Index (12.16) of polyethylene vs H_2/C_2 gas-phase ratio for both Union Carbide Cr-based catalysts [6].

Table 3.7 The effect of silica dehydration temperature on chromocene-based catalyst activity.

Silica Dehydration Temp. (°C)	Chromocene (mmol)	Catalyst Activity (g PE/mmol CR/100 psi ethylene)	Relative Activity
200	0.26	130	0.08
300	0.26	232	0.14
400	0.066	849	0.51
670	0.066	1,671	1.00

Note: 0.4 g Grade 56 silica; 15 psi H_2; 460 psi ethylene for 200–400°C and 185 psi ethylene for 670°C silica; polymerization temp. (60°C); one hour polymerization time.

Lunsford *et al.* reported a series of papers [27,30–32] in which the characterization of the chromocene on silica catalyst was investigated using infrared spectroscopy and a specifically designed infrared quartz cell. The chromocene catalyst was prepared by either a slurry impregnation method or a sublimation method directly in the quartz cell on a thin silica wafer using high vacuum techniques. Fu and Lunsford's data [30] showed that: (a) the chromocene reacted with both isolated and hydrogen-bonded surface hydroxyl groups in silica dehydrated at 200°C and 450°C and isolated hydroxyl groups on silica dehydrated at 900°C; (b) the chromium complexes were present in an agglomerated state or as clusters regardless of the Cr content in the catalyst; (c) the maximum amount of chromocene deposited on silica decreased as silica dehydration temperature was increased, and; (d) the cyclopentadienyl ligand on the chromium bonded to the silica retained its pi-bonded nature. In addition, the interaction of carbon monoxide [31] and nitrous oxide (NO) [32] were also investigated and the interaction of these oxides with the chromocene supported on silica verified the coordinately unsaturated nature of the chromium atom in this catalyst.

Figure 3.23(A) shows the O–H stretching region in silica dehydrated at 200, 450 and 900°C, and (B) the corresponding spectra after chromocene was reacted with the silica using a sublimation procedure. The sharp peak at 3746 cm^{-1} in (A) is due to isolated hydroxyl groups and the broad band at lower frequency is characteristic of hydrogen-bonded hydroxyl groups in the silica calcined at 200°C. This data shows that both types of hydroxyl groups react with chromocene. However, with silica dehydrated at 900°C, chromocene does not react sufficiently to eliminate all of the isolated hydroxyl groups, thus reaching a maximum concentration on the surface before isolated hydroxyl groups are completely eliminated (B). Interpretation of the C–H stretching region in Figure 3.23(C) suggested that the Cp ligands of the surface Cr complexes retained their pi-bonded nature and were thermally stable up to ca. 100°C.

Lunsford and Fu [30] also postulated, based on interpretation of infrared data, that chromocene reacted with hydroxyl groups to form mono-Cp chromium complexes (Figure 3.24) and surface Cr species (Figure 3.25) on which molecular chromocene was further adsorbed, giving an agglomerated species as illustrated in Figures 3.24 and 3.25. This data supports the conclusion of Karol *et al.* which showed that excess chromocene was deposited into silica in excess of the available surface hydroxyl groups. However, the silica siloxane group –Si-O-Si- was not involved, as proposed by Karol *et al.* The amount of such species was solvent dependent [25].

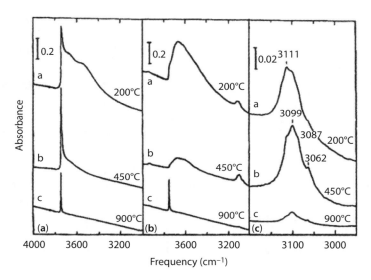

Figure 3.23 Infrared spectra of the O-H stretching region on (A) pure silica and (B) after chromocene deposition with silica dehydrated at (a) 200°C; (b) 450°C and (c) 900°C. The C-H stretching region of B is amplified in C after subtraction of silica background. Reprinted from [30] with permission from the American Chemical Society.

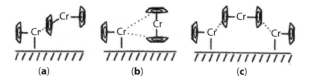

Figure 3.24 Proposed configuration of agglomerates of chromocene complexes on silica dehydrated at 900°C so that chromocene reacts with only one hydroxyl group to eliminate one cyclopentadienyl ligand. Reprinted from [30] with permission from the American Chemical Society.

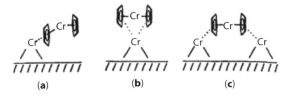

Figure 3.25 Proposed configuration of agglomerates of chromocene complexes on silica dehydrated at 200°C so that chromocene reacts with two hydroxyl group to eliminate both cyclopentadienyl ligands. Reprinted from [30] with permission from the American Chemical Society.

McDaniel reported that deposition of chromocene onto an aluminophosphate support that was previously dehydrated at 600°C, in place of silica, provided a catalyst that produced polyethylene with a significantly narrower molecular weight distribution than the silica-supported chromocene catalyst [33]. For example, he found that a polyethylene sample with a Melt Index ($I_{2.16}$) of 1.0 produced with the catalyst in which chromocene was supported on aluminophosphate exhibited a polydispersity (M_w/M_n) value of 4.2, while a similar polyethylene sample prepared by Karol *et al.* [28] with a catalyst in which the chromocene was deposited on silica exhibited a polydispersity value of 10.2, clearly showing a much more narrow molecular weight distribution for the polymer prepared with the aluminophosphate-supported catalyst. Moreover, a M_w/M_n value of 4.2 is comparable to polyethylene prepared with commercial Ziegler-type (titanium-based) catalysts that are used to provide polyethylene for applications that require a relatively narrow MWD.

3.2.5 Bis(indenyl) and Bis(fluorenyl) Chromium(II) Catalysts Supported on Silica

Karol and coworkers [34] reported silica-supported chromium-based catalysts based on Bis(indenyl) chromium (II), Bis(fluorenyl) chromium (II) and Bis(9-methylfluorenyl) chromium (II) in place of chromocene. These catalysts were prepared at room temperature by reacting the dehydrated silica with a hexane solution containing one of the Cr(II) compounds. The Bis(indenyl)-based catalyst exhibited good activity under some polymerization conditions, while the Bis(fluorenyl)-based catalyst was significantly less active than the chromocene-based catalyst. Table 3.8 summarizes some polymerization data for these catalyst systems.

The indenyl-based and fluorenyl-based experimental catalysts each required approximately three times as much hydrogen to produce a similar polymer molecular weight as the chromocene-based catalyst, while

Table 3.8 Activity of various silica-supported catalysts based on chromium(II) compounds [$R_2Cr(II)$], where R represent ligands containing a cyclopentadienyl group [34].

$R_2Cr(II)$ R=	Hydrogen/Ethylene vapor phase	Activity (g PE/ mmol Cr/hr/100 psi ethylene)	Melt Index ($I_{2.16}$)
C_5H_5 [a]	0.08	1,032	0.6
Indenyl	0.35	412	1.6
Indenyl	0.59	890	13.0
fluorenyl	0.15	185	0.16
fluorenyl	0.35	167	0.99
9-methylfluorenyl	0.29	346	no flow
9-methylfluorenyl	0.56	455	0.03

[a]Control catalyst

Note: Polymerization conditions: ethylene 170–210 psi; polymerization time 1.0–2.5 hr; 90°C.

the catalyst based on 9-methylfluorenyl produced very high molecular weight polymer even at a hydrogen/ethylene ratio of 0.56. However, chain transfer using molecular hydrogen remained the primary chain transfer reaction for all of the catalysts, as all polymer samples contained very low levels of terminal unsaturation. One unique feature of the Bis(indenyl)-based catalyst was its long-term polymerization stability, as indicated by a relatively stable polymerization rate even after five hours. The chromocene-based catalyst exhibits activity decay after 1–2 hours. Hence, under long-term polymerization experiments, the Bis(indenyl)-based catalyst is more active than the chromocene-based system. However, this is not an advantage under commercial polymerization conditions, where catalyst residence times in a commercial reactor are typically less than two hours.

3.2.6   Organochromium Compounds for Ethylene Polymerization Based on (Me)$_5$CpCr(III) Alkyls

Theopold [26] has reported a cationic chromium (III) complex with structural similarities to the chromocene/silica catalyst that polymerizes ethylene as a homogeneous catalyst in dichloromethane (CH_2Cl_2). This complex, [Cp*Cr(THF)$_2$CH$_3$]$^+$[BPh$_4$]$^-$(where Cp* is pentamethylcyclopentadienyl) is shown in Figure 3.26 undergoing an ethylene polymerization reaction [51].

Homogeneous polymerization was carried out at 0–25°C and 1.5 atm ethylene pressure to produce ethylene homopolymers with a M_n of ca. 14,000–20,000 and M_w of 23,000–77,000. Polydispersity values (M_w/M_n) of 1.6–2.3 in five of the six experiments reported, suggest that this catalyst is a single-site catalyst. Theopold produced commercially important molecular weights and molecular weight distributions. A M_w/M_n value of 2.0 (a Flory distribution) is considered a single-site catalyst for a polymerization process involving a chain transfer mechanism. Although Theopold reports one polydispersity value of 4.6, this higher value may have been due to nonsteady-state polymerization conditions. Polymerization experiments in which 1–10 equivalents of THF were added to the experiment showed the expected result of reducing the polymerization rate. This catalyst was similar to the Union Carbide chromocene/silica catalyst in that the homogeneous catalyst did not exhibit any significant reactivity with higher 1-olefins.

Additional homogeneous, single-site, chromium-based ethylene polymerization catalysts were reported by Theopold [35]. These complexes were [Cp*Cr(THF)$_2$Bz]BPh$_4$, Cp*Cr(THF)(Bz)$_2$, and Li[Cp*Cr(Bz)$_3$], which represent a cationic, neutral, and anionic Cr(III) complex, respectively. Table 3.9 summarizes the ethylene polymerization data for each of these three complexes.

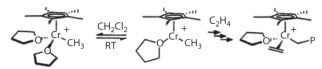

Figure 3.26 Polymerization process for the cationic complex [Cp*Cr(THF)$_2$CH$_3$]+[BPh$_4$]- in CH$_2$Cl$_2$. Complex becomes coordinately unsaturated by eliminating one THF ligand to Initiate the polymerization reaction. Anion omitted for clarity. Reprinted from [51] with permission from the American Chemical Society.

Table 3.9 Ethylene polymerization results for three Benzyl Chromium(III) complexes.

Complex	Solvent	PE Yield, mg (time)	M_w	M_n	M_w/M_n
Cp*Cr(THF)$_2$Bz] BPh$_4$	CH$_2$Cl$_2$	556 (45 min)	29,940	14,896	2.01
Cp*Cr(THF)(Bz)$_2$	Pentane	224 (45 min)	7,044	4,996	1.41
Li[Cp*Cr(Bz)$_3$]	Toluene	515 (120 min)	12,520	5,140	2.44

Conditions: 50 ml solvent, 2.2 mMol catalyst, 1.2 atm. Ethylene, room temperature; Bz is benzyl C$_6$H$_5$-CH$_2$ and Cp* is pentamethyl cyclopentadienyl.

Theopold proposed that the two complexes in Table 3.9 that contain THF dissociate a THF ligand similar to the Figure 3.26 complex in order to initiate the polymerization reaction. The anionic complex shown as a lithium salt in Table 3.9 that does not contain THF, dissociates a benzyl ligand as (benzyl)lithium in order to initiate ethylene polymerization.

The polydispersity values of 1.41–2.44 in Table 3.9 reported for the polyethylene produced with each of the three complexes, show that each complex acts as a single-site polymerization catalyst, similar to the polydispersity data for the polyethylene produced with $Cp*Cr(THF)_2 CH_3$. Finally, Theopold summarized [23] the results of a joint research program with Chevron Oil Corporation[4] in which additional chromium (III) alkyls were identified that polymerize ethylene. Two of these additional complexes are shown in Figure 3.27 for $[Cp*-Si(CH_3)_2-N-t-Bu]CrCH_2Si(CH_3)_3$ and Figure 3.28 for $Cp*Cr(Et_2O)_2CH_2Si(CH_3)_3$.

Examination of Figure 3.27 shows that the constrained geometry complex is a coordinatively unsaturated chromium(III) alkyl that polymerizes ethylene, however, it does not react with 1-hexene to provide an ethylene/1-hexene copolymer. In reactions with propylene, a head-to-tail dimerization reaction takes place without any homopolymerization of higher 1-olefins.

The complex in Figure 3.28 is very reactive in polymerizing ethylene. Theopold reports ethylene polymerization activity at $-104°C$ and the ability to react with higher 1-olefins to form oligomers with approximately 20 monomer units. Theopold postulates that the active complex may involve

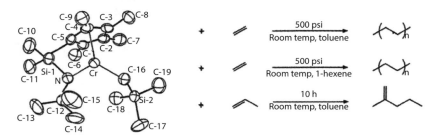

Figure 3.27 A constrained geometry chromium complex, [Cp*-Si(CH3)2-N-t-Bu] CrCH$_2$Si(CH$_3$)$_3$, and its reactivity with ethylene (top), ethylene/1-hexene mixture (middle) and propylene (bottom). Reprinted from [23] with permission from Prof. K. H. Theopold.

---

[4] Chevron Chemical Company was granted many U.S. Patents as a result of this joint Industrial/Acedemic research program, see, for example, U.S. 5,240,895; 5,302,674; 5,320,996; 5,393,720;5,399,634; and 5,418,200.

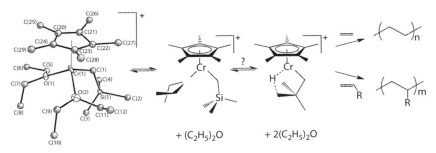

Figure 3.28 The complex Cp*Cr(Et$_2$O)$_2$CH$_2$Si(CH$_3$)$_3$ with very labile diethylether ligands in place of THF ligands. Reprinted from [23] with permission from Professor K.H. Theopold.

dissociation of both diethylether ligands and an agnostic interaction between one methyl group and the Cr center. This is the first complex that showed polymerization reactivity towards higher 1-olefins. Theopold [23] summarizes the large amount of data that his laboratory collected on Cp*Cr(III) alkyl complexes by suggesting that the high selectivity these complexes show towards polymerizing ethylene over 1-olefins is due to relative activation barriers, most likely due to the particular coordination environment around the chromium center. The difficultly of these complexes to react with higher 1-olefins is not a characteristic of chromium itself, as the Phillips catalyst exhibits relatively high reactivity with higher 1-olefins such as 1-butene and 1-hexene, which are used in commercial products.

3.2.7 Organochromium Complexes with Nitrogen-Containing Ligands for Ethylene Polymerization

Two additional chromium-based cationic complexes were reported by Gibson *et al.* that also exhibit high activity as homogeneous ethylene polymerization catalysts [36]. These cationic complexes were prepared from the Bis(imido) Chromium (VI) complex (R-N=)$_2$Cr(CH$_2$C$_6$H$_5$)$_2$, where R is t-butyl. The structure of this precursor complex is illustrated in Figure 3.29.

As stated by Gibson, the most striking feature in the structure of the Bis(imido) chromium(VI) precursor is the manner in which the two benzyl ligands are bonded to the chromium atom through η^1 and η^2 coordination modes. Figure 3.30 shows the two cationic complexes prepared from the precursor complex to form ethylene polymerization catalysts.

Interaction of solutions containing either of the two complexes in Figure 3.30 with 10 bar of ethylene for 60 minutes produced polyethylene with activities of 25,000 to 66,000 g PE/mol/hr/bar.

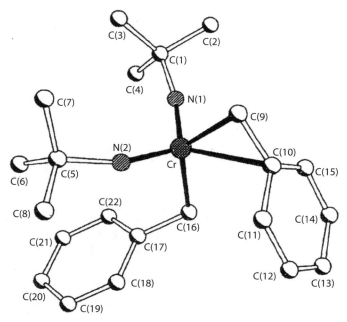

Figure 3.29 Molecular structure of (t-butyl-N=)$_2$Cr(CH$_2$C$_6$H$_5$)$_2$. Reprinted from [36]with permission from the Royal Society of Chemistry.

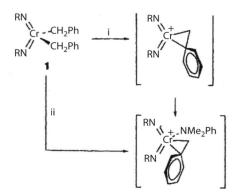

Figure 3.30 Treatment of 1 with (i) one equivalent of [Ph$_3$C][B(C$_6$F$_5$)$_4$] and (ii) one equivalent of [PhNMe$_2$H][B(C$_6$F$_5$)$_4$] in CH$_2$Cl$_2$, at room temperature for 30 min. Reprinted from [36]with permission from the Royal Society of Chemistry.

Gibson *et al.* [37] reported additional ethylene polymerization catalysts based on Cr(III) complexes bearing N,N-chelate ligands. These complexes were prepared from CrCl$_3$(THF)$_3$ and are shown in Figure 3.31.

Gibson points out in Figure 3.31 that complex 1 is a dimeric chromium species due to the more sterically demanding beta-diketimate ligand in

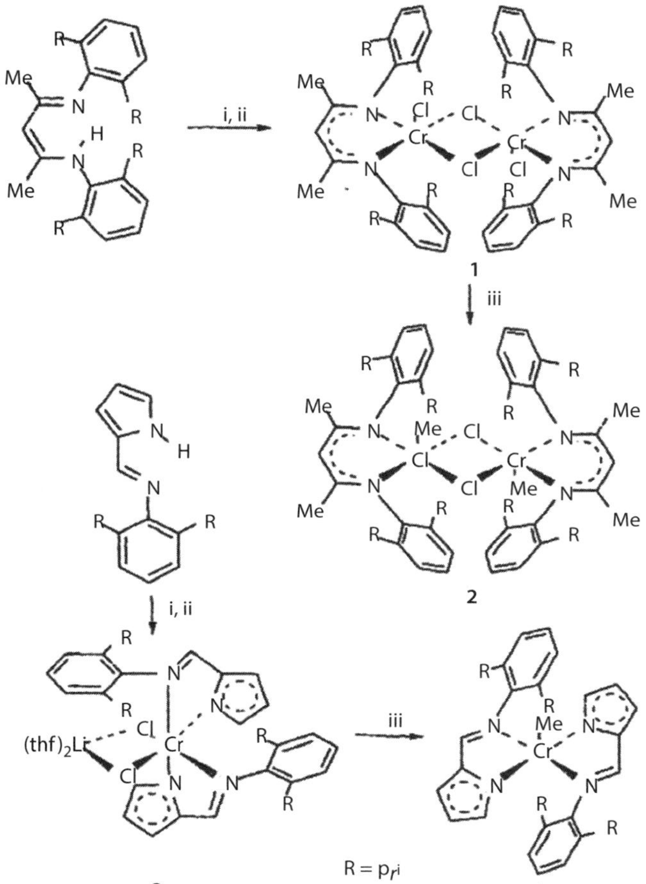

Figure 3.31 Preparation of chromium complexes 1–4. Reagents and conditions: (i) n-butyl-Li, −78°C, THF; (ii) [CrCl$_3$(THF)$_3$], −78°C, THF; (iii) AlMe$_3$, THF. Where R is isopropyl. Reprinted from [37] with permission from the Royal Society of Chemistry.

which only one beta-diketimate can be coordinated to each chromium center, while complex 3 is a monomeric chromium species because, with the less sterically demanding pyrrolide-imine ligand, two pyrrolide-imine ligands can coordinate to one chromium center. Complexes 1 and 3 can be converted into complexes 2 and 4 by reaction with trimethylaluminum (TMA). All four complexes are five-coordinate square pyramidal in structure and produce high molecular weight polyethylene with no branching. Polymerization data (1 bar ethylene, in toluene, 25°C for 60 min) for each complex was determined using methylaluminoxane (MAO), diethylaluminum chloride (DEAC) and 1/1 molar mixtures of DEAC and ethylaluminum dichloride. The most active catalysts with each of the four complexes were provided by DEAC [37]. Catalyst activity was 75, 54, 69 and 70 g PE/mmol/hr/bar for complexes 1 to 4, respectively.

Two additional ethylene polymerization complexes containing reduced Schiff-base N,O-chelate ligands were reported by Gibson et al. [38] in which the complexes were based on Cr(II) and Cr(III). These complexes were prepared from $CrCl_3(THF)_3$ and the bidentate N,O chelating ligand shown in Figure 3.32.

Complex 1 was prepared utilizing two equivalents of the ligand, which results in a four-coordinate, square planar Cr(II) complex with cis-coordinated oxygen and nitrogen atoms. Complex 2 was prepared with one equivalent of the ligand and produces a Cr(III) complex. Complex 2 was recrystallized using acetonitrile (CH_3CN) as solvent, which is coordinated to the chromium center in the isolated crystal. Complex 2 is a distorted octahedral with trans-chlorides and cis-acetonitriles. The polymerization activity of complexes 1 and 2 in Figure 3.32 was determined in toluene with 1 bar ethylene at 25°C for 60 minutes in the presence of MAO or DEAC as cocatalyst. As with the complexes shown in Figure 3.31, DEAC provided a

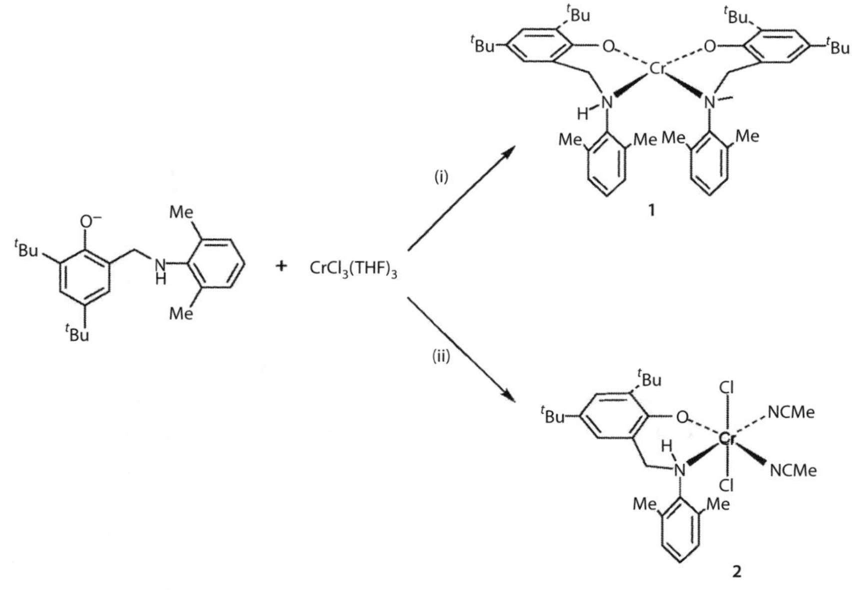

Figure 3.32 Preparation of Complexes 1 and 2 using two equivalents of the lithium salt of the anionic bidentate N,O ligand and $CrCl_3(THF)_3$ for (i) and one equivalent of the sodium salt of the anionic bidentate N,O ligand and $CrCl_3(THF)_3$ for (ii). Reprinted from [38] with permission from the Royal Society of Chemistry.

much higher activity of 60 and 130 g PE/mmol/hr/bar for complexes 1 and 2, respectively.

In a detailed investigation utilizing high throughput screening methodology, Gibson *et al.* [39] reported a comprehensive examination of bidentate (N,O) ligands and tridentate (O,N,N) ligands prepared from (p-tolyl) CrCl$_2$(THF)$_3$. The results show that bidentate (N,O)-containing ligands with small alkyl imine substituents generate highly-active chromium ethylene polymerization catalysts that produce high molecular weight polyethylene. The tridentate (O,N,N) ligand system produced an exceptionally active chromium-based catalyst that provided low molecular weight polyethylene containing approximately 20–40 ethylene units.

Esteruelas *et al.* [40] reported very high activity, single-site, ethylene polymerization catalysts prepared with tridentate (N,N,N) ligands based on Bis(imino)pyridyl chromium (III) complexes that were activated with tri-isobutyl aluminum (TIBA) or alkylaluminoxane (-Al(R)-O)$_n$, where R is methyl or isobutyl.

Figure 3.33 summarizes the preparation of chromium (III) complexes prepared by replacing the three THF ligands in the starting material, CrCl$_3$(THF)$_3$, with one tridentate (N,N,N) ligand to form six-coordinate chromium complexes in which the geometry around the chromium center

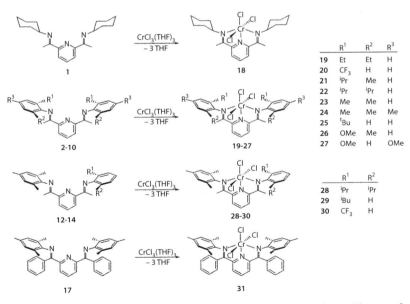

Figure 3.33 Preparation of Bis(imino)pyridyl Chromium (III) Complexes. The complex numbering was retained in order to provide some ethylene polymerization results in Table 3.10. Reprinted from [40] with permission from the American Chemical Society.

is described as a distorted octahedron with the Cl ligands in a mer position. The structure of complex 22 in Figure 3.33 that provided high molecular weight polyethylene with low extractables (waxes) and high catalyst activity with single-site characteristics is shown in Figure 3.34. Some polymerization data for the complexes shown in Figure 3.33 is shown in Table 3.10.

Most of the complexes examined as ethylene polymerization catalysts also produced a large amount of a low molecular weight polymer component (waxes), but if this polymer fraction was removed from the product, the remaining polyethylene exhibited a narrow MWD, indicating a single-site catalyst. Obviously, elimination of the species producing the low molecular weight polymer component would be necessary to develop a commercial catalyst.

However, data points 10 and 12 in Table 3.10 are particularly important. Data Point 10, in which complex 22 was tested with tri-isobutylaluminoxane (TIBAO) as cocatalyst, provided polyethylene without extractables (waxes) with commercially important molecular weight for certain applications such as high-density polyethylene for injection molding. Data Point 12 showed that complex 24 exhibited extremely high activity of 4.14×10^7 g/PE/mol Cr/bar/hr, which corresponds to a residual chromium content in the polyethylene of about 0.2 ppm, which most likely exceeds the activity of any other Cr-based catalyst.

In a copolymerization experiment carried out under the same polymerization conditions as Data Point 14, except in the presence of 15 ml of 1-hexene, the data show that complex 24 reacts with 1-hexene in an

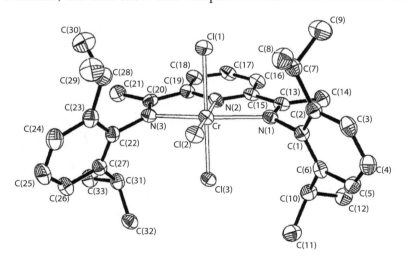

Figure 3.34 Molecular structure of Complex 22 in Figure 3.33. Reprinted from [40] with permission from the Amerian Chemical Society.

Table 3.10 Polymerization data for complexes prepared in Figure 3.33[a].

entry	complex	coactivator	Al/Cr coactivator	t (min)	total wt of polymer (g)	waxes (%)	activity[b]	M_w[c]	M_n	M_w/M_n
1	18	MAO	550	60	0.13	d	8.30×10^3	d	d	d
2	19	MAO	780	30	1.76	24.0	9.18×10^5	d	d	d
3	19	TIBA	810	30	34.69	10.0	3.80×10^6	2170	1450	1.50
4	19	MAO	800	30	15.13	15.0	1.66×10^6	2390	1450	1.86
5	20	TIBA	800	10	1.56	33.0	4.53×10^5	d	d	d
6	21	MAO	820	30	14.82	7.1	3.23×10^6	d	d	d
7	21	TIBA	830	30	34.98	4.4	7.66×10^6	5390	2480	2.17
8	22	MAO	400	30	12.90	0.9	8.57×10^5	d	d	d
9	22	TIBA	400	30	28.60	1.2	1.02×10^6	12900	8150	1.58
10	22	TIBAO	400	30	13.60	0.0	9.06×10^5	35900	27400	1.31
11	23	MAO	680	30	45.90	15.0	1.21×10^7	2130	1380	1.54
12	24	MAO	700	3.5	17.70	d	4.14×10^7	2100	1600	1.31
13[e]	24	MAO	1080	10	5.85	d	9.20×10^6	2660	1800	1.48
14	24	MAO	1110	30	46.60	d	2.51×10^7	1980	1540	1.29
15	24	TIBA	1050	30	29.00	d	1.52×10^7	1630	1100	1.48
16	24	TIBA	810	30	11.40	22.0	1.24×10^7	d	d	d

[a] Conditions: solvent heptane; temperature 70 °C; ethylene pressure 4 bar. The complexes were dissolved and preactivated in a MAO/toluene solution (ca. 10%); the actual concentration of Al_{MAO} for each solution was estimated from its ^{1}H NMR spectrum.
[b] Activity is expressed in units of g of PE (mol of Cr)$^{-1}$ bar^{-1} h^{-1}.
[c] Determined by GPC. Refers only to the polymer separated from the wax phase.
[d] Not recorded.
[e] Temperature 90 °C.

Source: Reprinted from [40] with permission from, the American Chemical Society.

ethylene/1-hexene copolymerization experiment to provide a method to reduce polyethylene crystallinity (density). Based on the polymerization conditions and the polymer characterization results described by Esteruelas *et al.* [40], an ethylene/1-hexene reactivity ratio (r_1) of approximately 100 can be estimated for complex 24, which is comparable to the value reported for some commercial Ziegler-type catalysts. However, unlike Ziegler-type catalysts that exhibit a significant increase in catalyst activity in the presence of 1-hexene as comonomer, complex 24 showed a 27% decrease in activity in the presence of 1-hexexe when 15 ml of 1-hexene was added to the polymerization experiment similar to Data Point 14 in Table 3.10.

The primary conclusions regarding the catalyst activity and the molecular weight of the polyethylene produced with these catalysts are that the substituents at the ortho positions of the N-aryl groups of the 2,6 bis(imino) pyridyl ligands affect both catalytic activity and the molecular weights of the resulting polyethylene. The most active catalysts are those with two substituents at the ortho position of the N-aryl groups. In terms of steric factors, complexes with small substituents provide very active systems but lower polymer molecular weight, while complexes with larger substituents exhibit less activity but higher polymer molecular weight.

Finally, from a commercial point of view, Esteruelas carried out an important experiment to determine if this chromium-based homogeneous catalyst could be transformed into a heterogeneous catalyst system (i.e., solid material), which is necessary in order to provide a catalyst system more easily adapted to commercial polyethylene processes. Most global polyethylene is manufactured in either a gas-phase reactor or a slurry system, where the particle size and particle size distribution of the catalyst particles are important in determining the particle size and particle size distribution of the granular polyethylene produced by these processes.

The heterogeneous catalyst was prepared by reacting 0.5 g of silica containing MAO within the silica pores with 25 mg of complex 24 and 10 ml of additional toluene and 5 ml of additional MAO/toluene solution. The contents of this slurry were stirred and ethylene was added (1 atm) to initiate a prepolymerization process that was carried out to increase the total solids to 2.06 grams which were isolated by solvent evaporation *in vacuo*. The silica-supported catalyst was evaluated in a 2-liter reactor containing one liter of isobutane at 80°C, 35 bar total pressure with TIBA as cocatalyst. Linear polyethylene was produced with an M_w of 1470 and M_w/M_n of 2.13 with a catalyst activity of 1309 Kg PE/g Cr/hr. These results clearly demonstrate that this particular single-site catalyst could be operated in commercial polyethylene manufacturing operations.

The ethylene polymerization behavior of both chromium (II) and chromium (III) complexes reported by B. L. Small and coworkers [41]were based on tridentate pyridine-based ligands with (N,N,N) [Type 1] as the donor set and similar complexes with (N,N,O) [Type 2] as the donor set and activated with methylaluminoxane (MAO) as cocatalyst. Fourteen ligand sets were investigated that were utilized to prepare Cr(II) and Cr(III) complexes. Figure 3.35 summarizes the composition of the tridentate ligands investigated.

Complexes 1–7 and 11–14 in Figure 3.35 are Cr(II) complexes (designated as Type 1) based on the tridentate ligand with the donor set (N,N,N), while complexes 8, 9 and 10 are Cr(II) complexes (designated as Type 2) based on the tridentate ligand with the donor set (N,N,O). Polymerization data for these chromium (II) complexes with MAO as cocatalyst is shown in Table 3.11.

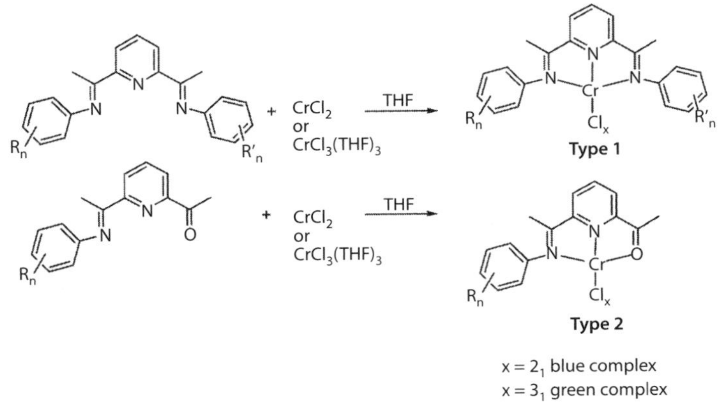

Complex	R_n	R'_n
1	Unsubst.	Unsubst.
2	2-Me	2-Me
3	2-Et	2-Et
4	2-iPr	2-iPr
5	2-tBu	2-tBu
6	2,6-Me$_2$	2,6-Me$_2$
7	2,5-tBu$_2$	2,5-tBu$_2$
8	2,6-Me$_2$	N.A.
9	2,6-iPr$_2$	N.A.
10	2-tBu	N.A.
11	2,4,6-Me$_3$	4-tBu
12	2,4,6-Me$_3$	2-Me
13	2,4,6-Me$_3$	2-Et
14	2,6-iPr$_2$	2,6-iPr$_2$

Figure 3.35 Preparation of Cr(II) and Cr(III) complexes with two tridentate ligand sets. Reprinted from [41] with permission from the American Chemical Society.

Table 3.11 Polymerization data for Cr(II) complexes summarized in Figure 3.35.

entry	complex	Al:Cr[a]	P_{C2}(bar), com.(amt.)[b]	T (°C)[c]	Yield (g)	Prod.[d] (g/g Cr complex)	major products/ notes[e]	GPC data (MWpeak x 10^3)[f]
1	1	500	1.0	35	n.d.	n.d.	2-butene	n.d.
2	2	500	1.0	35	n.d.	n.d.	1-butene	n.d.
3	3	500	1.0	35	n.d.	n.d.	1-butene	n.d.
4	4	500	1.0	35	n.d.	n.d.	1-butene	n.d.
5	5	500	1.0	25	1.9	370	wax/PE	0.72
6	6	500	1.0	35	10.4	1550	wax/PE	0.83
7	5	500	27	60	19.9	4000	wax/PE	0.17
8	7	500	27	100	57.3	11 500	PE	11.0
9	2	500	27	80	n.d.	n.d.	1-butene	40.0
10	8	160	1.0	25	6.0	330	PE	34.0
11	9	160	1.0	25	1.2	60	PE	68.2
12	10	500	1.0	25	1.1	180	PE	103
13	8	1150	27	80	101.0	20 200	wax/PE	1.54
14	8	1150	27	100	95.0	23 800	wax/PE	1.20
15	9	1150	27	45	4.9	1490	wax/PE	0.98
16	8	260	1.0, 1-hexene	25	22.2	1520	wax/PE	0.85
17	11	500	1.0	40	n.d.	n.d.	1-butene	0.35
18	12	1000	27	60	70.0	17 500	linear α-olefins	0.15
19	13	1000	27	60	45.0	11 300	linear α-olefins	0.16

[a]Molar ratio of Al to Cr in the reaction.
[b]Ethylene pressure and amount of comonomer present (entry 16). For entry 16, the 1-hexene was introduced as a 1:5 v:v mixture in *n*-heptane.
[c]Temperatures shown are the maximum values reached due to the heat generated by the reaction.
[d]Prod. = productivity, which was not determined for reactinos in which the major product was butene.
[e]All of the reactions produced small to moderate amounts of polyethylene byproducts. The reactions were run in *n*-heptane or cyclohexane.
[f]GPC data are for the isolated solid fraction of the product.
Source: Reprinted from [41] with permission from the American Chemical Society.

Examination of the data in Table 3.11 shows that Cr(II) complexes 1–7 and 11–13, that are based on the (N,N,N) donor set, produce mostly 1-butene, 2-butene, mixtures of linear 1-olefins or relatively very low molecular weight polyethylene and waxes. The catalysts that provide 1-butene or mixtures of linear 1-olefins may have important commercial applications for the manufacture of 1-olefins. For example, complexes 11, 12 and 13, acted as ethylene oligomerization catalysts producing 1-butene in 99% purity for complex 11 and mixtures of linear 1-olefins for complexes 12 and 13.

Data points 10–12 in Table 3.11 show that the Cr(II) complexes 8–10 in Figure 3.35, which are based on the donor set (N,N,O), provide much higher molecular weight polyethylene, as indicated by M_w values of 34,000 to 103,000.

Small *et al.* also reported limited ethylene polymerization for the analogous Cr(III) complexes based on ligand sets 6, 7 and 14 in Figure 3.35. The polymerization data for ligand sets 6 and 14 for the Cr(III) complexes were similar to those of the Cr(II) complexes. However, the Cr(III) complex with ligand set 14 provided significantly higher molecular weight polyethylene than the similar Cr(II) complex. In addition, the GPC curves of the polyethylene prepared with the Cr(III) complex and ligand set 14 showed a relatively broad MWD with polydispersity values of 7.9–12.5, indicating multiple polymerization sites. The amount of chain transfer to cocatalyst (Al) was rather significant and directly proportional to MAO concentration.

As pointed out by Small *et al.*, one of the more promising results in homogeneous chromium-based complexes was reported by Jolly *et al.* [42], who isolated neutral amine complexes in which the nitrogen atom was bonded to the chromium center through the electron pair on the nitrogen donor atom.

Figure 3.36 summarizes the general preparation scheme employed by Jolly to prepare a wide variety of chromium (III) complexes by primarily varying the substituents on the cyclopentadienyl ring and the amine ligand that coordinates to the chromium center, while Figure 3.37 illustrates the structure of two of the isolated complexes that were also evaluated as ethylene polymerization catalysts in the presence of methylalumoxane (MAO).

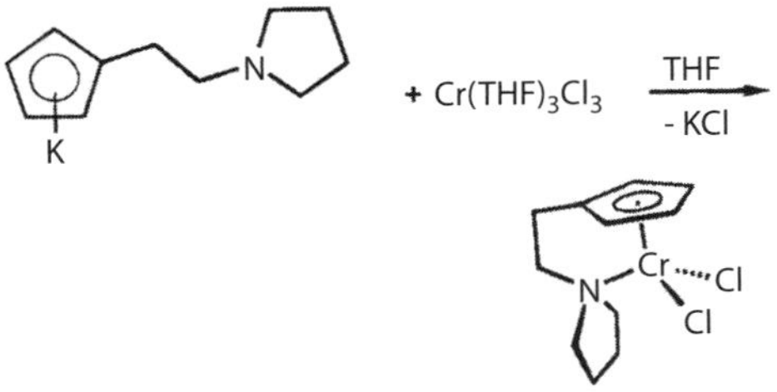

Figure 3.36 General preparation scheme employed by Jolly *et al.* Reprinted from [42] with permission from the American Chemical Society.

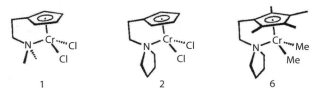

Figure 3.37 Structure of 1 ($Me_2NC_2H_4C_5H_4$)$CrCl_2$; 2 (Cyclo-$C_4H_8NC_2H_4C_5H_4$)$CrCl_2$; and the corresponding dimethyl derivative 6 (Cyclo-$C_4H_8NC_2H_4C_5H_4$)$CrMe_2$. Reprinted from [42] with permission from the American Chemical Society.

Polymerization data are summarized in Table 3.12 for some of the complexes reported [42] and the corresponding derivatives in which the $CrCl_2$ moiety has been replaced with the dimethyl derivative $CrMe_2$.

For the ethylene homopolymerization data summarized in Table 3.12, polymer melting points of 127–133°C and polymer crystallinity values of 67–74% were reported and show that the product is highly-linear polyethylene. Examination of the data in Table 3.12 shows that the polymer produced with each catalyst has a high molecular weight (1–3×10^6) with high catalyst activity. The polydispersity values of 1.56–4.05 for most samples indicate that the polyethylene has a relatively very narrow molecular weight distribution and that the catalyst is most likely a single-site species.

In an experiment to determine if this type of catalyst may have properties required for industrial applications, Jolly et al. carried out a polymerization at 80°C with catalyst 3. The results showed that this particular catalyst exhibited an activity of 50,750 Kg PE/mol Cr/hr with a constant polymerization rate for an extended period, thus indicating very beneficial polymerization properties for commercial applications. In addition, catalyst 6 was evaluated in ethylene/1-hexene copolymerization experiments. In neat 1-hexene an elastomeric material was isolated with an activity of 40,000 Kg PE/mol Cr/hr under polymerization conditions described in Table 3.12, while in toluene containing 20 vol% 1-hexene, an ethylene/1-hexene copolymer was isolated with a DSC melting point of 112°C, suggesting a LLDPE material (i.e., density of ca. 0.91–0.92 g/cc) was produced with a homogeneous branching distribution, which is expected based on the polydispersity value of 1.58 for the same catalyst in an ethylene homopolymerization experiment. This copolymerization data also shows that this particular catalyst has necessary copolymerization characteristics for industrial applications.

In two additional publications, Jolly et al. examined the low barrier to ethylene insertion for various homogeneous chromium (III) cationic complexes [43], which accounts for the preparation of high molecular weight polyethylene prepared from this type of Cr(III) species. They also reported [44] the isolation of chromium metallacycle complexes containing two or

Table 3.12 Ethylene polymerization data for selected complexes.

Compound	Activity (Kg PE/mol Cr/hr)	$10^{-6} M_w$	M_w/M_n
1 $(Me_2NC_2H_4C_5H_4)CrCl_2$	1,660	2.18	4.05
2 $(Cyclo-C_4H_8NC_2H_4C_5H_4)CrCl_2$	1,730	0.90	13.47
3 $(Cyclo-C_4H_8NC_2H_4C_5Me_4)CrCl_2$	7,170	1.91	2.11
4 $(Cyclo-C_4H_8NC_2H_4\text{-fluorenylH}_8)CrCl_2$	3,640	0.84	2.73
5 $(Me_2NC_2H_4C_5H_4)CrMe_2$	4,630	2.50	3.15
6 $(Cyclo-C_4H_8NC_2H_4C_5H_4)CrMe_2$	10,480	3.14	1.58
7 $(Cyclo-C_4H_8NC_2H_4\text{-fluorenylH}_8)CrMe_2$	10,680	1.89	1.56

Conditions: 250 ml toluene, 2 atm. Ethylene, 20°C, Cr:MAO = 1:100, 4 min. polymerization time.

three ethylene units if methylalumoxane is replaced by the lithium alkyl $Li(CH_2)_4Li$ or the Grignard reagent $ClMg(CH_2)_6MgCl$ in the presence of $(Me_2NC_2H_4C_5Me_4)CrCl_2$.

Jolly *et al.* also isolated the first chromium complex containing a coordinated ethylene molecule by reacting $CpCr(THF)Cl_2$ with magnesium in the presence of excess trimethylphosphine and ethylene.

Figure 3.38 shows the preparation and structures of the chromium metallacycles.

The metallacyclopentane complex is more stable than the metallacycloheptane complex, as shown by the thermal decomposition temperatures of 151°C and 56°C, respectively, as determined by a DSC method. The metallacycloheptane complex decomposes with the liberation of 1-hexene.

Figure 3.39 shows the preparation of the Cr(I) complex $CpCr(PMe_3)_2$[ethylene] by reducing $CpCr(THF)Cl_2$ with magnesium in the presence of PMe_3 and ethylene. Jolly *et al.* report that the ethylene complex in Figure 3.39 is extremely labile and that the complex decomposes slowly even at −60°C. However, in the presence of 50 bar of ethylene at room temperature the Cr(I) ethylene complex, $CpCr(PMe_3)_2$[ethylene], polymerizes ethylene.

In a paper that shows the significant effect of changing the ligand field around the Cr(III) center utilizing a tridentate ligand with the donor set (P,N,P) using the ligand $R_2P(CH_2)_2N[H](CH_2)_2PR_2$ (L-1), where R is phenyl, cyclohexyl or ethyl, Keim *et al.* [45] prepared ethylene trimerization catalysts that in the presence of MAO produce 1-hexene with very high selectivity. These complexes were prepared by reacting $CrCl_3(THF)_3$ with L-1 to produce $CrCl_3$(L-1)in which the three THF ligands are replaced

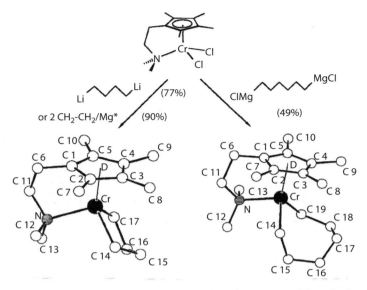

Figure 3.38 Preparation and structure of Cr metallacycles. Reprinted from [44] with permission from the American Chemical Society.

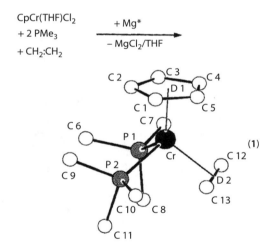

Figure 3.39 Preparation and structure of a chromium complex with coordinated ethylene. Reprinted from [44] with permission from the American Chemical Society.

with the tridentate ligand. The preparation is illustrated in Figure 3.40 and the experimental data are summarized in Table 3.13.

Examination of data points 1–3 in Table 3.13 shows that the catalyst productivity is very dependent on the substitution on the phosphorous

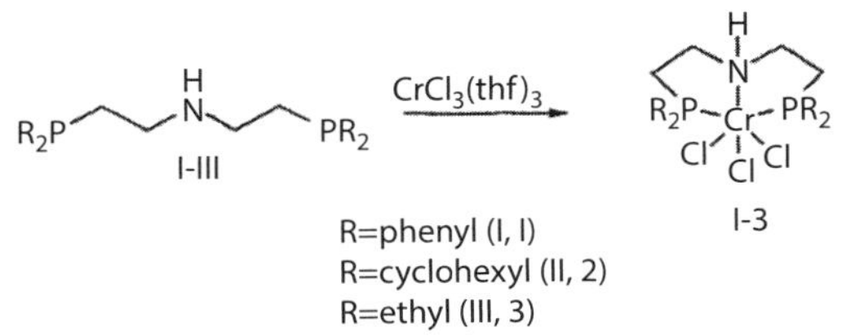

Figure 3.40 Preparation of ethylene trimerization catalysts. Reprinted from [45] with permission from the Royal Society of Chemistry.

Table 3.13 Ethylene trimerization results.

Data[a]	Catalyst	MAO (equivalents)	Temperature (°C)	PE (wt%)	Hexenes (wt%)	1-Hexene selectivity (wt%)[b]	Productivity (h⁻¹)[c]
1	1	120	100	0.1	98	99.2	8 670
2	2	120	100	85.7	14	80.0	580
3	3	100	100	0.2	98	99.2	17 300
4	1	680	100	10.2	83	99.1	17 620
5	3	340	100	0.3	98	98.9	48 060
6	3	690	100	0.7	98	98.9	58 570
7	3	850	100	2.1	94	99.1	69 340
8	3	630	100	0.3	92	99.2	31 220
9	3	600	50	21.9	78	98.8	2 950
10	3	680	80	0.4	97	99.2	50 770
11	3	640	80	0.3	97	99.3	39 890
12	3	660	120	0.2	93	99.3	49 620

[a]40 bar ethylene, 25mL toluene.
[b]Selectivity for 1-Hexene as a percentage of total hexenes.
[c]Average turnover frequency of ethylene conversion.
Reprinted from [45] with permission from the Royal Society of Chemistry.

atom on the ligand with productivity increasing in the order cyclohexyl < phenyl< ethyl, with the least bulky ethyl group providing the highest productivity, 1-hexene selectivity and least amount of polyethylene. In addition, the catalyst prepared with the ethyl substituent showed a significant increase in productivity with increasing levels of MAO (data points 3, 5, 6, and 7), while producing < 2.1 wt% polyethylene and ca. 99% 1-hexene selectivity. Kiem *et al.* found that the trimerization catalyst performed best at 80°C with good activity, less catalyst decay and high 1-hexene selectivity, as shown by a comparison of data points 8–12 in Table 3.13.

3.2.8   Catalysts for Ethylene Polymerization with *In-Situ* Formation of 1-Hexene

Chromium-based catalysts that polymerize ethylene while simultaneously producing a 1-olefin such as 1-hexene, provide a method of preparing polyethylene homopolymers that contain short-chain branching without adding a 1-olefin directly to the manufacturing process.

The deposition of tris(allyl)chromium, $Cr(allyl)_3$, onto silica forms an active catalyst for ethylene polymerization and was reported in the early 1970s [46–48].

However, more recently Bade and coworkers [49] reported that the deposition of $Cr(2\text{-Me-allyl})_3$ onto silica forms an active catalyst for ethylene polymerization in which the silica dehydration temperature used to dry the silica prior to the deposition of the chromium allyl affects the catalyst activity, the molecular weight of the polyethylene and the density of the polyethylene produced with the finished catalyst.

The deposition of $Cr(2\text{-Methyl-allyl})_3$ onto silica, previously dehydrated at 200, 400 or 800°C, was reported in which the stoichiometry of the deposition reaction was monitored by GC analyses of the pentane solution used to carry out the deposition reaction. Ethylene homopolymerization experiments with the catalyst based on the silica dehydrated at 800°C showed that the polyethylene contained a high concentration of short-chain branches, suggesting that this catalyst also produced 1-olefins (primarily 1-hexene) that were incorporated into the growing polymer chain. The catalyst prepared on 400°C silica produced polyethylene with a small amount of short-chain branches, while the catalyst prepared on 200°C silica exhibited no side chains.

Bade *et al.* postulate that the primary reaction products formed from the reaction of chromium allyls with the surface hydroxyl groups on dehydrated silica are shown in Figure 3.41A and B. In addition, chromium allyls may also react with a siloxane linkage with silica dehydrated at 800°C, and this reaction is shown in Figure 3.41C.

Species A in Figure 3.41 requires adjacent silica hydroxyl groups and would most likely be formed on silica dehydrated at 200°C, while species B requires isolated silica hydroxyl groups that are the predominate species on silica dehydrated at 400 and 800°C. Bade *et al.* found infrared evidence for the reaction illustrated in Figure 3.41C and suggested that this type of reaction takes place on silica dehydrated at 800°C where strained siloxane linkages may be present. The catalyst prepared on 800°C silica was unique in that it was the only catalyst that exhibited infrared bands associated with a η^3-bonded allyl group attached to the chromium center.

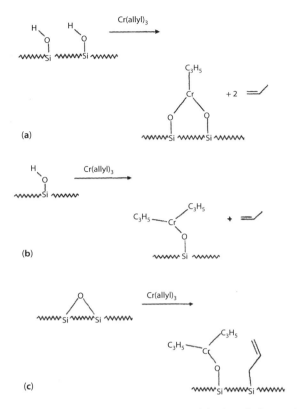

(a)

(b)

(c)

Figure 3.41 Postulated reactions of Cr(allyl)$_3$ between dehydrated silica with Si-OH groups (A and B) and a siloxane group (C). Reprinted from [49] with permission from the American Chemical Society.

Takashi Monoi and coworkers [50] reported a silica-supported organo-chromium compound that provided catalysts that produce polyethylene with a very broad molecular weight distribution.

In this catalyst, Cr[CH(SiMe$_3$)$_2$]$_3$ was supported on silica (surface area of 300 m^2/g and a pore volume of 1.6 cc/g) that was dehydrated at either 200 or 600°C using hexane as the slurry solvent. The catalyst prepared on silica dehydrated at 600°C also produces 1-hexene *in situ*, which results in polymer containing a relatively high number of short-chain branches that reduce the density of the polyethylene.

Figures 3.42 and 3.43 summarize the catalyst activity and the polyethylene density as a function of chromium content, respectively, for each type of catalyst. The polymerization data for each of the catalysts is summarized in Table 3.14.

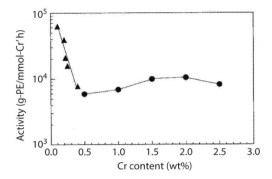

Figure 3.42 Effect of Cr content on catalyst activity density. Reprinted from [50] with permission from Nature Publishing Group.

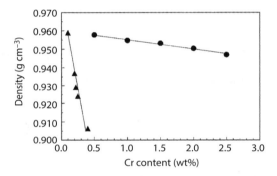

Figure 3.43 Effect of Cr content on density. (●) Silica dehydrated at 200°C; (▲) Silica dehydrated at 600°C. Polymerization conditions: 100°C, ethylene partial pressure 1.4 MPa, slurry solvent 700 ml isobutane, 1 h. Reprinted from [50] with permission from Nature Publishing Group.

Examination of Figures 3.42 and 3.43 clearly shows that the silica dehydration temperature has a very significant effect on the characteristics of a particular catalyst. For example, the catalyst prepared on silica dehydrated at 600°C undergoes a rapid decrease in activity by a factor of ten as the Cr content of the catalyst increases from 0.1 to 0.4 wt%. In addition, this same increase in Cr content increases the amount of 1-hexene generated *in situ,* as indicated by a rapid decrease in the density of the polyethylene produced by these catalysts. Monoi *et al.* rationalize this data by suggesting that three types of active centers are formed in this type of catalyst based on the amount of vicinal and isolated hydroxyl groups in the starting silica. The structures of these active sites are illustrated in Figure 3.14.

Active site 1 in Figure 3.44 is formed with silica dehydrated at 200°C due to the large number of vicinal hydroxyl groups. Active sites 2 and 3 are

Table 3.14 Effect of silica dehydration temperature and Cr content on ethylene polymerization characteristics [50].

Run	Calcination Temp (°C)	Cr Content to Silica (wt%)	Activity (g-PE/ mmol-Cr·h)	Density (g cm^{-3})	Mn ($\times 10^4$)	M_w ($\times 10^4$)	M_w/M_n
1	200	0.50	5930	0.9578	1.10	40.60	36.9
2	200	1.00	6920	0.9548	1.20	34.50	28.8
3	200	1.50	9980	0.9533	1.00	33.50	33.5
4	200	2.00	10500	0.9505	1.00	32.60	32.6
5	200	2.50	8260	0.9471	1.30	31.20	24.0
6	600	0.10	62900	0.9590	1.50	31.70	21.1
7	600	0.20	39300	0.9371	1.00	18.70	18.7
8	600	0.22	21000	0.9292	0.80	17.50	21.9
9	600	0.25	15800	0.9241	0.40	12.70	31.8
10	600	0.40	7800	0.9065	0.40	14.20	35.5

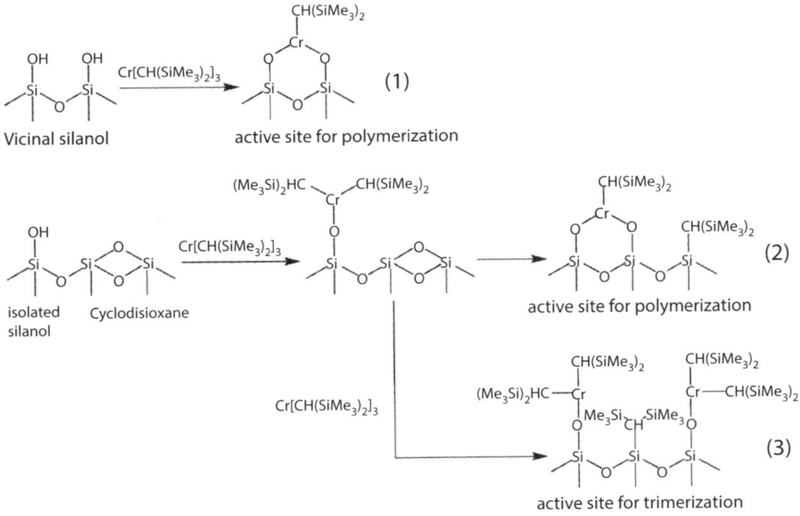

Figure 3.44 Proposed mechanism for active site formation. Reprinted from [50] with permission from Nature Publishing Group.ft

formed on silica dehydrated at 600°C due to the presence of only isolated hydroxyl groups, with active sites 2 and 3 responsible for ethylene polymerization and trimerization, respectively. Note that the interaction of surface siloxane linkages are required for the formation of active sites 2 and 3. This polymerization data is summarized in Table 3.14.

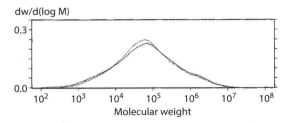

Figure 3.45 GPC curves for the polyethylene produced in data points 4 (dashed line) and 6. Reprinted from [50] with permission from Nature Publishing Group.

Examination of the data in Table 3.14 shows that the catalyst prepared on the silica dehydrated at 600°C (data points 6–10) is significantly more active on a chromium basis than the catalyst prepared of the silica dehydrated at 200°C (data points 1–5). However, as the data demonstrate, this catalyst also produces 1-hexene *in situ* and therefore the preparation of polyethylene with a density above 0.945 g/cc, which is necessary for many applications requiring polyethylene with a very broad MWD, can be prepared only with the catalyst containing 0.1 wt% chromium (data point 6). Importantly, the GPC curves for the polyethylene prepared with both types of catalysts indicate that the MWD of the polyethylene is relatively very broad with M_w/M_n values between 18.7–36.9, suggesting a high degree of shear-thinning during the polymer fabrication process, and therefore good polymer processability. A representative GPC curve for the polyethylene produced with the most active catalyst for each of the two silicas (data points 4 and 6 in Table 3.14) is shown in Figure 3.45.

The GPC curves in Figure 3.45 show an important feature in the high molecular weight region with a prominent shoulder in the molecular weight range 10^6–10^7, which may provide improved polymer properties in certain applications.

3.3 Next Generation Chromium-Based Ethylene Polymerization Catalysts for Commercial Operations

The research reported for chromium-based ethylene polymerization catalysts over the past 20 years clearly indicates that homogeneous catalysts with single-site characteristics can be prepared. Next generation chromium-based catalysts with commercial importance will require the combination of such single-site catalysts into a single, heterogeneous catalyst capable of producing polyethylene with unique molecular structures and a relatively broad molecular weight distribution as required for advanced polyethylene materials for blow-molding, high molecular weight high density polyethylene (HMW-HDPE) film and pipe applications.

For example, the combination of two or more single-site Cr-based catalysts that produce unique molecular weight polymer components is necessary. One single-site catalyst component would provide a very high molecular weight polymer component containing short-chain branches with a homogeneous branching distribution. A second single-site catalyst component would provide a very low molecular weight polymer component that does not contain short-chain branches (i.e., this catalyst component has very poor reactivity with higher 1-olefins). The catalyst components should both be present in each catalyst particle so that both polymer components are produced simultaneously during the polymerization process, hence producing polymer particles with an even distribution of polymer components. The amount of short-chain branching and the molecular weight and weight fraction of each polymer component need to be tailored for each specific application.

References

1. J.P. Hogan, R.L. Banks, U.S. Patent 2,825,721, issued on March 4, 1958 and assigned to Phillips Petroleum.
2. H.R. Sailors, and J.P. Hogan,History of polyolefins, *J. Macromol. Sci.Chem.*, Vol. A15, Iss. 7, pp. 1377-1402, 1981.
3. A. Zletz, (Standard Oil, Indiana), U.S. Patent 2,692,257, filed on April 28, 1951 and issued 1954.
4. M.P. McDaniel, and M.B. Welch, *J. Catal.*, Vol. 82, pp. 98-109, 1983.
5. J.P. Hogan, *J. Polym. Sci., Part A-1*, Vol. 8, p. 2637, 1970.
6. T.E. Nowlin, *Progress Polymer Science*, Vol. 11, pp. 29-55, 1985.
7. R.E. Hoff, T.J. Pullukat, and M. Shida, *J. Applied Polymer Science*, Vol. 26, p. 2927, 1981.
8. L.M. Baker, and W.L. Carrick, *J. Organic Chem.*, Vol. 33, No. 2, p. 617, 1968.
9. E. Groppo, C. Lamberti, S. Bordiga, G. Spoto, and A. Zecchina, The structure of active centers and the ethylene polymerization mechanism on the Cr/SiO2 catalyst: A frontier for the characterization methods, *Chem. Review*, Vol. 105, pp. 115-183, 2005.
10. J.P. Hogan, "Olefin copolymerization with supported metal oxide catalysts," in: *High Polymers Vol. XVIII: Copolymerization*, G.E. Ham, Ed., Interscience Publishers; John Wiley and Sons, Chap III, pp. 89-147, 1964.
11. T.J. Pullukat, R.E. Hoff, and M. Shida, *J. Polymer Sci., Polymer Chemistry Ed.*, Vol. 18, p. 2857, 1980.
12. M.P. McDaniel, and M.B. Welch, *J. Catal.*, Vol. 82, p. 98, 1983.
13. M.B. Welch, and M.P. McDaniel, *J. Catal.*, Vol. 82, p. 110, 1983.
14. M.P. McDaniel, M.B. Welch, and M.J. Dreiling, *J. Catal.*, Vol. 82, p. 118, 1983.
15. M.P. McDaniel, *Advances in Catalysis*, Vol. 33, p. 47, 1985.

16. M.P. McDaniel, "Review of Phillips chromium catalyst for ethylene polymerization," in: *Handbook of Transitional Metal Polymerization Catalysts*, R. Hoff, and R.T. Mathers, Eds., John Wiley & Sons, Inc., Chap. 10, pp. 291-446, 2010.

17. W.L. Carrick, et al., *J. Polymer Science: Part A-1*, Vol. 10, p. 2609, 1972.

18. K.J. Cann, et al., U.S. Patent application US2010/0291334 A1; published Nov. 18, 2010.

19. L.M. Baker, and W.L. Carrick, U.S. Patent 3,324,101, 1967.

20. N.P. Lesnikova, etal., *Kinet. Katal.*, Vol. 20, No. 6, p. 1533, 1979.

21. G.L. Goeke, B.E. Wagner, and F.J. Karol (Union Carbide Corporation), U.S. Patent 4,302,565, issued 1981.

22. F.J. Karol, etal., *J. Polymer Science: Part A-1*, Vol.10, p. 2621, 1972.

23. K.H. Theopold, Understanding chromium-based olefin polymerization catalysts, *CHEMTECH*, Vol. 27, No. 10, pp. 26-32, Oct. 1997.

24. J.P.Coleman, etal., *Principles and Applications of Organotransition Metal Chemistry*, University Science Books, Mill Valley, California, p. 585, 1987.

25. F.J. Karol, C. Wu, W.T. Reichle, and N.J. Maraschin, *J. Catal.*, Vol. 60, pp. 68-76, 1979.

26. K.H. Theopold, *Acc. Chem. Res.*, Vol. 23, pp. 263-270, 1990.

27. J.H. Lunsford, etal., *Langmuir,* Vol. 7, p. 1179, 1991.

28. F.J. Karol, etal., *J. Polymer Science: Polymer Chemistry Ed.*, Vol. 11, p. 413, 1973.

29. C. Wu, and F.J. Karol, *J. Polymer Science: Polymer Chemistry Ed.*, Vol. 12, p. 1549, 1974.

30. S. Fu, and J.H. Lunsford, *Langmuir,* Vol. 6, p. 1774, 1990.

31. S. Fu, and J.H. Lunsford, *Langmuir,* Vol. 6, p. 1784, 1990.

32. S. Fu, and J.H. Lunsford, *Langmuir,* Vol. 7, p. 1172, 1991.

33. M.P. McDaniel, *Ind. Eng. Chem. Res.*, Vol. 27, pp. 1559-1564, 1988.

34. F.J. Karol, etal., *J. Polymer Science: Polymer Chemistry Ed.*, Vol. 16, p. 771, 1978.

35. K.H. Theopold, etal., *Organometallics*, Vol. 14, p. 738, 1995.

36. V.C. Gibson, etal., *J. Chem. Soc., Chem. Commun.*, Iss. 16, p. 1709, 1995.

37. V.C. Gibson, etal., *Chem. Commun.*, Iss. 16, p. 1651, 1998.

38. V.C. Gibson, etal., *J. Chemical Society, Dalton Trans.*, Iss. 6, p. 827, 1999.

39. V.C. Gibson, etal., *Chem. Commun.*, Iss. 10, p. 1038, 2002.

40. M.A. Esteruelas, et al., *Organometallics*, Vol. 22, No. 3, pp. 395-406, 2003.

41. B.L. Small, et al., *Macromolecules*, Vol. 37, p. 4375, 2004.

42. P.W. Jolly, et al., *Organometallics*, Vol. 19, p. 388, 2000.

43. P.W. Jolly, etal., *Organometallics*, Vol. 19, p. 403, 2000.

44. P.W. Jolly, etal., *Organometallics*, Vol. 16, Iss. 8, p. 1511, 1997.

45. W. Keim, etal., *Chem. Commun.*, Iss. 3, p. 334, 2003.

46. F.J. Karol, etal., *J. Polymer Sci.*, Vol. 13, p. 1607, 1975.

47. F.J. Karol, U.S. Patent 4,054,538, filed Dec. 29, 1975, issued to Union Carbide Corp., March 29, 1977.

48. D.G.H. Ballard, *Adv. Catal.*, Vol. 23, p. 263, 1973.

49. O.M. Bade, *Organometallics*, Vol. 17, p. 2524, 1998.

50. T. Monoi, etal., *Polymer Journal*, Vol. 35, No. 7, p. 608, 2003.

51. K.H. Theopold, *J. Am. Chem. Soc.*, Vol. 113, p. 893, 1991

4

Single-Site Catalysts Based on Titanium or Zirconium for the Production of Polyethylene

4.1 Overview of Single-Site Catalysts

The introduction of single-site catalysts for the manufacture of polyethylene is the most recent catalyst innovation that has taken place in the polyethylene industry, with this type of catalyst introduced on a commercial scale in the late 1990s. Single-site catalysts have created new types of polyethylene for the polyethylene industry in which new grades of polyethylene have been introduced into new markets and applications. For the most part, single-site catalysts have not eliminated any other catalyst system from commercial production. These catalysts have primarily expanded the polyethylene industry into new markets.

4.1.1 Expanded Polyethylene Product Mix

Until the mid-1990s, the polyethylene product space consisted of the manufacture of ethylene-based homopolymers and copolymers over a density range of about 0.910–0.965 g/cc with Melt Index values ($MI_{2.16Kg}$) between about 0.05–100.

The introduction of single-site catalyst technology has expanded the polyethylene product space to densities as low as about 0.86 g/cc with traditional comonomers such as 1-butene, 1-hexene and 1-octene. In addition, single-site catalysts have provided new ethylene-based copolymers with comonomers based on styrene and cyclic olefins such as norbornene.

4.1.2 Types of Single-Site Catalysts

There are primarily three types of single-site catalysts used to manufacture polyethylene, each of which will be discussed in detail in this chapter.

1. The first type of single-site catalyst was identified by Walter Kaminsky and coworkers in 1976 at the University of Hamburg and is based on the interaction of a metallocene compound Cp_2MX_2, where Cp is a cyclopentadienyl ring ($C_5H_5^-$), M is a group 4 metal, primarily zirconium, and X may be an alkyl or halide. The cocatalyst is methylalumoxane, which is a mixture of $[-Al(Me)-O-]_n$ oligomers prepared by the slow hydrolysis of trimethylaluminum [1–4]. In this type of catalyst system, the Lewis acidic methylalumoxane acts to abstract an alkide species from the metallocene compound to form a cationic active center $[Cp_2MX]^+$, while the MAO becomes a weakly coordinated anion to form an ion pair. More recently, other methods have been found to form the metallocene-based cationic active center directly on a silica support, thus providing other reagents that can activate the metallocene catalyst [5–7].

2. The second type of single-site catalyst, designated as the constrained geometry catalyst (CGC), was developed by James C. Stevens and coworkers in the late 1980s at Dow Chemical Company in Freeport, Texas. This catalyst type contains one Cp ligand as one component in forming a metallocycle structure. The active site is based on titanium [8].

3. The third type of single-site catalyst is best described as a non-metallocene type of catalyst system, where a ligand system is used that is not based on a cyclopentadiene derivative. This third type of single-site catalyst may utilize a very wide variety of ligands and an early or late transition metal (i.e., iron, cobalt or nickel) may be utilized as the active center [9–13].

Many different variations of these types of catalysts have become commercial since the early 1990s.

This new generation of single-site catalyst has significantly changed the polyethylene industry. As with other technology breakthroughs that have taken place in this industry since 1933 with the development of low-density polyethylene in a high-pressure, free-radical process, these new single-site catalysts have broadened the polyethylene product mix by making available new compositions of polyethylene that possess unique properties. This has led to new applications and markets for polyethylene.

For example, prior to the discovery of this new single-site catalyst type, commercial grades of polyethylene were primarily manufactured over the compositional range of 0–4 mol% of comonomer (1-butene, 1-hexene or 1-octene) that provided ethylene copolymers over the density range of 0.915–0.970 g/cc. Commercial catalysts were primarily the Cr-based Phillips-type of catalyst or a Ti-based Ziegler catalyst with the understanding that both types of catalyst consisted of many different types of active sites. Each type of active site produced a different composition of polyethylene (different molecular weight and branching content) which resulted in a final polyethylene material with a complex molecular structure. These multi-site catalysts limited the composition of the polyethylene that was commercially available due to both process and product constraints imposed by such catalysts.

4.2 Polyethylene Structure Attained with a Single-Site Catalyst

Single-site catalysts primarily produce only one type of polymer structure, which is based on the polymerization characteristics of any particular type of catalyst. The polyethylene has a relatively very narrow molecular weight distribution with a polydispersity of 2.0 (defined as one Flory distribution with an M_w/M_n value of 2.0 due to one type of active center involving a chain transfer process in which each active site produces a large number of individual molecules). The comonomer content of the polyethylene is based on the reactivity of any particular single-site catalyst with the comonomer. The intramolecular and intermolecular comonomer distribution (comonomer/ethylene ratio) is constant and, therefore, the polyethylene molecules have a uniform branching distribution along the polymer backbone.

The uniform branching distribution found in this new type of polyethylene, and the relatively high reactivity of these catalysts with higher 1-olefins, has allowed the manufacture of ethylene/1-olefin copolymers containing very high levels of comonomer such as 1-butene and 1-hexene, i.e., 5–20 mol% comonomer. Consequently, the density range of the polyethylene product mix has been extended to lower densities of about 0.85 g/cc [8]. In fact, many of these new grades of polyethylene are elastomers in which there is sufficient comonomer in the polymer chain to eliminate crystalline regions in the polyethylene structure, making these materials completely amorphous. Such products were not readily attainable with Ti/Mg-based Ziegler catalysts due to the multi-site nature of these catalysts that contain a wide variety of active sites that each have a very different reactivity with higher 1-olefins [14]. Hence, polyethylene produced with Ziegler-type, Ti/Mg-based multi-site catalysts contain molecular species with widely different ethylene/comonomer ratios.

Single-site ethylene polymerization catalysts based on titanium or zirconium as the active center are primarily used in the low-pressure manufacture of polyethylene with a relatively very narrow molecular weight distribution to include high-density polyethylene (HDPE), linear low-density polyethylene (LLDPE) and very low density LLDPE to include amorphous material. Table 4.1 summarizes a few examples of end-use applications for this type of polyethylene in terms of the density, molecular weight distribution and molecular weight for various applications. Note that in Table 4.1 a Melt Flow Ratio (MFR) value of 16 approximately corresponds to an M_w/M_n polydispersity value of 2.0, which is indicative of a single-site catalyst.

Table 4.1 Examples of commercial products manufactured with single-site catalysts.

Product Type	Melt Index (MI) $(I_{2.16})$[a]	Flow Index (FI) $(I_{21.6})$[b]	MFR (FI/MI)[c]	Density (g/cc)
Film[d]	0.5–2	8–32	16	0.910–0.930
Elastomers/Film	1–5	16–80	16	0.85–0.910
Injection molding	5	80	16	0.916–0.928
	30	480	16	0.916–0.928
Rotational molding	10	160	16	0.925–0.950

[a] Melt Index is an inverse relative measure of polymer MW.

[b] Flow Index is a measure of the polymer processability.

[c] Melt Flow Ratio of the FI and MI values is a relative measure of the polymer MWD where an MFR value of 16 indicates that the MWD is very narrow due to the single-site catalyst that produces PE with a single Flory MWD with a Mw/Mn value of 2.0. Melt Index, Flow Index, MFR and density are only approximate values for each application.

[d] Film products contain a comonomer (1-butene, 1-hexene or 1-octene) which is used to reduce polymer crystallinity. The type of comonomer used has a significant effect on film toughness properties; 1-butene is used for general purpose film applications and 1-hexene or 1-octene used for more demanding applications.

The elastomers and film products highlighted in Table 4.1 represent examples of the introduction of new products into the polyethylene markets that were the direct result of the development of single-site catalysts. The injection molding and rotational molding markets remain dominated by Ti-based Ziegler catalysts.

4.2.1   Product Attributes of Polyethylene Manufactured with Single-Site Catalysts

Polyethylene manufactured with a single-site catalyst possesses several advantages relative to a similar material produced with a titanium/magnesium-containing, second-generation Ziegler catalyst. These include:

- Lower heat seal temperature useful in food and other packaging applications.
- High film clarity due to the homogeneous branching distribution in which the branch frequency (i.e., branches/1000 carbons) is independent of polymer molecular weight.
- Balanced film mechanical strength values (i.e., tear strength in machine direction, MD, and transverse direction, TD, are comparable) which are due to the very narrow MWD of this type of polyethylene.
- The ability to introduce LLDPE products with lower densities (ca. 0.80–0.915 g/cc) due to the more homogeneous branching distribution and higher reactivity with higher 1-olefins such as 1-butene, 1-hexene and 1-octene. Titanium-based catalysts are usually limited to providing LLDPE materials with a density of about 0.915 g/cc.
- High reactivity of other comonomers such as styrene for the manufacture of ethylene/styrene copolymers, providing new compositions of polyethylene.

4.2.2   Processing Disadvantage of Polyethylene Manufactured with Single-Site Catalysts

The very narrow molecular weight distribution characteristic of single-site catalysts makes the processing of this type of polyethylene more difficult and is, therefore, considered a disadvantage over other grades of polyethylene with a slightly broader MWD. Polyethylene with a very narrow molecular weight distribution results in less shear-thinning (a reduction of the melt viscosity with increasing shear) during the fabrication process.

Consequently, these materials are more difficult to process into finished goods, which may lower manufacturing rates and increase production costs.

Consequently, one objective for scientists working with polyethylene produced from single-site catalysts has been to find methods to improve the processability of polyethylene manufactured from these catalysts. Scientists at Dow Chemical Company, for example, have found that by introducing some long-chain branching into the polymer structure, which is possible with Dow's constrained geometry single-site catalyst, provides a polyethylene with improved processability. Dow's INSITE catalyst, for example, produces polyethylene with a low degree of long-chain branching that improves the shear thinning properties (processability) of INSITE-based polyethylene. Chevron Phillips scientists have improved processability in a similar manner by using advanced metallocene catalysts to also provide a low degree of long-chain branching.

Materials scientists have taken other steps to improve processability such as blending high-pressure LDPE or more advanced resins such as ethylene/norbornene-based polyolefins into the base resin to improve shear thinning behavior.

This chapter will be limited to single-site catalysts important only in the polyethylene industry. There has been an enormous amount of research carried out with this type of catalyst, which is capable of polymerizing a wide variety of monomers other than ethylene. Other monomers extensively investigated include propylene, higher 1-olefins, norbornene, styrene, cyclic olefins, linear conjugated dienes and acrylate. This research has been summarized in the book *Stereoselective Polymerization with Single-Site Catalysts* [15]. Appendix 4.1 summarizes some books and review articles that have been published since the early 1990s that pertain to ethylene polymerization and copolymerization with single-site catalysts.

4.3   Historical Background

Scientists as early as the mid-1950s recognized the importance of using metallocene compounds in conjunction with aluminum alkyls for olefin polymerization.

The interaction of bis(cyclopentadienyl)titanium dichloride [$(Cp)_2TiCl_2$] and diethylaluminum chloride in toluene was reported in 1957 by Breslow [16–18] as a weakly active ethylene polymerization catalyst if the DEAC was allowed to react with the titanium compound for more than one hour. Much more active catalysts were found if oxygen was introduced with the ethylene into the polymerization system. These oxygen-modified homogeneous catalysts were comparable in activity to Ziegler's Mülheim heterogeneous catalyst system based on $TiCl_4$ and triethylaluminum (TEAL). The activity of these catalysts was ca. 100–200 g PE/g Ti/hr/atm ethylene.

One interesting feature of Breslow's research with these homogeneous catalyst systems was the type of polyethylene which was isolated. Polyethylene prepared with these catalysts was more linear, as indicated by a lower methyl content (melting point of 137°C), and had a narrower molecular weight distribution than the polymer provided by the Mülheim catalyst. It is possible that some of these homogeneous systems may have been "single-site" catalysts, in which case this type of polyethylene would have had a MWD as indicated by the Mw/Mn value of 2 or one Flory component in the polymer.

However, almost twenty years later, Breslow [19] reported in 1975 a similar polymerization system with very high ethylene polymerization activity if Cp_2TiCl_2 was added to toluene in the presence of dimethylaluminum chloride (DMAC) previously treated with water. The combination of Cp_2TiCl_2 and only dimethylaluminum chloride (DMAC) in the absence of water had low activity as an ethylene polymerization catalyst. The water/Al molar ratio of 0.2–0.5 provided a highly active catalyst, while a higher water/Al ratio > 0.5 provided a very low activity catalyst. Some of Breslow's data is shown in Table 4.2.

Long and Breslow postulated at the time that the interaction of water with dimethylaluminum chloride provided an intermediate material $[Me-Al(Cl)-0]_n$ that promoted the catalyst activity by interacting with the Cp_2TiCl_2. Breslow proposed a structure shown in Figure 4.1 where an alumoxane dimer underwent a methyl exchange reaction with Cp_2TiCl_2.

$$Cp_2TiCl_2 + Me-Al(Cl)-O-Al(Cl)-Me \rightarrow Cp_2Ti(Cl)Me + Al(Cl)_2-O-(Cl)Me$$

A similar catalyst water-activation affect was reported by Reichert and Meyer in 1973, which was two years earlier than the Breslow results, in which the catalyst system Cp_2Ti EtCl/EtAlCl$_2$ and water was investigated [20].

Table 4.2 The effect of water/AL molar ratio on catalyst activity.

$(H_2O/Al)_{molar}$	Activity (g PE/g Ti/hr/bar ethylene)
0.0	< 2 50
0.13 – 0.40	2,500
0.53 – 0.66	< 2 50

Conditions: 0.05 mmol Cp_2TiCl_2; 50 ml toluene, 30°C, 0.7 bar ethylene, 0.9 mmol DMAC/liter.

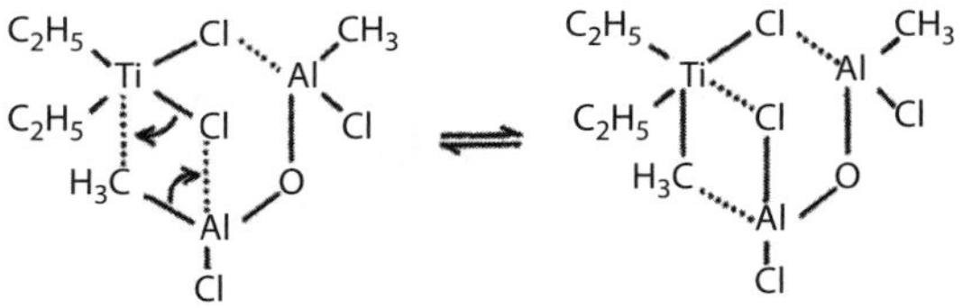

Figure 4.1 Breslow and Long's proposed structure. Reprinted from [19] with permission from Wiley-VCH Journal.

4.3.1   First Single-Site Catalyst Technology – Canadian Patent 849081

Perhaps the first report on the preparation of single-site catalysts that lead to the discovery of ethylene/1-olefin copolymers with a homogeneous branching distribution was provided in Canadian Patent 849081. In what can be considered as a classic patent in the polyethylene industry, due to the relative importance of the discovery disclosed in the patent, the Canadian patent issued to Clayton T. Elston and assigned to DuPont of Canada on August 11, 1970, disclosed homogeneous vanadium-based catalysts activated with diethylaluminum chloride/ethylaluminum dichloride mixtures for the preparation of ethylene/1-octene copolymers with a homogeneous branching distribution.

This patent disclosed a series of single-site, vanadium-based coordination catalysts for the preparation of ethylene/1-olefin copolymers with an intermolecular uniform branching distribution. The patent describes this preferred branching distribution as a homogeneous branching distribution in which the comonomer is randomly distributed within a given molecule, while the copolymer molecules contain the same ethylene/comonomer ratio. In addition, the copolymers produced with these coordination catalysts possess a narrow molecular weight distribution.

Elston defined homogeneously branched copolymers as those in which the comonomer is randomly distributed within a given molecule with all copolymer molecules having the same ethylene/comonomer ratio. Elston's data showed that such copolymers exhibit a narrow molecular weight distribution, provide film products with reduced haze levels and film products with higher impact strength and a better balance of physical properties in the machine and transverse direction, when compared with similar polyethylene products that exhibit a heterogeneous branching distribution.

Elston used copolymer melting point values to determine the branching homogeneity of copolymers noting that ethylene/1-olefin copolymers with a heterogeneous branching distribution show crystalline melting points at a given comonomer content, which are significantly higher than the melting points of homogeneously branched copolymers with the same comonomer content.

At the time, DuPont operated a 275 million pound/year solution-polymerization process in Canada (Sclair Process) that was commercialized in the 1960s for the manufacture of LLDPE. However, DuPont sold the business to Canadian-based Nova Corporation in the mid 1970s.

The coordination catalysts described by Elston were based on vanadium compounds and activated with ethyl aluminum dichloride (EADC), mixtures of EADC and diethyl aluminum chloride (DEAC) or isoprenyl-aluminum (IPRA). The type of branching distribution obtained in the ethylene copolymers were a function of the specific vanadium compound and cocatalyst used in the polymerization experiment. Some typical examples from this patent are summarized in Table 4.3.

Elston used polyethylene melting point data to readily ascertain the type of branching present in any particular polyethylene sample. It should be noted that samples containing a similar amount of comonomer are required in order to identify the type of branching present. A polyethylene sample with a heterogeneous branching distribution contains relatively higher molecular weight polymer components with low comonomer content. Such a polymer component contains crystalline regions with a higher melting point, which is readily apparent from the DSC melting point data and is displayed in the data in Table 4.3 by a melting peak at 119.6°C for sample 1. The polyethylene (sample 2) with a homogeneous branching distribution, and a similar amount of comonomer relative to the heterogeneous branching sample 1, displayed a much lower melting point of 105.6°C. Polyethylene film fabricated from polymer samples similar to

Table 4.3 Vanadium-based catalysts described in Canadian Patent 849081.

Catalyst/ Cocatalyst	Comonomer Type	Branching per 1000 Carbon	Branching Type	Density (g/cc)	Melting (°C)
VOCl$_3$/ isoprenylAl[a]	1-butene	17.0	Heterogeneous	0.9201	116.5;119.6
VO(O-n-butyl)$_3$/ EADC	1-butene	14.8	Homogeneous	0.9193	105.6
VO(O-n-decyl)$_3$/ EADC DEAC mixture 1:1	1-octene	47.9	Homogeneous	0.8770	63.0

[a] See J.J. Ligi and D.B. Malpass, *Encyclopedia of Chemical Processing and Design*, Marcel Dekker, NY, Vol. 3, p. 32, 1977, for a detailed discussion of isoprenylaluminum (IPRA).

samples 1 and 2, also demonstrated that the film obtained from sample 2 with a homogeneous branching distribution possessed very good film clarity (13.9 Haze) compared to the film obtained from sample 1 (31.1 Haze) with a heterogeneous branching distribution.

The ethylene/1-octene polymer in sample 3 contained a high level of comonomer (9.4 mol%) and provided a very low density material (0.8770 g/cc) and possessed elastomeric properties and a very low melting point, 63.0°C.

Additional data provided in this patent also showed that the branching content of polyethylene samples with a homogeneous branching distribution was independent of molecular weight. This was demonstrated by fractionating the polymer into eight separate fractions with increasing molecular weight and determining the branching content and molecular weight of each fraction.

4.3.2   Discovery of Highly Active Metallocene/Methylalumoxane Catalysts

In the mid 1970s, Professor Kaminsky and coworkers were investigating the reduction of Cp_2TiMe_2 with trimethylaluminum (TMA) when they discovered that the Cp_2TiMe_2 compound became a very active ethylene polymerization catalyst by the addition of small amounts of water added to the TMA before reaction with the metallocene [21].

As is often the case with important scientific discoveries, Kaminsky outlined the serendipitous events that took place in the laboratory in the mid 1970s that lead to this discovery [4]. These events are briefly summarized.

A Ph.D. student in Professor Kaminsky's laboratory was investigating the reduction of Cp_2TiMe_2 with TMA by NMR methods. Samples were prepared in NMR tubes at −78°C and the glass tubes were then sealed by melting the glass tube with a torch. Next, the tubes were placed into the NMR machine and allowed to slowly warm to room temperature. In order to save time, the student prepared solutions of Cp_2TiMe_2 in toluene in NMR tubes sealed with only a plastic lid without the TMA added. Because these tubes became slightly contaminated with small amounts of water vapor, future NMR experiments with the TMA added showed unexpected $-CH_2$ bonds in the NMR spectrum. Ethylene polymerization experiments carried out in an autoclave with these solutions exhibited slow ethylene polymerization. When the autoclave was opened to take a sample for NMR analysis and then reclosed, it was noted that the ethylene polymerization rate increased significantly. After opening and then closing the autoclave several times, it was noted that the ethylene polymerization rate increased each time. After eliminating oxygen and traces of chloride as the possible contaminants responsible for the ethylene polymerization behavior, it was discovered that water was the critical reagent. The polymerization rate was dependent on the amount of water added with the highest polymerization rate found at a water/TMA ratio of one.

Professor Kaminsky reported on the Zirconocene/MAO catalyst system for the first time at the IUPAC Polymer Conference in Florence, Italy, in 1980 and at the Macromolecular Meeting in Midland, Michigan, in 1981. Kaminsky reported that the Zirconocene/MAO catalyst system was 10 to 100 times more active than the initial Ziegler catalyst and that the zirconium-based activity was 40×10^6 g PE/g Zr/h at 95°C and 8 bar ethylene pressure.

4.3.2.1   Early Publications of Kaminsky

In order to circumvent the results of Breslow and Reichert that had both reported the increase in ethylene polymerization rate by addition of water to titanium-based metallocene compounds with other aluminum alkyls, Kaminsky and coworkers continued their research with mostly zirconium-based metallocenes activated by TMA/water mixtures [1–3].

These early results were obtained by investigating the ethylene polymerization behavior of zirconium and aluminum containing metallocene complexes shown as structures 1 and 2 in Figure 4.2 using trimethylaluminum or triethylaluminum (TEAL) treated with water as activators [1]. At the time, structures 1 and 2 were unique in terms of the $Zr-CH_2-CH_2$ bond angle of only 76° exhibited in 1 and 2, while compound 2 contained an ethylene bridging group. Both halogen-containing compounds were very active homogeneous catalysts in ethylene polymerization. In a second paper [2,3], Kaminsky examined the ethylene polymerization behavior of several additional metallocenes based on titanium and zirconium and reported extremely high activity of 400,000 to 600,000 g polyethylene/g titanium for Cp_2TiMe_2 at 21°C, ethylene pressure of 8 bar and a polymerization time of 1.5 hr. Maximum catalyst activity required an Al/H_2O ratio of 2 to 5. In these early polymerization experiments, the alumoxane material was prepared *in situ* by the slow addition of water to the aluminum alkyl at 12°C using toluene as the polymerization solvent followed by the addition of the metallocene compound.

The 1980 paper by Sinn, Kaminsky, Vollmer and Woldt [3] is probably the classic paper that initiated enormous interest in the metallocene/alumoxane catalyst system in both industrial and academic laboratories worldwide. In this paper it became clear that the addition of a metallocene compound

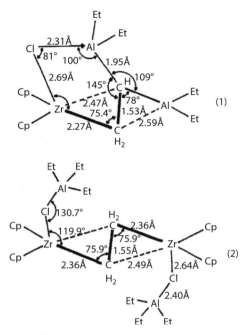

Figure 4.2 Structures of Zr/Al complexes. Reprinted from [1] with permission from John Wiley and Sons.

and a separate alumoxane material provided a living polymerization system with extremely high catalyst activity. The methyl alumoxanes were synthesized as separate reagents to various degrees of oligomerization as shown below, thus eliminating the *in situ* hydrolysis step that was used previously.

$$Al(CH_3)_3 + nH_2O \rightarrow [-Al(CH_3)-O-]_n + 2nCH_4$$

Catalyst activity increases with higher degrees of oligomerization with the highest activity obtained with methylalumoxane (MAO), where n was between 5–12. In the case of $Cp_2Zr(CH_3)_2$ an extremely high catalyst activity of 10^7g polyethylene/g Zr was reported with a methylalumoxane that had an n value of 12 and an Al/Zr molar ratio of approximately 15,000. Polymerization was carried out at 70°C, 8 bar ethylene in toluene for two hours. Polyethylene molecular weight was easily controlled with polymerization temperature without hydrogen added as a chain transfer agent.

In a later paper, Herwig and Kaminsky [22] were able to isolate by fractional precipitation one methylalumoxane compound that was shown to have a cyclic structure containing five $-[Al(CH_3)-O-]$ units, as shown in Figure 4.3 with the corresponding mass spectrum which was used to postulate the cyclic structure. Note the mass peaks at 275, 217, 159, 101 and 43 correspond to the elimination of one $Al-(CH_3)-O$ unit, sequentially from the cyclic structure.

4.3.2.2   Kinetic Parameters of the Homogeneous Cp_2ZrCl_2/MAO Catalyst

Chien and Wang [23] determined the initial active site concentration of this homogeneous catalyst system using a radiolabeling technique based on CH_3OT and found that 80% of all the Zr atoms are catalytically active at 70°C and an Al/Zr molar ratio of 1,000, and that 100% of the Zr atoms are active at an Al/Zr ratio of 10,000. Lowering the polymerization temperature to 50°C and the Al/Zr ratio to 550 reduces the Zr active site content to 20%. The ethylene propagation rate constant at 70°C and an Al/Zr ratio of 11,000 was calculated to be 1.6×10^3 M/sec with the maximum ethylene polymerization rate obtained in about one minute at 1.06×10^{-3} M/sec and the rate decayed to about one-half the maximum rate of polymerization in about 40 minutes.

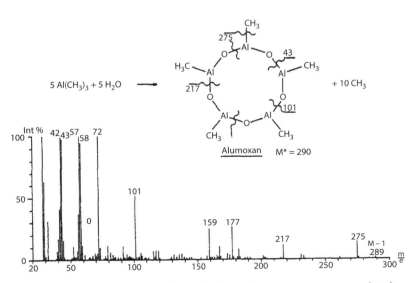

Figure 4.3. J. Herwig and W. Kaminsky isolated a cyclic alumoxane compound with five Al-O- unit, data from [22]. Reprinted from [22] with permission from Springer Publishing.

As pointed out by Kaminsky and Arndt [24], this Cp_2ZrCl_2/MAO catalyst is about 10–100 times more active in ethylene polymerization than the conventional Ziegler systems, as an activity of 4×10^7 g polyethylene/g Zr/hour are obtained, which translates to about 46,000 polymer molecules/hour or one ethylene insertion in 3×10^{-5} seconds.

4.3.3   Alkylalumoxanes – Preparation, Structure and Role in Single-Site Technology

4.3.3.1   Background

Methylalumoxanes (MAO) have become important commercial products in the preparation of single-site catalysts for the manufacture of polyethylene. Therefore, a brief discussion on the preparation, structure and role of MAO is necessary.

Alkylalumoxanes are oligomeric organometallic compounds with the general formula:

$$R_2\text{-Al-O-}[\text{Al-O}]_n\text{-AlR}_2,$$

where the degree of oligomerization,(n), may vary over a wide range. However, because of the strong Lewis acidity of the Al atom, the representation of the alumoxane species as a three-coordinate aluminum center is not an accurate description of the structure of alumoxane compounds, which will be discussed in more detail below.

Alkylalumoxane compounds became readily attainable in the late 1950s when alkylaluminum compounds were produced in relatively large quantities to support the Ziegler-Natta catalyst technology that had been discovered to provide a low-pressure route to polyethylene. Therefore, alkylalumoxane compounds were extensively studied in the 1960s as catalysts in the polymerization of epoxides, aldehydes and olefins [25].

4.3.3.2   Preparation

The preparation of alumoxanes is carried out by the slow, controlled hydrolysis of an aluminum alkyl with water to eliminate an alkane. An example of one such reaction is shown below utilizing an Al/water ratio of two, which results in a dimeric species.

$$8AlR_3 + CuSO_4{:}5H_2O \rightarrow R_2Al\text{-O-AlR}_2 + CuSO_4{:}H_2O + 8RH$$

Because of the high reactivity of water with aluminum alkyls, and the difficulty in adding water in a controlled manner, the use of hydrated metal salts such as cupric sulfate pentahydrate, $CuSO_4 \cdot 5H_2O$ or the aluminum sulfate $Al_2(SO_4)_3 \cdot 14H_2O$, has become a widely used method in the preparation of methylalumoxanes [26]. The work of Kaminsky and coworkers since the late 1970s, when methylalumoxane was found to be a unique cocatalyst in conjunction with metallocene compounds, led to renewed interest in alumoxanes and, more particularly, methylalumoxane.

Pasynkiewicz [27] summarizes a wide variety of other methods that may be used to synthesize alumoxanes using reagents such as CO_2, $RCONR_2$, $RCOOH$, PbO and R_4B_2O in place of water.

The hydrolysis of alkylaluminum compounds is discussed in considerable detail by A. R. Barron [28], who also provides some alternative routes to synthesize alkylalumoxane compounds. Reaction temperature (ambient to low temperature) and Al/water ratio are the primary variables to control the specific composition of the alkylalumoxane mixture obtained. Note that Barron strongly recommends, as a safety precaution, that any alkylalumoxane synthesis NOT be carried out at an elevated temperature.

4.3.4   Structure of Alumoxanes

Commercial samples of methylalumoxane are available as 30 wt% (MAO) in toluene which contain approximately 14 wt% Al. This solution is thrixotopic and on standing forms a gelatin material due to the dipole-dipole intramolecular and intermolecular interactions between the Al and oxygen atoms present in these materials. The solution is a complex mixture of linear, branched and cyclic structures that coexist in solution. In addition, various equilibrium processes between different structures, such as the reaction illustrated in Figure 4.4, demonstrate the dynamic behavior of these complex solutions [27]. As will become clear in the structures shown in Figure 4.4, aluminum does not have a simple coordination number of three because it is common for aluminum to achieve a coordination number of four due to the strong Lewis acidity of the highly positive Al atom.

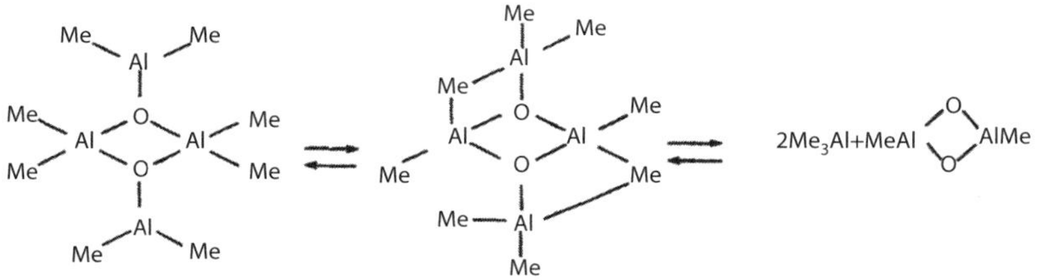

Figure 4.4 Possible equilibrium processes in MAO solution. Reprinted from [27] with permission from Elsevier Publishing.

As pointed out by Pasynkiewicz [27], because of the complex nature of these MAO solutions, and the wide range of structures possible, relatively few structures have been determined, and the structures that have been established usually involve bulky alkyl groups or are anionic species that are more readily isolated due to their propensity to precipitate from solution.

Pasynkiewicz [27] and Barron [28] show the structure of the $(Al_7O_6Me_{16})^-$ anion, which was obtained by Atwood et al. as a potassium salt with a molecule of coordinated benzene [29], which was used as solvent for the hydrolysis reaction, and was retained in the lattice with the molecular formula $K(Al_7O_6Me_{16}):C_6H_6$. This anion, shown in Figure 4.5, demonstrates the complexity of these materials and the propensity of the aluminum atom to achieve a coordination number of 4. The paper of Atwood et al. is thought to be the first crystallographic evidence showing the four-coordinate nature of Al in these compounds. The anion consists of an Al_6O_6 ring capped by a seventh aluminum atom which is bonded to three alternate oxygen atoms in the ring. The six-ring aluminum atoms are each bonded to two terminal methyl groups, while the unique aluminum atom in the center of the structure is attached to only one methyl group.

Atwood and coworkers [30] have also obtained the most basic anion, which results from the dimeric species $[Me_2Al-O]_2^{2-}$ with a molecule of $AlMe_3$ coordinated to each of the two bridging oxygen atoms. This anion is illustrated in Figure 4.6.

Probably the best description of the actual structure of alumoxane solutions, as pointed out by Barron [31], is to consider the $Al-O-(Al-O)_n-Al$ oligomeric backbones to stack as three-dimensional clusters or cages [32], as illustrated in Figure 4.7.

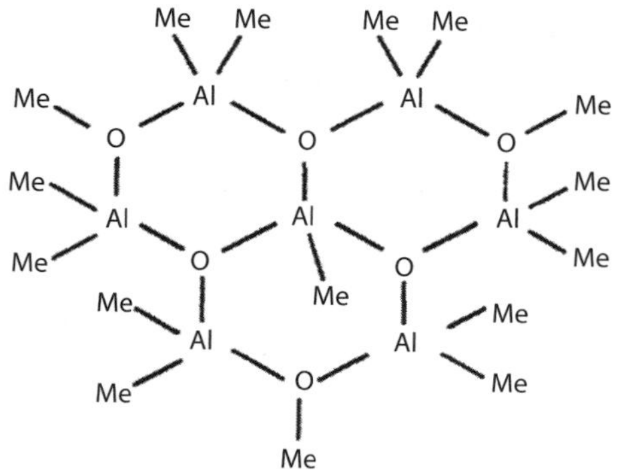

Figure 4.5 Structure of the anion $K(Al_7O_6Me_{16}):C_6H_6$ reported by Atwood et al. Reprinted from [29] with permission from the American Chemical Society.

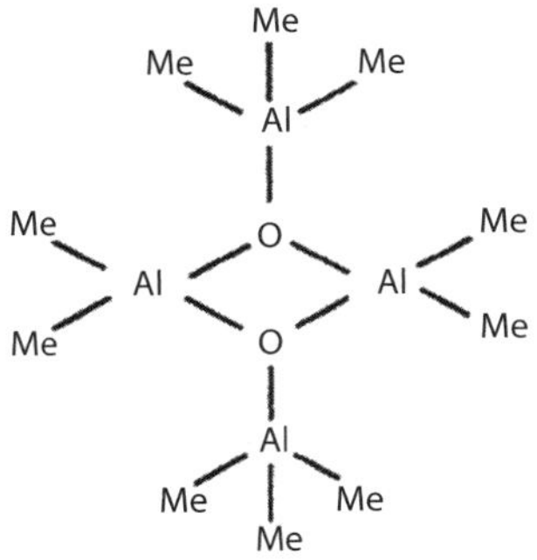

Figure 4.6 Structure of the anion [Me$_2$Al-O]$_2^{2-}$ reported by Atwood and coworkers. Reprinted from [30] with permission from Elsevier Publishing.

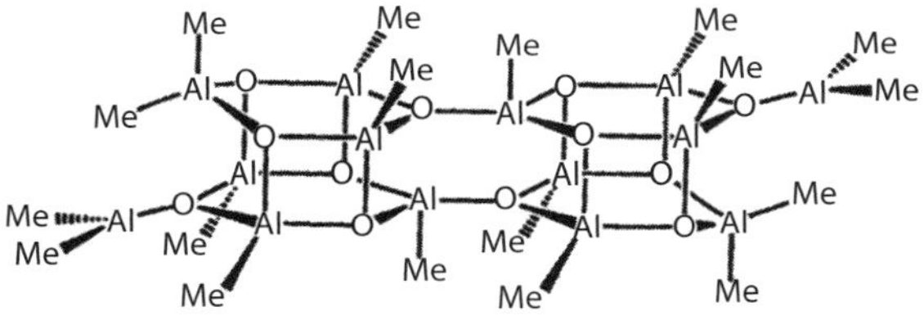

Figure 4.7 Description by Barron of the three-dimensional stacked structure of MAO which would explain the thixotropic behavior of these solutions. Reprinted from [32] with permission from John Wiley and Sons.

4.3.4.1   Role of Methylalumoxane in Single-Site Catalysts

The role of methylalumoxane (MAO) as a cocatalyst to activate zirconocene compounds such as Cp$_2$ZrCl$_2$ to create a single-site ethylene polymerization catalyst is similar, in some respects, to the role simple aluminum alkyls (AlR$_3$) play in activating Ziegler catalysts. For example, MAO acts as an alkylating agent to form the initial Zr-carbon bond (Zr-CH$_3$) necessary to initiate the polymerization process. However, experimental evidence obtained by a variety of methods clearly has shown that the MAO reacts with the zirconium center to form a zirconium cation of the type [Cp$_2$Zr-CH3]$^+$ in which the zirconium is not reduced to a lower oxidation state, but remains as a d^0 metal and Zr(IV) oxidation state. The MAO, therefore, forms an anion moiety to complete the ion pair necessary to create the active species, as illustrated in Equation 4.1.

$$Cp_2ZrCl_2 + MAO \rightarrow [Cp_2Zr-CH_3]^+ [MAO]^- \text{ (Formation of Zirconium/}$$
MAO ion pair) (4.1)

The ion pair shown above suggests that the more important role of MAO in activating the metallocene compound as an ethylene polymerization center is to stabilize the zirconium cation by providing a bulky non-coordinating moiety similar to the well-established bulky anion tetraphenyl borate $[(C_6H_6)_4B]^-$.

4.3.4.2 Supporting Evidence for Cationic Active Site for Ethylene Polymerization

In a classic paper in the field of Ziegler catalysis, experimental evidence into the insertion of an unsaturated monomer into a titanium-methyl bond associated with a titanium cation was first reported by Eisch *et al.* [33].

Eisch and coworkers (Figure 4.8) investigated the reaction, at −20°C, of equimolar amounts of Cp_2TiCl_2 (2) and methylaluminum dichloride (3) in chloroform with trimethyl(phenylethynyl)silane (1) to yield the ionic insertion product shown in Figure 4.8 as compound (4). Eisch shows that two equilibria take place in which the titanocene and aluminum compound form an intermediate species (5) which is in equilibrium with the (6) titanium-containing cation, which is the active species for the insertion reaction.

The solution reaction to form (4) was monitored by ^1H, ^{13}C and ^{27}Al NMR spectroscopy and these data conclude that (4) is the only insertion product

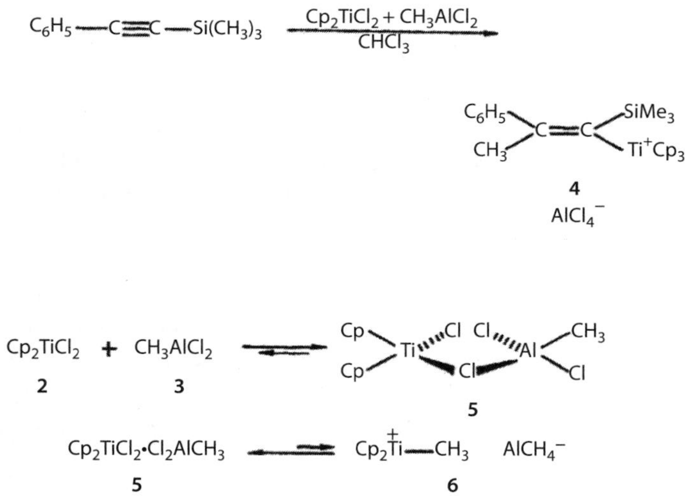

Figure 4.8 Proposed formation of $[Cp_2TiMe]^+$ cation by J.J. Eisch. Reprinted from [33] with permission from the American Chemical Society.

formed from this catalyst system, demonstrating that the unsaturated monomer (1) undergoes insertion into the Ti-carbon in a regiospecific and syn-stereospecific manner. Consequently, the unsaturated monomer coordinates to the titanium cation in only one of two possible directions (regiospecific) and produces only one insertion product in which the trimethylsilyl group and phenyl groups are in a cis configuration with respect to the carbon-carbon double bond. The structure of the insertion product as reported by Eisch is shown in Figure 4.9.

Important features of the above structure are bond angles, (°): Ti-C-Si, 88.9; Ti-C=C, 144.9; Si-C=C 125.6; C=C-CH$_3$, 123.5; C=C-Phenyl, 122; and CH$_3$-C-phenyl, 114.3. In addition, the Ti, Si, vinylic carbons, CH$_3$ and first carbon atom in the phenyl ring are approximately planar in configuration.

Eisch and coworkers conclude that the crucial step in the activation of this catalyst is the isomerization of species (5) into (6) with this conclusion possible only based on the prior work of Long [34] and what Eisch terms the elegant kinetic studies of Fink [35]. Fink's data was obtained with the homogeneous solution of Cp$_2$Ti(R)Cl, where R is propyl and ethylaluminum dichloride in ethylene polymerization, and found that the first of the two equilibria shown in Figure 4.8 is very much shifted to the right, while the second equilibrium can only be shifted to the right by a mass action effect using high Al/Ti ratios.

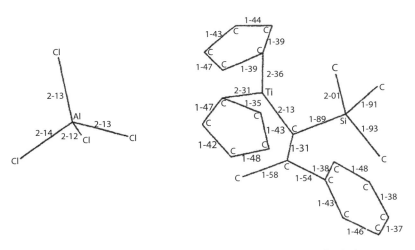

Figure 4.9 Structure of the single insertion product [C$_6$H$_5$(CH$_3$)C=C(TiCp$_2$) Si(CH$_3$)$_3$]$^+$[AlCl$_4$]$^-$. Note: Not all lines represent chemical bonds. For example, the distances between Ti and Cp rings are merely the distances of closest approach. Reprinted from [33] with permission of the American Chemical Society.

Jordan and coworkers [36] reported the reactive cationic complexes based on dicyclopentadienylzirconium (IV), $[Cp_2Zr(CH_3)(THF)][BPh_4]$, which was prepared from the corresponding acrylonitrile complex $[Cp_2Zr(CH_3)(CH_3CN)][BPh_4]$, by displacing the CH_3CN ligand with THF. Both complexes isolated by Jordan *et al.* contain the $[Cp_2Zr(CH_3)]^+$ cation. Both complexes were shown to easily react in CH_3CN as solvent with carbon monoxide (CO) to insert the CO between the Zr-carbon bond to yield the corresponding complex $[Cp_2Zr(COCH_3)(CH_3CN)][BPh_4]$. Note that carbon monoxide is isoelectronic with ethylene. In a following paper [37], Jordan *et al.* reported the structure of $[Cp_2Zr([CH_3)(THF)][BPh_4]$ shown in Figure 4.10.

As pointed out by Jordan *et al.*, some key structural features of the cation are the $Zr-CH_3$ and Zr-Cp ring bond distances that are 0.02 and 0.04 angstroms shorter than the same distances in the similar neutral complex $Cp_2Zr(CH_3)_2$ [38]. The THF ligand is oriented nearly perpendicular to the plane formed by the C_{35}-Zr-O atoms.

Jordan *et al.* investigated the ethylene polymerization behavior of the $[Cp_2Zr(CH_3)]^+$-containing complex. In dichloromethane at 25°C, the cationic complex produced linear polyethylene with a minimum activity of 0.2 g PE/mmol catalyst/minute/bar ethylene pressure. Characterization of the polyethylene produced from the catalyst showed an M_w value of 33,000 and an M_w/M_n value of 2.58, suggesting approximately single-site behavior.

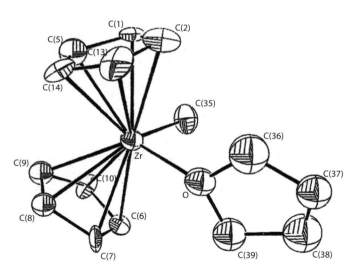

Figure 4.10 Structure of $[Cp_2Zr([CH_3)(THF)]^+[BPh_4]^-$ anion omitted for clarity. Reprinted from [37] with permission of the American Chemical Society.

The polymerization did not require an additional aluminum alkyl, which shows that the Zr cation is an ethylene polymerization active site.

Consequently, the work reported by Jordan and coworkers supports the role of the methylalumoxane postulated above in which the MAO forms the initial Zr-CH$_3$ bond, abstracts a X$^-$ counter ion from the Zr center, and acts as a bulky, weakly coordinated anion which stabilizes the ion pair shown above in Equation 4.1.

4.3.5   Additional Methods for Activating Metallocene Single-Site Catalysts

The work of Jordan and his coworkers was important because it supported the hypothesis that the zirconium cation was the active site in these single-site catalysts and led to the search for finding other methods to form the zirconium cation without using MAO as the activating agent.

Kissin *et al.* reported new cocatalyst systems based on sterically hindered organoaluminates [40] or mixtures of dialkylaluminum chlorides (R$_2$AlCl) and dialkylmagnesium compounds (MgR$_2$) [39,41] at an Al/Mg molar ratio >2 for activating metallocene complexes for ethylene polymerization. Catalysts were prepared with and without silica. The silica-supported catalyst was prepared [39] in two steps. First, a toluene solution of the metallocene compound was mixed with toluene solutions of AlEt$_2$Cl and MgBu$_2$. The Al/Mg molar ratio in the mixtures was 3 and the Al/Zr molar ratio varied from 20 to 30. In the second step, the mixtures were combined with Davison Grade 955 silica (previously calcined at 600°C) at an Al/silica ratio of 3 mmol Al/g, and the toluene was removed with a nitrogen flow at 50–55°C to obtain a free-flowing powder. Unsupported catalysts were prepared *in situ* by adding toluene solutions of the catalyst components directly to the polymerization vessel. The catalysts were evaluated in ethylene/1-hexene copolymerization experiments and some of these results are shown in Table 4.4.

The copolymer melting point data shown in Table 4.4 indicate that the copolymers prepared with this novel cocatalyst system possess uniform branching distribution consistent with single-site catalysts. However, gel permeation chromatography (GPC) data of the copolymers found that the copolymers have a relatively broad molecular weight distribution as indicated by M_w/M_n values in the range of 10–15. Consequently, each of the catalysts investigated consists of more than one type of active site, each producing a copolymer with a different molecular weight, thus producing a polymer with a broader molecular weight distribution.

Table 4.4 Polymerization of zirconocene compounds activated with mixtures of dialkylaluminum chlorides (R_2AlCl) and dialkylmagnesium compounds (MgR_2) [39].

Catalyst Type	Temp (°C)	Silica (yes/no)	Ethylene (psi)	1-hexene Molar	Activity (kg PE/g Zr(hr))	1-hexene mol% (DSC mp)
$(n\text{-}BuCp)_2$ $ZrCl_2$	80	no	140	1.38	65.2 (0.5)	0.9
$C_2H_4(Ind)_2$ $ZrCl_2$	80	no	180	1.66	134.8 (0.5)	5.3 (99.6)
$C_2H_4(Ind)_2$ $ZrCl_2$	80	no	180	1.38	195.6 (2.0)	2.0 (118.7)
$C_2H_4(Ind)_2$ $ZrCl_2$	90	yes	180	1.75	84.8 (1.0)	4.4 (105.1)
$Me_2Si(Cp)(Flu)$ $ZrCl_2$	90	no	180	1.75	72.8 (1.0)	3.4 (106.9)
$Me_2Si(Cp)(Flu)$ $ZrCl_2$	90	yes	180	1.75	8.1 (1.0)	–

In a follow up paper, Kissin and coworkers [41] postulate that the reaction of the aluminum alkyl and the magnesium alkyl produce an ionic material of the general formula shown below, which acts as a Lewis acid to activate the metallocene complex.

Cocatalyst ion pair formation [39,41]:

$$MgR'_2 + R_2AlR' \rightarrow [MgR']^+[R_2R'_2Al]^-$$

Formation of Zr cation active center by cocatalyst ion pair [39,41]:

$$Cp_2Zr(R)Cl + [MgR']^+[R_2R'_2Al]^- \rightarrow [Cp_2Zr\text{-}R]^+[R_2R'_2Al]^- + MgR'Cl$$

Kissin also explains that the silica-supported catalysts are not suitable for a high-temperature (80–90°C) hydrocarbon slurry polymerization process, as the zirconium catalyst component is not chemically fixed to the silica surface and the zirconium catalyst component is extracted from the silica due to its solubility in the hydrocarbon medium resulting in poor polymer particle morphology. However, the catalyst may be retained in the silica pores during a gas-phase polymerization process and, therefore, may provide good particle morphology necessary for a gas-phase process. Consequently, this unique cocatalyst activator may be able to provide a polyethylene product with a uniform branching distribution and a broader MWD.

4.3.6   Characterization Methods that Identify Polyethylene with a Homogeneous Branching Distribution Obtained with Single-Site Catalysts

By the early 1980s it became clear that the zirconocene/methylalumoxane catalyst system produced high-density polyethylene (HDPE) and linear low-density polyethylene (LLDPE) with a very narrow molecular weight distribution. And in the case of LLDPE, the polyethylene possessed a homogeneous branching distribution which was in contrast to the LLDPE provided with highly-active Ti/Mg containing Ziegler catalysts, which were commercialized in the late 1970s. Figure 4.11 illustrates a representation of the key differences in the branching distribution of ethylene/1-olefin copolymers produced by each type of catalyst.

Figure 4.11(A) shows the structural features of LLDPE prepared with a high activity Ti/Mg containing Ziegler catalyst where the relatively

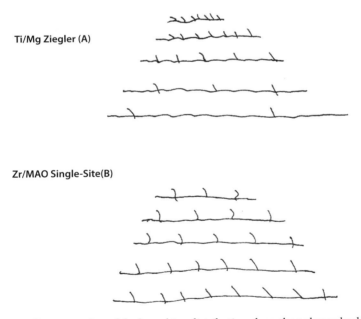

Figure 4.11 Representation of the branching distribution along the polymer backbone with a heterogeneous (non-uniform) branching distribution shown as (A) and a homogeneous branching distribution (uniform) shown as (B).

low molecular weight polymer component is highly branched, e.g., butyl branches with 1-hexene as the comonomer and the relative branching frequency (i.e. branches per 1000 carbon atoms) decreasing as the polymer component molecular weight increases. The molecular weight distribution, as determined by gel permeation chromatography (GPC), of this type of LLDPE depends on the particular catalyst used and on polyethylene composition such as comonomer content and average molecular weight of a particular sample. However, the M_w/M_n values are typically in the range of 3.5–6. Figure 4.11(B) illustrates LLDPE prepared with a Zr/MAO single-site catalyst where the branching frequency is independent of the polymer component molecular weight and the molecular weight distribution as determined by GPC has an M_w/M_n value of approximately 2.0.

Figures 4.12, 4.13 and 4.14 illustrate analytical data that quantify the important structural differences between polyethylene prepared with a high activity Ti/Mg Ziegler-type catalyst and the polyethylene prepared with a zirconocene/methylalumoxane single-site catalyst.

Examination of Figure 4.12 shows the significantly higher amount of 1-hexene required to lower the density (reduce crystallinity) in ethylene/1-hexene copolymers prepared with a high-activity Ti/Mg catalyst that produces LLDPE with a heterogeneous branching distribution compared to a zirconocene/methylalumoxane single-site catalyst. For example, the single-site catalyst requires only about 3 mol% 1-hexene to produce a LLDPE material with a density of approximately 0.915 g/cc, while the Ti/Mg-type catalyst requires approximately twice as much 1-hexene (ca. 6 mol%) to

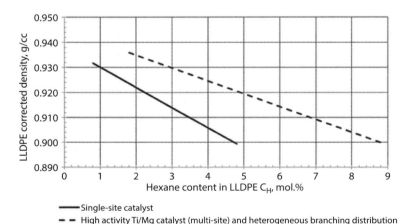

——— Single-site catalyst

– – High activity Ti/Mg catalyst (multi-site) and heterogeneous branching distribution

Figure 4.12 Density and 1-hexene content of ethylene/1-hexene copolymers with a heterogeneous branching distribution (----) and homogeneous (solid line) branching distribution [41].

produce a LLDPE material with a density of 0.915 g/cc. Because the single-site catalyst incorporates the same amount of 1-hexene into each molecule (ethylene/1-hexene ratio), this results in a more efficient use of comonomer. Because the Ti/Mg catalyst produces LLDPE in which the relatively higher molecular weight polymer molecules contain less comonomer, higher comonomer levels are needed in the copolymer to reduce the polyethylene crystallinity to the same degree to produce the same density.

Figure 4.13 shows the melting point data obtained by DSC analysis for a wide range of polyethylene samples containing 0.5 - 5.5 mol% 1-hexene and prepared with the Zr/MAO single-site catalyst. Because of the homogeneous branching distribution in this type of polyethylene, the melting point decreases linearly with increasing 1-hexene content in the polymer. In addition, the DSC melting point curve exhibits only a single peak.

Figure 4.14 shows a typical DSC melting point scan of LLDPE prepared with a Ti/Mg Ziegler catalyst that produces polyethylene with a heterogeneous branching distribution. The relatively high molecular weight polymer component, with a lower branching frequency, forms crystalline units with a relatively high and well-defined melting point near approximately 123°C. The relatively highly-branched lower molecular weight polymer component melts over a broad range from about 80–120°C and is due to the wide range of comonomer in various lower molecular weight polymer components.

Kissin and coworkers found [42] that the lowest molecular weight polymer component contains as much as 25 mol% 1-hexene; while the least branched, relatively very high molecular weight polymer component contains only 1.2 mol% 1-hexene in a LLDPE sample, with a 1-hexene content of 3.6 mol%.

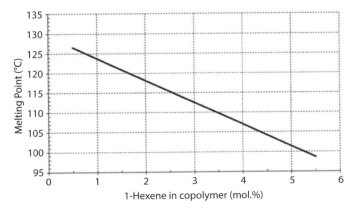

Figure 4.13 DSC determined melting point data for polyethylene prepared with single-site zirconocene/methylalumoxane catalyst [41].

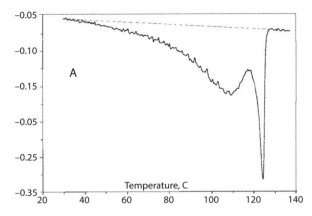

Figure 4.14 DSC melting point data for ethylene/1-hexene copolymer with a hetergeneous branching distribution [41].

Due to the important structural differences outlined above in these two types of polyethylene prepared with each catalyst type, industrial scientists began working in the early 1980s to translate this Zr/MAO single-site catalyst into practical innovations that would adapt this technology to the commercial polyethylene industry. There were significant technical barriers that needed to be solved in order for this catalyst to operate in commercial polyethylene process reactors. A Zr/MAO single-site catalyst needed to be developed that could:

a. Operate in slurry and gas-phase reactors at approximately 75–100°C while producing the necessary high molecular weight polyethylene needed for commercial polyethylene products.

b. A Zr/MAO catalyst needed to be supported on a solid material so that the catalyst particle could be easily added to a commecial reactor as a free-flowing powder and produce polyethylene particles with a suitable particle size for either a slurry or gas-phase process, utilizing particle form technology in which one catalyst particle produces one polymer particle.

4.3.7   Control of Polymer Molecular Weight

As outlined by J. A. Ewen [43], the first priority in industrial research programs in the early 1980s was to find zirconocene compound(s) that provided suitable polymer molecular weights for industrial applications. This was reported by Ewen and Welborn [44], who found that replacing one hydrogen on each cyclopentadienyl ring with an alkyl group (methyl

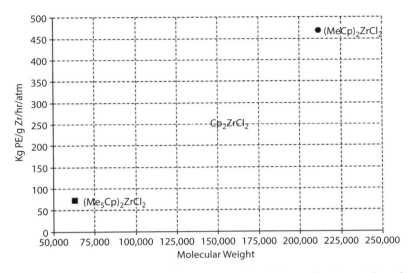

Figure 4.15 Correlation between catalyst activity, polyethylene molecular weight and alkyl-group substitution on Cp ring in various metallocene compounds [44].

or ethyl) increased the ethylene propagation rates and also increased the polyethylene molecular weight. This was an extremely important result for industrial scientists working to commercialize a single-site catalyst system. The activity of these mono-alkyl substituted metallocene compounds and corresponding polymer molecular weights is shown in Figure 4.15.

This electronic ring substitution effect was further examined by other scientists that found that bis(n-butylcyclopentadienyl) zirconium dichloride/MAO catalyst provided a commercially important catalyst system for the gas-phase process used in the commercial manufacture of polyethylene [45,46].

4.4 Single-Site Catalyst Based on (BuCp)$_2$ZrCl$_2$/MAO and Silica for the Gas-Phase Manufacture of Polyethylene

In 1988 Howard W. Turner reported the preparation of a solid catalyst component based on the reaction of (BuCp)$_2$ZrCl$_2$ and methylalumoxane at room temperature. The solid was isolated by cooling the reaction vessel to −30°C for one hour and then decanting the liquid layer and washing the solid material with pentane to yield a glassy solid. The solid with an Al/Zr ratio of 20 was examined as a homogeneous ethylene polymerization catalyst using toluene as a solvent. Catalyst activity was 164 g PE/g catalyst at 80°C, 35 psi ethylene in 10 minutes [45].

Lo and coworkers reported [46] a silica-supported catalyst that was utilized in a gas-phase fluidized-bed reactor process for the preparation of HDPE and LLDPE using $(BuCp)_2ZrCl_2$/MAO and a Davison Grade 955 silica (pore volume of 1.5 cc/g) that was previously calcined at 600°C for 12 hours. A typical catalyst formulation is shown in Table 4.5.

$$Al = 13.3 \text{ wt\%}; Zr = 0.225 \text{ wt\%}; (Al/Zr)_{molar} = 200$$

The key feature of this catalyst formulation is the low Al/Zr molar ratio of 200 that was required to produce a high-activity, silica-supported catalyst which is readily adaptable to the gas-phase polymerization process. In addition, this silica-supported, single-site catalyst also provided polyethylene with acceptable molecular weight for commercial applications. The amount of catalyst that can be impregnated into the silica is limited by the pore volume of the silica, so that the relatively low Al/Zr ratio in this catalyst system could be completely impregnated into the silca pore structure.

The catalyst described by Lo *et al.* was prepared at ambient temperature by adding a toluene solution of $(BuCp)_2ZrCl_2$ dissolved in the MAO solution, dropwise to the calcined silica such that the solution volume was slightly greater than the total pore volume of the silica. After the addition of the toluene solution to the dry silica, toluene was removed by heating the impregnated silica at about 45°C for six hours under a nitrogen flow to obtain a dry, free-flowing powder. The finished catalyst was sieved to remove any catalyst particles larger than 150 microns. The catalyst was evaluated in a gas-phase fluidized-bed pilot unit that produced approximately 25 lbs of polyethylene/hour in a continuous operation. Table 4.6 summarizes the process data.

The data in Table 4.6 show that catalyst activity increases with polymerization temperature from 1,150 g PE/g catalyst at 65°C to 5,100 g PE/g catalyst at 77.5°C. However, very importantly, the polyethylene molecular weight may be controlled over the range necessary for the manufacture of commercial grades of polyethylene for industrial applications, as shown by Melt Index ($MI_{2.16 kg}$) values from 4.1 to 0.6. Isopentane or isopentane containing 0.3 ppm oxygen was needed to produce the relatively higher molecular weight products.

Table 4.5 Typical silica-supported catalyst formulation [46].

Reagent	Quantity
Silica	1.00 g
$(BuCp)_2ZrCl_2$	0.014 g
Methylalumoxane	0.406 g*

*MAO was added as a 30 wt% solution in toluene.

Table 4.6 Gas-phase process data for single-site catalyst based on $(BuCp)_2ZrCl_2$/MAO and silica.

Temp (°C)	Isopentane psi	(hexene/ethylene) gas phase molar ratio	ethylene psi	Zr ppm	MI $I_{2.16\,kg}$	Density g/cc	Activity g PE/g
65	0	0.0088	261	1.95	4.1	0.931	1,150
77.5	0	0.0030	180	0.44	2.6	0.943	5,100
77.5	0	0.0210	206	0.30	4.1	0.918	7,500
77.5	44	0.0160	185	0.46	1.0	0.918	4,900
77.5	52 (0.3ppm Oxygen)	0.0180	195	1.0	0.6	0.919	2,250

Note: Catalyst activity was estimated from the amount of Zr in the catalyst (0.225 wt%) and the Zr found in the polyethylene. Catalyst residence time in a continuous pilot unit (bed weight/production rate) is approximately three hours [46].

The catalyst described above was supported on silica by first dissolving the bis(n-butylcyclopentadienyl) zirconium dichloride compound into the methylalumoxane before this solution was added to the calcined silica. In this method, the MAO is chemically attached to the silica by reacting the silica hydroxyl group (Si-OH) with the Al-CH$_3$ group present in the MAO as shown below:

$$Si\text{-}OH + -Al(CH_3)\text{-}O- \rightarrow Si\text{-}O\text{-}Al\text{-}O- + CH_4$$

Kaminsky and Renner [47] reported another method in which a metallocene compound was first reacted with the calcined silica to chemically attach the zirconium compound to the silica and then MAO was added as a second step. In this procedure the Al/Zr ratio was only 39 and the catalyst had an activity of 121 kg PE/mol Zr/hr at 50°C and 2 bar of ethylene.

Kamfjord, Wester and Rytter [48] reported a method in which only the methylalumoxane material is reacted with silica that was previously calcined at 450°C for six hours. For example, 7 ml of 30 wt% MAO in toluene solution was added to 2.0 g of calcined silica at 20°C using an incipient wetness method in which the volume of the MAO solution used corresponds to the total pore volume of the dried silica. Next, the toluene was removed by evaporation to isolate a free-flowing silica/MAO intermediate reaction product. Then a solution of (BuCp)$_2$ZrCl$_2$ (0.0076 g/0.019 mmol) in 3 ml of toluene is added to 2 g of the silica/MAO intermediate. The finished dry catalyst is obtained after the toluene is removed by evaporation. Analytical data found that the finished catalyst contains 0.09 wt% Zr and 22.0 wt% Al to provide an Al/Zr molar ratio of 870. The activity of this catalyst was 5,025 kg PE/mol Zr/hr using triethylaluminum as an external cocatalyst during a slurry ethylene homopolymerization process carried out at 70°C and 4 bar ethylene pressure.

In another approach to anchor metallocene compounds to a silica surface, Lee et al. [49] attached various functional groups to the silica surface which were designated as spacer groups that contained a Cp ligand as part of the end group which was utilized to chemically bond a zirconium/indenyl group to the spacer group. One example of this approach is illustrated in Figure 4.16

This silica/zirconium intermediate was then activated with modified methylalumoxane (MMAO) as cocatalyst. This approach produced various catalysts with an activity of 300–1300 kg PE/mol Zr/h/atm. The polyethylene had a relatively high molecular weight, i.e., > 200,000, and exhibited M_w/M_n values of 3–4.8, suggesting a somewhat broader molecular weight distribution than one single-site catalyst. This catalyst probably contained two or more similar single-site catalysts.

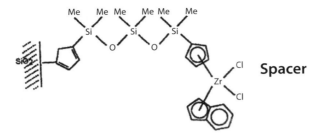

Figure 4.16 Reaction of a zirconium compound to a silica containing a Cp ligand attached to silica. Reprinted from [49] with permission from John Wiley and Sons.

4.5   Activation of the Metallocenes Cp_2ZrCl_2 or $(BuCp)_2ZrCl_2$ by Solid Acid Supports

In a novel approach to activate the metallocene compound with a strong Lewis acid supported on a solid support such as silica or alumina, scientists at Chevron-Phillips Chemical Company reported a series of high-activity catalysts suitable for the slurry process [5,7]. This method does not employ an alumoxane compound to activate the catalyst.

The inorganic supports shown in Table 4.7 were impregnated with various Lewis acids using aqueous solutions following an incipient wetness technique in which the total pore volume of the support was approximately equal to the pore volume of the aqueous solution. Supports were initially dried at 110°C to isolate a free-flowing powder intermediate material [5].

The intermediate material was placed into a quartz tube contained in a furnace and a stream of dry air or dry nitrogen was used to fluidize the material in the quartz tube. Then the fluidized material was calcined for several hours at 200–700°C, after which the column was cooled to ambient conditions and the calcined material was stored in an inert atmosphere until needed.

In one set of experiments, silica was fluoride-treated with aqueous solutions of ammonium fluoroborate, NH_4BF_4, from 0.3 to 3 mmol/g, and then calcined from 200–700°C to provide a finished support material. The surface area of each finished support was determined and the results showed that some supports underwent a high degree of sintering due to the collapse of the internal pore structure, especially evident at relatively higher loadings of NH_4BF_4 and higher calcining temperatures, i.e., 600–700°C. In a second set of experiments, silica was treated with an aqueous solution of aluminum nitrate containing tetrafluoroboric acid, HBF_4, and then calcined over a similar range of temperatures.

Table 4.7 Supports investigated for activation of metallocenes by solid acid supports.

Support	Grade	Surface Area (M²/g)	Pore Volume (cc/g)	Source
Silica	952	300	1.6	W.R. Grace
Alumina	Ketjen Grade B	400	1.5	Akzo-Nobel
Silica/Alumina	MS 13-110 (13% alumina)	450	1.1	W.R. Grace

The alumina support was also treated with a variety of Lewis acids, but the most active supports were treated with chlorinating agents such as CCl_4 or $TiCl_4$, which were added to the alumina in a fluidized column maintained at 400–600°C. Alumina was treated with an aqueous solution of ammonium sulfate, $(NH_4)_2SO_4$ or H_2SO_4.

Each type of support was evaluated in a 2.2-liter stainless steel autoclave fitted with a temperature control jacket and a marine stirrer. Each supported, finished catalyst was prepared *in situ* by sequentially adding to the polymerization reactor 10–100 mg of finished support, 2.0 ml of a toluene solution containing 0.5 wt% (n-butyCp)$_2$ZrCl$_2$, 0.6 liter of isobutane, 1.0 mmol triethylaluminum and a second addition of 0.6 liter of isobutane. The contents of the reactor were stirred at 400 rpm while the reactor was heated to 90°C, after which ethylene was added to the reactor system to maintain total pressure at 550 psig. After the polymerization experiment, the ethylene flow to the reactor was stopped, the reactor was cooled to ambient temperature and solvents were removed by evaporation, and the reactor was opened and the polymer was collected. Note that no reactor fouling was detected and the granular polyethylene had acceptable morphology important for a commercial process.

The activity of the finished catalysts varied over a wide range, but the highest activity systems with each type of support are shown in Table 4.8.

McDaniel and coworkers postulate that a possible mechanism for the interaction of the metallocene compound, in the presence of an aluminum alkyl, with a Lewis acid site is illustrated in Figure 4.17.

However, the scheme shown in the figure does not explain the relatively more important role of the aluminum alkyl (such as TEAL) that is required to achieve high catalyst activity. For example, starting with a dialkyl metallocene compound provides a supported acid catalyst with significantly lower activity than the activity of the same catalyst in the presence of an excess of AlR_3. In addition, the type of AlR_3 compound used to prepare the finished acid-supported catalyst (e.g., TMA, TEAL or TIBA) can produce a finished catalyst with very different activity. One possible explanation offered by McDaniel to rationalize these experimental findings is that the AlR_3 compound is coordinated to the active site in a manner illustrated in Figure 4.18.

Some important conclusions by McDaniel and coworkers about these metallocene catalysts activated by the acidity introduced into the various solid supports are as follows:

Table 4.8 High-activity, silica-supported catalysts treated with a Lewis acid.

Support Type	Lewis Acid		Calcining Temp (°C)	Activity g PE/g/h
	Type	Amount (mmol/g)		
Silica	NH_4BF_4	0.5	500	1140
Silica	$Al_2(NO_3)_2/HBF_4$	2.0 each	550	2300
Alumina	CCl_4	2.0	400–600	2800
Alumina	$(NH_4)_2SO_4$	3.0	500–600	2600
Alumina/Silica	NH_4HF_2	3.5	450	18000
Alumina/Silica	Trific acid[a]	2.0	350	12000

[a] Trifluoromethanesulfonic acid

Figure 4.17 Formation of zirconium cation by reaction of Cp_2ZrClR with acid support. Reprinted from [5] with permission from Wiley-VCH Verlag GmbH & Co. KGaA.

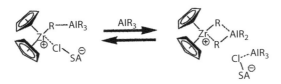

Figure 4.18 Formation of zirconium cation by reaction of Cp2ZrClR with acid support. Reprinted from [5] with permission from Wiley-VCH Verlag GmbH & Co. KGaA.

a. These solid acid supports activate the metallocene compound (either Cp_2ZrCl_2 or $(BuCp)_2ZrCl_2$ used in this study) to produce a single-site catalyst, as indicated by the very narrow molecular weight distribution of the polyethylene obtained by these catalysts. Polydispersity values, M_w/M_n, found were approximately 2.0.

b. The results found using various analytical procedures employed in this study to better characterize these catalyst systems, indicate that these solid acid catalysts contain a relatively high number of acid sites that fix the metallocene compound to the solid surface, a relatively low number of these active sites (ca. 2%) become ethylene polymerization sites, suggesting that very few surface acid sites possess some unknown characteristic that is capable of providing the unique ion pair necessary for formation of an active center.

c. Catalysts prepared by these solid acid supports chemically fix the metallocene active site to the inorganic surface so that metallocene active sites remain fixed to the solid surface during the polymerization process. This provides the particle-form technology required of a commercial catalyst employed in the ethylene polymerization slurry process.

It should be noted that these catalyst systems would most likely operate very well in gas-phase commercial reactors.

4.5.1   Activation of Bridged Metallocenes by Solid Acid Supports

In an important follow-up study to the activation of the non-bridged metallocene compounds by solid acid supports discussed above, scientists at Chevron-Phillips Chemical Company reported [7] the activation of bridged metallocenes, designated as ansa-metallocenes, using the same solid acid supports investigated above [5].

One significant disadvantage of the polyethylene produced with a single-site catalyst is the very narrow molecular weight distribution (M_w/M_n = 2) of the polyethylene obtained with such a catalyst. Polyethylene with a very narrow MWD undergoes much less shear thinning (i.e., a reduction in the viscosity of the molten polyethylene) as the polymer melt is conveyed through the polymer fabrication equipment used to form a finished product. Hence, manufacturing rates may decline, and the energy required to manufacture these end-use products increases.

One method of improving the processability of polyethylene produced with a single-site catalyst is to introduce moderate amounts of long-chain branching (LCB) into the polyethylene molecular structure. However, relatively higher levels of LCB may improve polymer processability but may decrease the mechanical strength characteristics of the finished fabricated product. For example, in injection molding applications, the finished article may undergo warpage in the finished product. In a film product manufactured using a blown film fabrication method, the mechanical strength properties of the finished film product may undergo a higher level of film orientation in which the film strength properties are diminished in either the machine direction or the transverse direction which are present during the film fabrication process.

The activation of certain ansa-metallocenes to solid acid supports produces polyethylene with relatively low levels of long-chain branching, which is sufficient to moderately increase the molecular weight distribution of the polyethylene, and improve the polymer processability, but not be detrimental to the properties of the finished fabricated product. Consequently, low levels of LCB improve the shear-thinning behavior of the molten polyethylene, which resolves the processing disadvantage of polyethylene with a polydispersity index of 2.

One example of a bridged metallocene compound that was activated using a fluorinated Davison silica-alumina support calcined at 500°C that provided polyethylene with the desired low levels of LCB is shown in Figure 4.19.

The catalyst prepared with this particular bridged metallocene provided an ethylene/1-hexene copolymer at 80°C with an M_w of 283,000 and a polydispersity index value (M_w/M_n) of 3.1 utilizing commercial polymerization conditions.

Polyethylene products produced with the Phillips loop process using catalysts prepared with metallocene compounds activated with solid acid supports are available commercially.

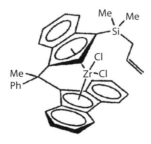

Figure 4.19 U.S. Patent 7,148,298 B2 [7].

As of 2009, resins with Melt Index ($I_{2.16 \text{ kg}}$) values from about 0.5 to 20 and densities of 0.912–0.945 g/cc are available from Chevron Phillips Chemical Company for film, rotomolding and injection molding applications. The film applications would be for high strength and/or high clarity requirements [6].

4.6 Dow Chemical Company Constrained Geometry Single-Site Catalysts (CGC)

Scientists at the Dow Chemical Company discovered [50] a remarkable single-site catalyst in the late 1980s based on titanium as the active center. The catalyst development research was primarily credited to James C. Stevens and David R. Neithamer, who were listed as coinventors on U.S. Patent 5,064,802 issued on November 12, 1991. Stevens described the structure of this catalyst system in the general terms [51,52] shown in Figure 4.20.

One important feature of this catalyst system, which most likely accounts for the extremely high reactivity of this catalyst with 1-olefins such as 1-hexene and 1-octene and other olefins such as styrene, is due to the short bond length of the R'₂Si-bridging group attached to the amide ligand and the Cp ligand. This opens up one side of the complex (hence, the term constrained geometry) to reduce steric hindrance effects and allows olefins ample room for coordination in the first coordination sphere around the metal (usually titanium) active site. However, as pointed out by Stevens [51,52], this constrained geometry feature of the catalyst does not allow steric control of the polymerization reaction. Therefore, stereoregular polyolefins such as isotactic polypropylene are not produced by this

Where: M = Ti, Zr, Hf
R = halide or CH₃
R' and R" + alkyl, aryl
Note: Cp ring may also have alkyl substitution

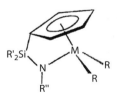

Figure 4.20 General structure of CGC catalyst [50].

catalyst, but polyolefins such as ethylene/styrene and ethylene/1-octene copolymers with very high 1-octene content are produced because of the constrained geometry characteristic.

4.6.1 Cocatalyst Activation of Constrained Geometry Catalyst

The catalyst as illustrated in Figure 4.20 requires a cocatalyst in order to generate the metal-based cation, similar to the metallocene/MAO system, which is the catalytic active site. Stevens reports [51,52] an activity of 1.5–7.5 x 10^5 g PE/g Metal when the complex is activated with a large excess of MAO, with the specific activity depending on the structure of the specific catalyst and polymerization conditions employed.

In addition, the catalyst system is uniquely adaptable to Dow's solution polymerization process which operates at a relatively high polymerization temperature (110–160°C) in order that the polymer product remains completely soluble in the polymerization solvent. For a catalyst based on the complex $(C_5Me_4)SiMe_2N(t-Bu)TiCl_2$/MAO operating at an ethylene partial pressure of 450 psi with 1-octene as a comonomer and a 10 minute polymerization time, relatively high molecular weight polyethylene is obtained, which is necessary for a commercial catalyst system. Figure 4.21 illustrates this molecular weight/polymerization temperature data.

Note that other key features of Dow's continuous solution process are the relatively high ethylene conversion rates and the relatively short catalyst

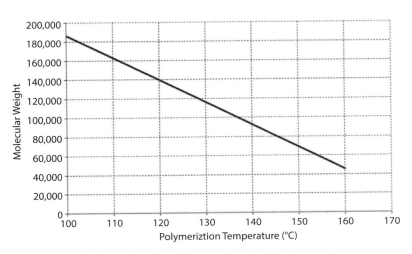

Figure 4.21 Polyethylene molecular weight at polymerization temperatures 110–160°C [51,52].

residence times of usually less than 0.2 h. An important conclusion from these findings is that Dow's single-site catalyst can most likely be employed in all three of the standard commercial polyethylene processes which are used worldwide to manufacture polyethylene (i.e., the gas-phase, slurry and solution processes). The same conclusion cannot be reached for the metallocene-based, single-site catalyst system. Metallocene single-site catalysts are not adaptable to the solution process because polyethylene molecular weight would be very low at these relatively high polymerization temperatures. However, because the gas-phase and slurry-phase processes operate at a lower polymerization temperature (70–110°C), the Dow catalyst system would require a chain transfer agent such as hydrogen to control polyethylene molecular weight.

Although MAO is an effective cocatalyst in activating the Dow catalyst, overall activity (g PE/g Metal) is relatively lower than the metallocene/MAO catalyst, and a higher overall activity is preferred for a commercial catalyst system.

Stevens found that activating the constrained geometry catalysts with fluorinated phenyl borates, $B(C_6F_5)_3$, produced a much higher catalyst activity exceeding 10^7 g PE/g Metal, and that these borates are particularly useful in the solution process because they are soluble in the polymerization solvent. An example illustrating $B(C_6F_5)_3$ as a cocatalyst which forms the necessary ion-pair is shown in Figure 4.22.

4.6.2 Processability of Polyethylene Manufactured with Dow's CGC System

As discussed above with Chevron-Phillips metallocene-based catalyst which is activated with solid acid catalyst supports, the introduction of low levels of long-chain branching is an important structural feature of polyethylene manufactured for commercial applications with single-site catalysts. Dow's CGC system is able to incorporate low levels of long-chain branching into the polyethylene due to the high level of vinyl-terminated polymer molecules, the relatively high polymerization temperature that is

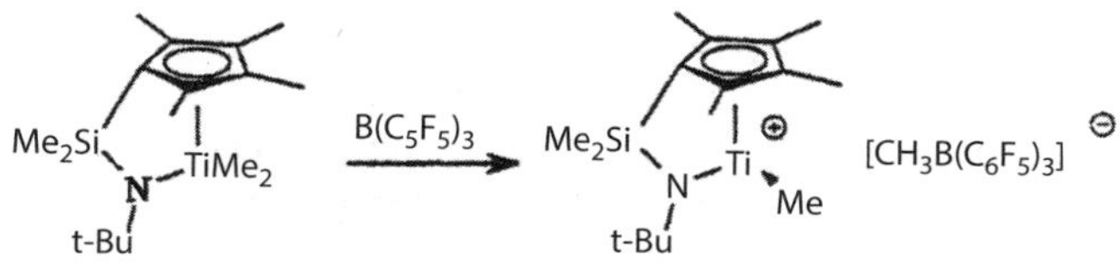

Figure 4.22 Activation of Dow's CGC with B(C6F5)3 [51,52].

responsible for the polymer remaining in solution and the relative ly high reactivity of the CGC with 1-olefins [52].

Hence, the polyethylene manufactured with Dow's catalyst system possesses sufficient levels of long-chain branching to impart higher levels of shear-thinning in the polymer melt, low melt fracture and high melt strength during the polyethylene fabrication process.

4.7 Novel Ethylene Copolymers Based on Single-Site Catalysts

Table 4.1 outlines a very brief summary of the types of polyethylene that are available commercially due to the introduction of single-site technology in the 1990s. However, a more detailed discussion of the novel materials that have been introduced based on these new catalysts is necessary to demonstrate how the polyethylene industry has expanded outside the previous product mix. Prior to the development of single-site catalysts, the last significant catalyst-related change in the polyethylene industry took place in the late 1970s with the introduction, on a larger scale, of linear low-density polyethylene (LLDPE) as a result of new higher activity Ziegler catalysts.

Although ethylene/1-olefin copolymers were well documented in the late 1950s with the discovery of the chromium-based Phillips catalyst and the titanium-based Ziegler catalyst, it was the discovery of metallocene-based single-site catalysts and the constrained geometry catalyst system that significantly increased the various types of new ethylene-based copolymers that are available for commercial applications. These new catalysts created new products, applications and markets for the polyethylene industry.

These new products were primarily due to three characteristics that were unique to this new catalyst family. These are: (1) homogeneous or uniform branching distribution of the 1-olefin employed in the copolymerization, as illustrated in Figure 4.11; (2) relatively much higher reactivity of the 1-olefins such as 1-butene, 1-hexene and 1-octene, increasing the amount of comonomer that could be produced with commercial reactors, and; (3) the introduction of new olefins such as styrene and cycloolefins that reacted with these single-site catalysts, leading to entirely new polyethylene compositions.

Some approximate reactivity ratio data with single-site catalysts compared to a commercial Ziegler-type catalyst is shown in Table 4.9. This data shows that single-site catalysts react approximately 2–50 times better with 1-hexene than the Ziegler-type catalyst which requires relatively

Table 4.9 Approximate reactivity of 1-hexene with various catalysts.

Catalyst	Reactivity ratio r_1 [a]
Ti/Mg/Cl	100
Metallocene/MAO (BuCp)$_2$ZrCl$_2$	40
Bridged Metallocene/MAO	5–30
Constrained Geometry	<5

[a]$r_1 = k_{ee}/k_{eh}$, where k_{ee} is the rate in which an ethylene molecule inserts into a growing polymer chain in which the last insertion involved an ethylene molecule, while k_{eh} is the rate in which a 1-hexene molecule inserts following an ethylene insertion. Therefore an r_1 value of 100 implies relatively low 1-hexene reactivity, while a value of 5 represents extremely high 1-hexene reactivity.

high concentrations of 1-hexene in the polymerization reactor to produce ethylene/1-hexene copolymers with sufficient branching to lower the polymer density to the 0.916 g/cc range, which is a common density of LLDPE resins used in film applications.

The significant increase in the compositional range of ethylene/1-octene copolymers available based on the discovery of the Dow constrained geometry single-site catalyst is illustrated by the data reported by G. W. Knight and coworkers at Dow Chemical [53] and is shown in Figure 4.23.

The Dow catalyst introduced new commercial products with densities between about 0.85 g/cc to about 0.91 g/cc.

Ethylene/1-octene copolymers containing approximately 10–50 wt% 1-octene with corresponding polyethylene densities of 0.91–0.85 g/cc and Melt Index ($I_{2.16 kg}$) values of 0.2–30 dg/min, are new compositions that are now available in the polyethylene industry as a result of single-site catalyst technology. It is important to note that the Melt Index range covers a very wide range of polyethylene molecular weights in such compositions. Dow Chemical Company, for example, identifies these compositions with the unique trade name ENGAGE™. ENGAGE also includes other 1-olefins such as 1-butene and propylene. This type of new polyethylene composition bridges a gap between conventional rubber, which is an amorphous (containing little or no crystallinity) polymeric network that is often crosslinked, and conventional polyethylene which existed before the introduction of single-site catalyst technology. Dow uses the AFFINITY™ trade name for other polyethylene compositions which include very low density (high 1-olefin content) and very low molecular weight materials (MI values of 500 or 1,000) employed as hot melt adhesives. These materials have a low degree of crystallinity and a low melting point of about 68–70°C.

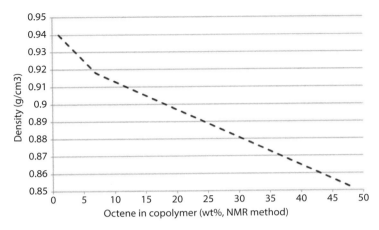

Figure 4.23 Density of ethylene/1-octene copolymers with a homogeneous branching distribution at various 1-octene levels. Products below a density of about 0.91 g/cc represent new polyethylene compositions available for commercial applications [53].

It is also necessary to point out that the metallocene single-site catalyst system can also produce very similar polymer compositions as the Dow constrained geometry technology. However, many of these products may be more easily produced using the solution polyethylene process rather than the gas-phase or slurry processes. The slurry process is limited to polymer compositions that are insoluble in the slurry solvent (usually isobutane or other very volatile organic solvent) because soluble polymer components may foul the slurry process and cause production problems. The gas-phase process may also be limited due to very low polymer softening temperatures which may cause polymer particle agglomeration in the gas phase process. The solution process, although a higher cost manufacturing process, is most likely the preferred method for manufacturing many of these new grades of polyethylene.

Table 4.10 summarizes some important dates in the commercialization of single-site catalysts during the 1990s and the use of specific trade names by the manufacturers of these new types of polyethylene in order to distinguish these new polyethylene compositions from previous materials.

4.8   Non-Metallocene Single-Site Catalysts

More recent catalyst developments have identified additional catalyst compositions that are not based on a metallocene complex or the Dow constrained geometry system. Nova Chemicals Inc. has most likely

Table 4.10 Commercialization of single-site technology.

Company/Date	Trade name	Comments
Exxon/1991	Exact	Exxpol high pressure process 0.87–0.92 g/cc Ethylene/1-butene Ethylene/1-hexene
Exxon/1996	Exceed	Gas Phase UNIPOL Process
Dow/1993	Insite/Engage	0.85–0.92; MI 0.2–1,000
	Insite/Affinity	Solution Process
Mitsui/1995	Evolue	Gas Phase Process Ethylene/1-hexene 0.90–0.94 g/cc MI 1.0–3.8
BASF/1995	Luflexen	Slurry Loop Process 0.90–0.92 g/cc Ethylene/1-butene
Chevron-Phillips/1996	Marflex/mPact	Slurry Loop Process 0.91–0.94 g/cc; MI 0.5–20; Excellent film clarity; Roto- and Inject-molding grades
Univation[a]		Gas Phase Process Film, Injection Molding
Borealis	Borecene	Film and Rotomolding
Innovene/Ineos[b]		Gas Phase and Slurry Process
Nova Chemicals/2001[c]	SURPASS	Advanced Sclairtech Solution Process (either single- or multi-solution reactors) Film, Injection and Rotomolding Grades

[a]Joint venture of Dow and ExxonMobil Chemical; initially Union Carbide.
[b]Former BP gas-phase process; Dow has formed a joint venture for using CGC system in gas-phase process.
[c]Proprietary non-metallocene single-site catalyst.

commercialized such a catalyst using the Advanced Sclairtech process, which is a solution polymerization process.

Based on U.S. Patent 6,649,558 issued on November 18, 2003 to Stephen J. Brown *et al.* and assigned to Nova Chemicals International S.A., the Nova single-site catalyst is based on a bis-phosphinimine ligand to synthesize the catalyst precursor bis(tri-t-butylphosphinimine)titanium dichloride, which is then converted to the corresponding dimethyl derivative using methylmagnesium bromide and then activated with an ionic activator to produce the polymerization catalyst shown below.

$$[(t\text{-butyl})_3P=N]_2\text{-TiMe}_2 + [CPh_3]^+[B(C_6F_5)_4]^- \rightarrow [(t\text{-butyl})_3$$
$$P=N]_2\text{-TiMe}]^+ [B(C_6F_5)_4]^- + MeCPh_3$$

Examination of this catalyst system under solution polymerization conditions carried out at 160°C in 216 ml of cyclohexane as solvent containing 10 or 20 ml 1-octene as comonomer, 140 psig total reactor pressure, the catalyst exhibited a relatively high activity of 2430 g PE/mmol catalyst/hr. Characterization of the polyethylene showed a molecular weight (Mw) of 114,000 with a polydispersity index value of 2.6, indicating near single-site MWD.

Additional patents [10] awarded to Nova Chemicals such as U.S. 6,339,161 issued to W. Xu et al. on January 15, 2002, show similar ligand systems in which late transition metal complexes were isolated using halides of Fe(II), Fe(III), Co(II) and Ni(II). One example of these additional ligands is the 2,5 substituted thiophene ligand also containing the bis-phosphinimine moiety (Figure 4.24).

The interaction of $FeCl_3$ with the ligand 2,5-(t-Bu$_2$P=N-SiMe$_3$)$_2$thiophene provided the corresponding Fe(III) complex, which was examined as an ethylene homopolymerization catalyst in toluene at 140°C using methylalumoxane as activator and 286 psig of ethylene. The initial polymerization rate was reported as very high with a very rapid decay in activity after only 1.5 minutes, after which the polymerization experiment was terminated. The polyethylene isolated was characterized and found to have an Mw of 470,000 and a polydispersity index value of 1.9, indicating a single-site catalyst.

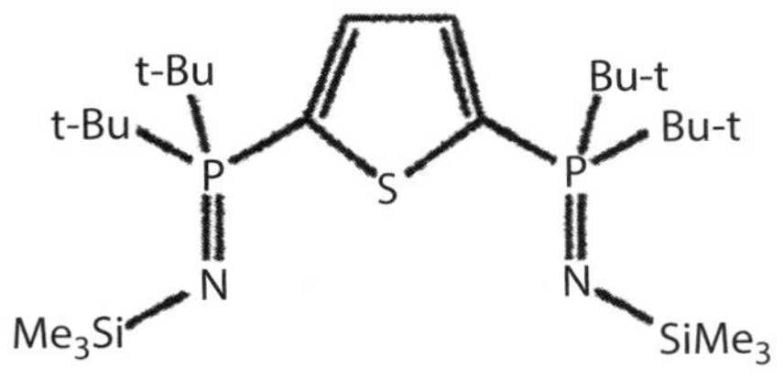

Figure 4.24 2,5-(t-Bu$_2$P=N-SiMe$_3$)$_2$thiophene.

4.8.1 LyondellBasell Petrochemical

Scientists at Occidental Petroleum[1] patented [11] two additional non-metallocene single-site catalysts based on a six-membered borabenzene complex and a nitrogen-containing azametallocene or half metallocene complexes [12,13]. These complexes were activated with both methyl-alumoxane and boron-containing activators such as $B(C_6F_5)_3$. Lyondell described a wide range of ethylene homopolymers and copolymers prepared with these complexes over a wide range of molecular weights.

Table 4.11 summarizes some polymerization data for the catalyst systems shown in Figure 4.25.

Table 4.11 Polymerization data for catalysts 1–4 in Figure 4.25.

Data Point	Temp (°C)	Catalyst number	Comonomer	Al/Zr (molar)	Activity kg PE/g/hr	Melt Index $I_{2.16\,kg}$
1	80	1	none	3300	690	0.04
2	80	1	1-butene	3300	680	3.44
3	110	1	none	3300	270	1.40
4	80	2	none	1630	100	0.04
5	80	3	1-butene	2900	200	0.14
6	80	4	1-butene	2200	380	25.8

Conditions: toluene diluent; methylalumoxane activator; 1.03 MPa ethylene pressure [12,13].
Data points 1–4 reference: S. Nagy, R. Krishnamurti and B.P. Etherton, PCT Int. Appl. WO 96/34021 (1996); and data points 5,6 reference: S. Nagy, R. Krishnamurti and B.P. Etherton, U.S. Patent 5,554,775 issued to Occidental Chemical Corp., on Sept. 10, 1996.

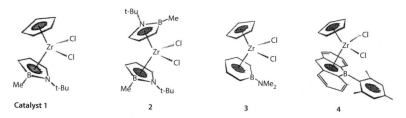

Figure 4.25 Complexes patented by Occidental Chemical Corp. (now LyondellBasell) as single-site catalysts [12,13].

---

[1] Lyondell Petrochemical purchased the polyethylene unit of Occidental Petroleum and later Lyondell Petrochemical merged with Basell, which became LyondellBasell.

4.9   New Ethylene Copolymers Based on Single-Site Catalysts

Prior to the discovery of single-site catalysts the polyethylene industry was limited to conventional 1-olefins, primarily 1-butene, 1-hexene and 1-octene, as comonomers that were used to control polyethylene properties. In this regard, the amount and type of 1-olefin in the finished polymer were the important variables.

This limitation changed with the introduction of single-site catalysts. Two commercial examples of new ethylene-based copolymers with novel olefins are ethylene copolymers based on norbornene and styrene.

Kaminsky and coworkers found that copolymerization of cyclopentene or norbornene with ethylene, using various metallocene compounds activated with MAO, produced new thermoplastic amorphous materials [55] that were transparent and had high stiffness and excellent stability.

4.9.1   Ethylene/Norbornene

Ethylene/norbornene copolymers are presently available commercially from TOPAS Advanced Polymers, Inc., a company formed in 2006 after early development efforts by Hoechst AG and Mitsui Petrochemical. Hoechst initiated the commercialization of this cyclic olefin copolymer (COC) in 1990, and Mitsui Petrochemical became involved in evaluating the manufacture of these copolymers in a continuous pilot-scale operation in 1993. TOPAS Advanced Polymers, Inc., is a joint venture of Daicel Chemical Industries Ltd., and Polyplastics Co., Ltd., with headquarters in Frankfurt, Germany, and a United States subsidiary in Florence, Kentucky. The structure of this copolymer is shown in Figure 4.26.

Unlike typical ethylene-based copolymers available in the polyethylene industry, which are semicrystalline materials, the ethylene norbornene copolymers are amorphous thermoplastic materials with a high heat distortion temperature of up to 170°C. They have a glossy, crystal clear appearance, with a high modulus (stiff material) and low shrinkage, which are important in the processability of the molten material. The high modulus of this copolymer is not unexpected. The polymer backbone due to the

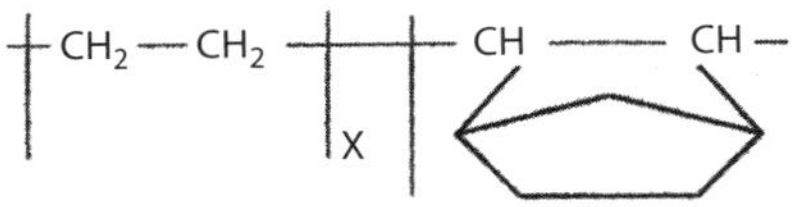

Figure 4.26 Structure of cyclic olefin copolymer. Data from www.topas.com.

bicyclic norbornene branch is a rigid structure with little mobility. Various grades of this copolymer contain 30–60 mol% norbornene and are processed into intermediate or finished products as extruded sheet, cast film, blown film and molding equipment such as injection or blow molding.

Because these copolymers are compatible with other grades of polyethylene, TOPAS may be used as blends with other grades of polyethylene to enhance certain mechanical properties or as processing aids with other single-site-derived products to improve processability. Typical properties for this copolymer are shown in Table 4.12.

Such a copolymer serves as an excellent example of the introduction of new ethylene-based polymer compositions as a direct result of the development of new catalyst technology.

Ethylene/styrene copolymers are presently available commercially from Dow Chemical Company using Dow's constrained geometry catalyst (INSITE) technology. These novel, random, semicrystalline copolymers contain 20–80 wt% styrene. The copolymer melting points decreased from about 90°C to 45°C with increasing styrene content from 20 to 45 wt%, respectively [56,57].

These copolymers behave in a similar manner to ethylene/1-olefin copolymers as the amount of styrene is increased over the range of 20–50 wt%. Melting point temperatures and copolymer crystallinity decrease with increasing amounts of styrene, with the copolymer becoming an amorphous material at about 50 wt% (21 mol%) styrene.

At this level of styrene, the chain-folding mechanism of ethylene sequences is completely interrupted by the phenyl branch, thus eliminating ethylene-containing crystalline regions from the solid lattice.

However, copolymer density does not continue to decrease with higher levels of styrene incorporated into the copolymer. This effect is clearly demonstrated with copolymers containing 50–75 wt% styrene as the copolymer

Table 4.12 Typical properties for TOPAS.

Density (g/cc)	1.02
Melt Index (190°C)	0.1–1.8
Haze	< 1%
Water vapor permeability	
4 mil thickness	
M²/day	0.8–1.3 g
Glass transition Temp. (T_g)	65–178°C

Data from www.topas.com.

density steadily increases from about 0.95 g/cc to a density of 1.02 g/cc with about 75 wt% styrene, with higher levels of styrene content approaching the density of amorphous polystyrene of about 1.05 g/cc.

The tensile stress-strain behavior of these copolymers measured at 23°C was also reported by the Dow scientists [56,57]. Copolymers exhibit a lower modulus as styrene content is increased over the range of 20–60 wt% styrene. The copolymer with 60 wt% styrene was highly elastic.

More specifically, the copolymers investigated contained from 20 wt% styrene (ES20) to 73 wt% styrene (ES73). These samples exhibited strain rupture greater than 200% with samples ES20 and ES30 showing tensile properties similar to ethylene/1-olefin copolymers. The ES43 copolymer and particularly the ES60 copolymer demonstrated increased elastomeric behavior with corresponding lower modulus. The ES73 copolymer was more unique in that it displayed properties of a glassy material.

In the examination of potential applications for these unique materials that possess a wide variety of properties depending on copolymer composition, the Dow group examined finished articles formed by injection and blow molding, blown and cast film and melt extrusion. Potential applications for these new materials would be as substitute materials for flexible PVC, styrenic block copolymers, ethylene/vinyl acetate copolymers and ethylene/propylene-based elastomers. These new ethylene/styrene copolymers once again demonstrate that new catalyst technology creates new markets and applications for the polyethylene industry by competing with materials outside of the polyethylene product mix.

4.9.2   Ethylene/Styrene Copolymers Using Nova Chemicals Catalyst

In U.S. Patent 6,579,961 issued to Q. Wang and coworkers on June 17, 2003, Nova Chemicals International reported using a different catalyst system to prepare ethylene/styrene copolymers over a broad compositional range of styrene. With this catalyst system, Nova scientists report that styrene is inserted into the growing polymer chain in a unique manner in which the sequence distribution along the polymer backbone is different than similar copolymers prepared with the Dow constrained geometry catalyst system.

Nova scientists refer to the unique incorporation of styrene as a double reverse styrene incorporation in which one styrene molecule inserts in a 1-2 addition, followed by two styrene molecules that insert in a 2-1 addition, providing a styrene triad unit that is not present in the ethylene/styrene copolymers produced with the Dow constrained geometry catalyst.

The Dow ethylene/styrene copolymers have at least two methylene groups (-CH$_2$CH$_2$-) between phenyl branches [58].

Nova scientists describe an ethylene/styrene copolymerization using a toluene soluble catalyst system based on CpTi-N=P-R$_3$(Me)$_2$, where R is isopropyl. The catalyst is activated with MAO and the ionic activator [CPh$_3$]$^+$[B(C$_6$F$_5$)$_4$]$^-$, which are both added to the polymerization vessel. In a typical experiment, 136 ml of dry cyclohexane and 85 ml of dry styrene were added to the reactor; the temperature was set at 90°C and ethylene was added to maintain a total pressure of 70 psig. Next 2.59 mmol of CpTi-N=P-R$_3$(Me)$_2$ in toluene were added to the vessel followed by 2.72 mmol of [CPh$_3$]$^+$[B(C$_6$F$_5$)$_4$]$^-$ and 0.518 mmol of Al (MAO), and polymerization was carried out for 0.25 hr. The copolymer was isolated to provide 30.8 g of copolymer containing 35 wt% styrene. Polymer characterization found a MI of 15.1, T$_g$ = -23°C, T$_m$ = 65°C, M$_w$ = 74,000 and the polydispersity index was 3.2, indicating a relatively narrow MWD, but slightly broader than the MWD of a single-site catalyst.

A range of ethylene/styrene copolymers were reported over a styrene content of 33–65 wt% and MI values of 0.2 to 460 g/10 minutes, showing a very wide range of polymer molecular weights. The Nova scientists disclose a more complex polymer backbone based on ^{13}C NMR data. These microstructures, designated I–V in Figure 4.27, are illustrated in the figure with the ^{13}C NMR chemical shift values marked accordingly.

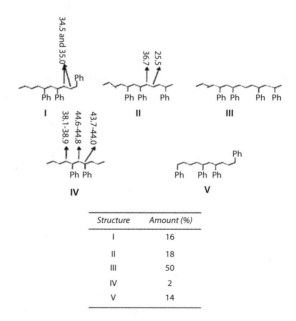

Structure	Amount (%)
I	16
II	18
III	50
IV	2
V	14

Figure 4.27 Polymer backbone based on ^{13}C NMR data microstructures with corresponding ethylene styrene copolymer containing 35 wt% styrene microstructure composition [59].

4.10   Compatible Metallocene/Ziegler Catalyst System

The global polyethylene business involves two types of polyethylene [60] broadly defined in terms of the polyethylene molecular weight distribution (MWD). Polyethylene with a relatively narrow MWD is manufactured by Ziegler or single-site catalysts, usually in a single polymerization reactor, and the MWD is defined by a polydispersity index (Mw/Mn) in the range 2–5. Polyethylene with a relatively broad MWD is manufactured with various Cr-based catalysts in a single reactor or may be manufactured in a multi-reactor process that usually involves two or more polymerization reactors in series. Both single-site and Ziegler-type catalysts are utilized with multi-reactor process technology. This second type of polyethylene is defined by polydispersity index values (M_w/M_n) in the range 8–20. A few examples of multi-reactor processes utilized to manufacture polyethylene with a broad MWD are shown in Table 4.13.

Because multi-reactor process technology involves higher capital investment and operating costs than single reactor processes, advanced catalyst technology in which polyethylene with a broad MWD can be manufactured in a single reactor is more cost effective.

One example of this approach was reported by scientists at Mobil Chemical Company in the late 1980s in order to prepare high molecular weight, high density polyethylene (HMW-HDPE) with a bimodal molecular weight distribution in a single reactor [61,62]. The HMW-HDPE with a bimodal MWD was initially developed by scientists at Conoco Oil Corporation in the 1970s. They used a multi-reactor process where the polymer component with the relatively low molecular weight was produced in one reactor under certain process conditions to provide low molecular weight polymer, and then the contents of this first reactor (polymer and catalyst) were transferred to a second polymerization reactor under suitable process conditions to provide a relatively high molecular weight polymer component. The final polyethylene possessed a bimodal MWD as shown by the GPC data.

Table 4.13 Examples of multi-reactor processes used to manufacture polyethylene with a broad MWD.

Company	Reactor Configuration
Mitsui	Tandem slurry reactors
Univation UNIPOL II/Dow ExxonMobil	Tandem gas phase
Borstar/Borealis	Gas phase/slurry phase
Advanced Sclairtech/Nova	Tandem solution reactors

The HMW-HDPE with a bimodal MWD is primarily used in film applications to fabricate grocery bags (T-shirt bags) and other merchandise bags. However, polyethylene with a bimodal MWD is also used in premium blow-molding markets and high-pressure pipe applications such as natural gas distribution pipe.

The catalyst developed by the Mobil scientists [61,62] was based on a finished bimetallic catalyst that contained two different active sites, a Ti-based Ziegler catalyst component and a metallocene (Zr)/MAO catalyst component. The catalyst was prepared in two steps. The first catalyst preparation step involved reacting Davison Grade 955 silica (previously dried at 600°C to provide a silica that contained approximately 0.72 mmol silica hydroxyl groups/g) sequentially with dibutylmagnesium, 1-butanol and titanium tetrachloride to isolate a dry free-flowing powder containing the Ziegler catalyst component. The second catalyst preparation step was carried out by adding a toluene solution of $(BuCp)_2ZrCl_2$ and methylalumoxane to the titanium-containing catalyst component using an incipient wetness impregnation technique in which the volume of the toluene solution containing the Zr/MAO was approximately equal to the total pore volume of the titanium catalyst component. Toluene was removed at about 45°C using a dry nitrogen purge through the catalyst preparation vessel that contained a helical stirrer operating at about 40–60 rpm. The final catalyst contained 1.85 wt% Ti and 0.30 wt% Zr. The Al (MAO)/Zr molar ratio was 100.

The bimetallic catalyst was evaluated in a continuous, gas-phase fluidized-bed pilot reactor at 95°C that produced about 20lb/hr of HMW-HDPE at an ethylene partial pressure of 180 psi, hydrogen/ethylene molar ratio of 0.005–0.008, 1-hexene/ethylene molar ratio of 0.015 and trimethylaluminum (TMA) as cocatalyst. A comparison of the product properties of the HMW-HDPE produced with the bimetallic catalyst in a single polymerization reactor to a similar product available commercially using two polymerization reactors operated in series is shown in Table 4.14. The gel permeation chromatogram (GPC) of the bimodal HMW-HDPE produced in a single reactor with the bimetallic catalyst is shown in Figure 4.28.

Examination of the GPC data in Figure 4.28 shows the bimodal MWD of the polyethylene produced with the bimetallic (Ti/Zr) catalyst in a single gas-phase reactor and illustrates that the relatively LMW polymer component is produced by the Zr catalyst component, while the relatively HMW polymer component is provided by the Ti-based catalyst component. This GPC data is very similar to a similar product produced with tandem reactors. As discussed previously, the Conoco scientists developed this polymer composition to improve the mechanical properties of the film fabricated from this type of polyethylene. The relatively high MW polymer

Table 4.14 Comparative properties of bimodal HMW-HDPE prepared in a single reactor and tandem slurry process.

	Single Reactor (Ti/Zr Catalyst)	Tandem Slurry Process Ti-based Catalyst
$I_{21.6}$ (Flow Index)	5.3	8.0
MFR (FI/MI)	113	160
Density (g/cc)	0.949	0.950
Fabrication Rate (lbs/hr)	98	120
Dart Drop, 1 mil (g)	565	325
Dart Drop, 0.5 mil (g)	410	420
MD Elmendorf Tear, 0.5 mil; g/mil	37	25

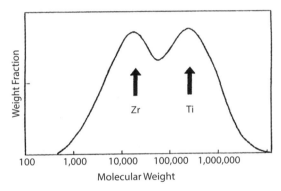

Figure 4.28 GPC curve of a HMW-HDPE sample prepared in a single reactor with a bimetallic (Ti/Zr) catalyst, with the polyethylene component provided by each of the two metals identified [61,62].

component is responsible for the toughness (measured using a dart impact test) of the film, while the relatively very low MW polymer component improves the processability of the polyethylene during the film-forming process. The polymer component with an intermediate MW was eliminated from the polymer composition because this particular polymer component was relatively difficult to process and did not contribute a sufficiently high molecular weight component to enhance the polymer film properties.

The important physical properties shown in Table 4.14 indicate that the dart drop values of the two HMW-HDPE samples are approximately the same, with the single-reactor property exhibiting a higher machine direction (MD) Elmendorf tear, which is an important property in that it relates to less film failure in the finished item (i.e., grocery T-shirt bag).

4.11   Next Generation Catalysts

Attempts to forecast long-term future technology in any area of science is an impossible task left to individuals with relatively high egos. However, it is possible to state some very general trends which the polyethylene industry may follow over the course of the next 10–20 years based on the present day status of ethylene polymerization catalyst technology.

Catalyst developments for the polyethylene industry in the near future will continue to expand the types of molecular structures available for future commercial applications. These catalyst developments can be classified as follows:

a. **Mixtures of one or more single-site catalysts with different 1-olefin reactivity:** The combination of two or more single-site catalysts [63] with significantly different reactivity with a comonomer such as 1-hexene will allow the preparation of polyethylene with a bimodal branching distribution in which one single-site catalyst provides a relatively higher molecular weight polymer component containing short-chain branching, while the second single-site catalyst, that exhibits very poor reactivity with 1-olefins, will provide a second lower molecular weight polymer component. The polymer properties will be controlled by the difference in branching and molecular weight in each of the two polymer components.

b. **New catalysts with unique ligand systems based on early and late transition metals:** Catalysts based on late transition metals such as nickel, iron and cobalt for olefin polymerization have been reported over the past 10–20 years and two excellent reviews of such catalysts were published by Vernon C. Gibson and coworkers in 1999 [64] and 2003 [65]. The later review is a remarkable extensive manuscript containing 513 references which is highly recommended for any scientist seeking a summary of the research in emerging catalyst technology that will most likely lay the groundwork for future catalysts for the commercial manufacture of new grades of polyethylene. It is important to note that the nickel (II) and palladium (II) complexes developed by Brookhart and coworkers based on square-planar cationic compounds utilizing bulky diimine ligands were the first examples of late transition metal complexes that provided high molecular weight polyethylene for possible commercial applications. Some complexes identified, produced very highly-branched (70–100 branches/1000 carbon atoms) polymers in ethylene homopolymerization experiments. Such polyethylenes would be amorphous materials with applications as thermoplastic elastomers.

c. **Catalysts for the manufacture of ethylene/polar comonomer polyolefins:** Ethylene/vinyl acetate and ethylene vinyl alcohol are two important commercial polymers that are derived from the high-pressure polyethylene process which would be suitable candidates for manufacture with a low-pressure process with the development of a suitable catalyst. Some palladium (II) complexes identified by Brookhart and coworkers are able to copolymerize 1-olefins with comonomers such as methyl acrylate, resulting in highly-branched copolymers with ester groups on some chain ends (e.g., see ref. [66]).

References

1. W. Kaminsky, J. Kopf, H. Sinn, and H. Vollmer, *Angew. Chem. Int. Ed. Engl.*, Vol. 15, No. 10, p. 629, 1976.
2. W. Kaminsky, et al., *Angew. Chem. Int. Ed. Engl.*, Vol. 15, No. 10, p. 630, 1976.
3. H. Sinn, W. Kaminsky, H. Vollmer, and R. Woldt, *Angew. Chem. Int. Ed. Engl.*, Vol. 19, No. 5, pp 390-392, 1980.
4. W. Kaminsky, The discovery of metallocene catalysts and their present state of the art, *J. Polymer Science, Part A: Polymer Chemistry*, Vol. 42, pp. 3911-3921, 2004.
5. M.P. McDaniel, et al., "Metallocene activation by solid acids," in: *Tailor-Made Polymers*, J.R. Severn, and J.C. Chadwick, Eds., Wiley-VCH Verlag GmbH & Co., Weinheim, Germany, Chap. 7, pp. 171-210, 2008. ISBN: 978-3-527-31782-0.
6. Personal communication with M.P. McDaniel.
7. M.D. Jensen, J.L. Martin, M.P. McDaniel, D.C. Rolfing, Q. Wang, M.G. Thorn, A.M. Sukhadia, Y. Yu, and J.T. Lanier, U.S. Patent 7,148,298 B2, issued to Chevron-Phillips Chemical Co., Dec. 12, 2006.
8. J.P. Stevens, *Studies in Surface Science and Catalysis*, Vol. 101, p. 11, 1996.
9. G.G. Hlatky, et al., U.S. Patents 5,554,775 and 5,539,124, originally filed by Occidental Petroleum which merged with Lyondell Petrochemical in May 1995.
10. U.S. Patents: Gao, et al., 6,300,435, Oct. 9, 2001; Stephan, et al., 6,342,463, Jan. 29, 2002; von Haken Spence, et al., 6,355,744, March 12, 2002; Xu, et al., 6,380,333, April 30, 2002; McMeeking, et al., 6,420,300, July 16, 2002; Estrada, et al., 6,867,160, March 15, 2005; and Gao, et al., 7,001,962, Feb. 26, 2006; all issued to Nova Chemicals (Internatonal) S.A. (Calgary, CA).
11. R. Krishnamurti, et al. U.S. Patent 5,554,775, issued Sept. 10, 1996, original assignee Occidental Chemical Corp.; B.P. Etherton, et al., 5,539,124, issued July 23, 1996, original assignee Occidental Chemical Corp.; and S. Nagy, et al., 5,637,660, June 10, 1997, original assignee Lyondell Petrochemical Company.

12. S. Nagy, R. Krishnamurti, and B.P. Etherton, PCT Int. Appl. WO 96/34021, 1996;

13. S. Nagy, R. Krishnamurti, and B.P. Etherton, U.S. Patent 5,554,775, issued to Occidental Chemical Corp., Sept. 10, 1996.

14. T.E. Nowlin, Y.V. Kissin, and K.P. Wagner, *J. Polymer Science, Part A: Polymer Chemistry*, Vol. 26, p. 755, 1988.

15. L.S. Baugh, and J.A.M. Canich, Eds., *Stereoselective Polymerization with Single-Site Catalysts*, CRC Press; Taylor & Francis Group, Boca Raton, Fl., 2008.

16. D. Breslow, and N.R. Newburg, *J. Am. Chem. Soc.*, Vol 79, p. 5072, 1957.

17. D. Breslow, and N.R. Newburg, *J. Am. Chem. Soc.*, Vol 81, p. 81, 1959.

18. W.P. Long, and D. Breslow, *J. Am. Chem. Soc.*, Vol 82, p. 1953, 1960.

19. W.P. Long, and D.S. Breslow, *Liebigs Annalen der Chemie*, Iss. 3, pp. 463-469, 1975.

20. K.H. Reichert, and K.R. Meyer, *Makromol. Chem.*, Vol. 163, No. 169, 1973.

21. J. Scheirs, and W. Kaminsky, Eds., "A brief historical perspective," in: *Metallocene-Based Polyofefins, Vol. 1*, John Wiley and Sons, p. XIX, 2000.

22. J. Herwig, and W. Kaminsky, *Polymer Bulletin*, Vol. 9, pp. 464-469, 1983.

23. J.C.W. Chien, and B.P. Wang, *J. Polymer Science, Part A: Polymer Chemistry*, Vol. 27, p. 1539, 1989.

24. W. Kaminsky, and M. Arndt, "Metallocenes for Polymer Catalysis," in: *Advances in Polymer Science, Vol. 127*, Springer-Verlag, p. 143, 1997.

25. A.R. Barron, et al., *J. Am. Chem. Soc.*, Vol. 115, p. 4971, 1993, and references therein.

26. G.A. Razuvaev, et al., *Izv. Akad. Nauk SSSR, Ser. Khim.*, p. 2547, 1975.

27. S. Pasynkiewicz, Alumoxanes: Synthesis, structures, complexes and reactions, *Polyhedron*, Vol. 9, No. 2/3, p. 435, 1990.

28. A.R. Barron, in: *Alkylalumoxanes: Synthesis, Structure and Reactivity, Vol. 1*, J. Scheirs, and W. Kaminsky, Eds., John Wiley and Sons, Chap. 2, pp. 33-67, 2000.

29. J.L. Atwood, et al., *Organometallics*, Vol. 2, p. 985, 1983.

30. J.L. Atwood, et al., *J. Chem. Soc., Chem. Commun.*, p. 302, 1983.

31. A.R. Barron, *J. Am. Chem. Soc.*, Vol.115, p. 4971, 1993.

32. A.R. Barron, *Alkylalumoxanes: Synthesis, Structure and Reactivity, Vol. 1*, J. Scheirs, and W. Kaminsky, Eds., John Wiley and Sons, Chap. 2, p. 58, 2000.

33. J.J. Eisch, et al., *J. Am. Chem. Soc.*, Vol. 107, No. 24, p. 7219, 1985.

34. W.P. Long, *J. Am. Chem. Soc.*, Vol. 81, p. 5312, 1959.

35. G. Fink, and W. Zoller, *Makromol. Chem.*, Vol. 182, p. 3265, 1981.

36. R.F. Jordan, W.E. Dasher, and S.F. Echols, *J. Am. Chem. Soc.*, Vol. 108, p. 1718, 1986.

37. R.F. Jordan, et al., *J. Am. Chem. Soc.*, Vol. 108, p. 7410, 1986.

38. J.L. Atwood, et al., *Organometallics*, Vol. 2, p. 750. 1983.

39. Y.V. Kissin, et al., *Macromolecules*, Vol. 33, No. 12, p. 4599, 2000.

40. Y.V. Kissin, *Macromolecules*, Vol. 36, p. 7413, 2003.

41. Y.V. Kissin, R.I. Mink, A.J. Brandolini, and T.E. Nowlin, *J. Polymer Sci., Part A: Polymer Chemistry*, Vol. 47, p. 3271, 2009.

42. T.E. Nowlin, Y.V. Kissin, and K.P. Wagner, *J. Polymer Science, Part A: Polymer Chemistry*, Vol. 26, p. 755, 1988.

43. J.A. Ewen, in: *Metallocene-Based Polyofefins, Vol. 1*, J. Scheirs, and W. Kaminsky, Eds., John Wiley and Sons, Chap. 1, pp. 3-31, 2000.

44. H.C. Welborn, Jr., and J.A. Ewen, U.S. Patent 5,324,800 issued to Exxon Chemical Patents, Inc., June 28, 1994.

45. H.W. Turner, U.S. Patent 4,791,180 issued to Exxon Chemical Company, Dec. 13, 1988.

46. F.Y. Lo, et al., U.S. Patent 5,608,019, issued to Mobil Oil Corp., March 4, 1997.

47. W. Kaminsky, and F. Renner, *Makromol. Chem., Rapid Commun.*, Vol. 14, p. 239, 1993.

48. T. Kamfjord, T. Wester, and E. Rytter, *Makromol. Chem., Rapid Commun.*, Vol. 19, p. 505, 1998.

49. D.-H. Lee, K. Yoon, and S.K. Noh, *Macromolecular Rapid Communications*, Vol. 18, p. 427, 1997.

50. J.C. Stevens, and D.R. Neithamer, U.S. Patent 5,064,802, issued to Dow Chemical Co., Nov. 12, 1991. Additional Dow Patents include: See for example, U.S. Patents 5,132,380; 5,350,723; 5,399,635; 5,494,874; 5,532,394; 5,539,068; 5,665,800.

51. J.P. Stevens, "Studies in surface science and catalysis," in: *11th International Congress on Catalysts – 40th Anniversary*, J.W. Hightower, W.N. Delgass, E. Iglesia, and A.T. Bell, Eds., Elsevier, Vol. 101, pp. 11-20, 1996.

52. S.Y. Lai, J.R. Wilson, G.W. Knight, and J.C. Stevens, U.S. Patent 5,665,800, issued to Dow Chemical Co., Sept. 9, 1997.

53. G.W. Knight, et al., *Structure, Properties and Preparation of Polyolefins Produced by Single-Site Catalyst Technology, Vol. 1*, J. Scheirs, and W. Kaminsky, Eds., John Wiley and Sons, Chap. 12, pp. 261-286, 2000.

54. W. Xu, Q. Wang, and R.P. Wurz, U.S. Patent 6,339,161 B1, issued to Nova Chemical International, Jan. 15, 2002.

55. W. Kaminsky, A. Bark, and M. Arndt, *Makromol. Chem. Macromol. Symp.*, Vol. 47, p. 83, 1991.

56. M.J. Guest, Y.W. Cheung, C.F. Diehl, and S.M. Hoenig, in: *Metallocene-Based Polyofefins, Vol. 2*, J. Scheirs, and W. Kaminsky, Eds., John Wiley and Sons, Chap. 12, pp. 271-292, 2000.

57. F.J. Timmers, U.S. Patent 5,703,187, issued to Dow Chemical Company, Dec. 30, 1997.

58. K.B. Sinclair, "New polyolefins from emerging catalyst technologies," in: SPE Polyolefins VIII International Conference, Houston, Texas; Feb 21-24, 1993.

59. Q. Wang, P. Lam, Z. Zhang, G. Yamashita and L. Fan, U.S. Patent 6,579,961, issued to Nova Chemicals International, June 17, 2003.

60. T.E. Nowlin, Low pressure manufacture of polyethylene, *Progress in Polymer Science*, Vol. 11, No. 1/2, pp.29-55, 1985.

61. T.E. Nowlin, et al., U.S. Patent 5,332,706, issued to Mobil Oil Corporation, July 26, 1994.

62. T.E. Nowlin, et al., U.S. Patents 5,473,028; 5,539,076; 5,614,456; and 6,740,617. The last patent in the group issued to Univation Technologies, LLC, Houston, TX.

63. J.A. Ewen, and H.C. Welborn, U.S. Patent 4,530,914, issued to Exxon Research & Engineering Co., July 23, 1985.
64. G.J.P. Britovsek, V.C. Gibson, and D.F. Wass, The search for new-generation olefin polymerization catalysts: Life beyond metallocenes, *Angew. Chem. Int. Ed.*, Vol. 38, pp. 428-447, 1999.
65. V.C. Gibson, and S.K. Spitzmesser, Advances in non-metallocene olefin polymerization catalysis, *Chem. Rev.*, Vol. 103, pp. 283-315, 2003.
66. M. Brookhart, et.al., *J. Am. Chem. Soc.*, Vol. 117, p. 6414, 1995; *J. Am. Chem. Soc.*, Vol. 118, p. 267, 1996; *J. Am. Chem. Soc.*, Vol. 120, p. 888, 1998.

Appendix 4.I

Review articles pertaining to ethylene polymerization with single-site catalysts.

W. Kaminsky, and M. Arndt, "Metallocenes for Polymer Catalysis," in: *Advances in Polymer Science, Vol. 127*, Springer-Verlag, pp. 143-187, 1997.

A.A. Montagna, and J.C. Floyd, Single-sited catalysis leads next polyolefin generation, *Hydrocarbon Processing*, p. 57-62, March 1994.

A.D. Horton, Metallocene catalysis: Polymers by design, *TRIP*, Vol. 2, No. 5, pp. 158-166, May 1994.

J.A. Ewen, New chemical tools to create plastics, *Scientific American*, pp. 86-91, May 1997.

M. Hackmann, and B. Rieger, Metallocene catalysis, *CATTECH*, Baltzer Science Publishers, pp, 79-92, Dec. 1997.

O. Olabisi, M. Atiqullah, and W. Kaminsky, Group 4 metallocenes: Supported and unsupported, *Rev. Macromol. Chem. Phys.*, Vol. C37, No. 3, pp. 519-554, 1997.

J. Scheirs, and W. Kaminsky, Eds., *Metallocene-Based Polyofefins, Vol. 1 and 2*, John Wiley and Sons, 2000.

W. Kaminsky, and H. Sinn, Eds., *Transition Metals and Organometallics as Catalysts for Olefin Polymerization*, Springer-Verlag Berlin, Heidelberg, 1988.

P.C. Mohring, and N.J. Coville, Homogeneous group 4 metallocene catalysts: The influence of cyclopentadienyl-ring substituents, *J. Organometallic Chemistry*, Vol. 479, pp. 1-29, 1994.

S. Pasynkiewicz, Alumoxanes: Synthesis, structures, complexes and reactions, *Polyhedron*, Vol. 9, No. 2/3, pp. 429-453, 1990.

G.J.P. Britovsek, V.C. Gibson, and D.F. Wass, The search for new-generation olefin polymerization catalysts: Life beyond metallocenes, *Angew. Chem. Int. Ed.*, Vol. 38, pp. 428-447, 1999.

V.C. Gibson, and S.K. Spitzmesser, Advances in non-metallocene olefin polymerization catalysis, *Chem. Review*, Vol. 103, pp. 283-315, 2003.

M.P. McDaniel, J.D. Jensen, K. Jayaratne, K.S. Collins, E.A. Benham, N.D. McDaniel, P.K. Das, J.L. Martin, Q. Yang, M.G. Thorn, and A.P. Masino, *Tailor-Made Polymers*, J.R. Severn, and J.C. Chadwick, Eds., Wiley-VCH Verlag GmbH & Co., Weinheim: Germany, Preface XV, Chap. 7, pp. 171-210, 2008. ISBN: 978-3-527-31782-0.

5

Commercial Manufacture of Polyethylene

5.1　Introduction

Polyethylene is a semicrystalline, thermoplastic material classified as a synthetic organic polymer. Depending on manufacturing process conditions, polyethylene melts between approximately 110–135°C to form a non-Newtonian liquid that may be molded into a wide variety of shapes utilizing various fabrication techniques. One of the first questions many scientists and engineers ask when beginning a career in the polyethylene industry is: How can such a simple polymer, represented by the formula $-(CH_2-CH_2)_n-$, result in such a complex business? The answer lies in the enormous variety of molecular structures that are possible in the polymerization and, more importantly, copolymerization of ethylene with a wide variety of other 1-olefins.

Polyethylene is manufactured by either a high-pressure process or a low-pressure process and each process is discussed in detail in this chapter. However, the two methods may be briefly summarized as follows:

(a) **Low-pressure process:** There are a wide variety of low-pressure manufacturing processes with two common features, (1) all low-pressure methods require a transition metal catalyst in which ethylene is inserted into a transition metal-carbon bond to produce a polyethylene molecule, and (2) each transition metal active site undergoes a chain termination step that provides a continuous active site that produces additional polyethylene molecules. The low-pressure method was discovered in the early 1950s.

(b) **High-pressure process:** There are two process methods for the high-pressure manufacture of polyethylene, (1) a tubular process where the polyethylene is formed in a long, small diameter tube, and (2) an autoclave process where polyethylene is produced in larger reaction vessels in series. Both methods require a free-radical source to initiate the process in which the polyethylene is produced by the reaction of the free radical with an ethylene molecule in a free-radical, growth mechanism. The high-pressure method was discovered in the early 1930s.

5.1.1 First Manufacturing Facility

The first commercial manufacture of polyethylene, using high-pressure ethylene gas in the presence of trace amounts of oxygen that acted as a free-radical source for the initiation of the polymerization process, was due to the important physical properties of polyethylene for electrical applications at the beginning of World War II. In 1940 about 90 metric tons of polyethylene was produced by Imperial Chemical Industries (ICI) in Great Britain in support of the war effort.

The early history of polyethylene is especially important because the first practical uses of polyethylene were in electrical applications that played a significant role in the outcome of World War II. Polyethylene exhibits low dielectric loss and great water resistance, which are important product attributes for wire and cable applications. Therefore, polyethylene was used by the British in submarine wire and cable, but perhaps more importantly, polyethylene-based wire and cable were used to make extremely rapid progress in the introduction of radar [1]. Radar played a significant role in the Battle of Britain (July25 – September 24, 1940), allowing the Royal Air Force to monitor enemy aircraft long before the aircraft reached airspace over Great Britain, thus providing the RAF with a significant tactical advantage.

Over the past 80 years, this synthetic polymer has played an important role in providing various types of products that have contributed to improving the standard of living around the globe.

This chapter will discuss the various methods that have been developed between 1940 and 2010 to manufacture polyethylene in a cost-effective manner to become the most important plastic material in use today.

5.1.2 Early Documentation of Manufacturing Processes

In order to document the enormous amount of literature that has been published concerning ethylene polymerization and the characterization of polyethylene in the molten and solid state, it is important to highlight several books that were published between 1956 and 1965 discussing the technology that created the rapidly growing polyethylene industry.

The references listed below are some important texts for anyone interested in understanding the chronological events that shaped the polyethylene industry since the discovery of polyethylene utilizing high-pressure ethylene polymerization. These sources, and references cited within these sources, document most of the early research during the rapid growth period of the polyethylene industry.

- 1956:*Polyethylene*, R. A. V. Raff and J. B. Allison, Koppers Company, Inc., Pittsburgh, Pennsylvania; High Polymers Series, Volume XI, Interscience Publishers, Inc., New York. Comment: This 551 page text deals with the high-pressure process and has chapters dedicated to ethylene, polymerization of ethylene, molecular structure, properties and testing of polyethylene, and fabrication and applications of polyethylene.
- 1957: *Polyethylene*, Theodore O. J. Kresser, Spencer Chemical Company, Orange, Texas, Rinhold Plastics Applications Series; Reinhold Publishing Corporation, New York. Comment: This 212 page text provides an overview of the polyethylene industry as a business and contains an excellent discussion of the growth of the industry and the product attributes of polyethylene that provided the stimulus for this growth. It does not contain detailed references, but is a good text for anyone that wants to understand the early growth of the polyethylene industry.
- 1964:*Copolymerization*, George E. Ham, Editor; Spencer Chemical Company, Merriam, Kansas, High Polymers

Volume XVII, Interscience Publishing, Division of John Wiley & Sons, New York. Comment: These 15 chapters by individual authors deal with low-pressure ethylene copolymers over a broad range of compositions from elastic materials, block copolymers, and ethylene copolymers with styrene, and copolymers prepared with cationic and anionic methods. Each chapter includes references.

- 1965:*Crystalline Olefin Polymers*, High Polymers Series Volume XX; R. A. V. Raff, Koppers Company, and K. W. Doak, Rexall Chemical Company, Paramus, New Jersey; Editors; Interscience Publishing, Division of John Wiley & Sons, New York. Comment: A two-part series, Part I includes 16 authored chapters covering ethylene copolymers with higher 1-olefins, low-pressure transition metal catalysts, stereospecific polymerization, high-pressure free-radical technology, polyethylene structure/property relationships and process technology. Part II of the series covers important polyethylene properties such as stress-cracking, permeability, degradation, and stabilization. Each chapter is referenced very well.
- 1966: "Polymerization by Organometallic Compounds," Leo Reich, Picatinny Arsenal, Dover, New Jersey, and A. Schindler, Camille Dreyfus Laboratory, Research Triangle Institute, Durham, North Carolina;in: *Polymer Reviews*,H. F. Mark and E. H. Immergut, Editors; Volume XII, Interscience Publishing, Division of John Wiley & Sons, New York. Comments: Nine chapters with Chapters III and IV a good source for a detailed discussion on research reported on Ziegler-Natta catalysts from about 1955–1965.

5.2   Commercial Process Methods

The physical state of the polyethylene during the polymerization process varies depending on the type of process utilized in the manufacturing process and the molecular structure of the polyethylene produced. The polyethylene contained in the polymerization reactor may be in the molten state, a solid particle of polyethylene (referred to as granular polyethylene) or polyethylene which is dissolved in an organic solvent. In some cases, polyethylene exists as a solid particle with a significant amount of dissolved polyethylene such as a low molecular weight polymer component.

The specific polymerization conditions utilized in the manufacturing process determine the physical state. The polymerization conditions in the process are primarily the polymerization temperature, ethylene pressure, polyethylene composition (amount of branching, molecular weight and molecular weight distribution) and type of solvent employed in the process.

Commercial polyethylene is manufactured by either a high-pressure process or a low-pressure process summarized in Table 5.1.

The low-pressure polymerization processes were originally used in a single reactor configuration, but since the 1970s more complex polymerization systems have been developed that utilize two or more reactors that are in parallel and/or in series. Consequently, the polymerization conditions in each reactor may be varied over a wide range so that the final polyethylene material manufactured has a complex molecular structure, which is designed to provide various premium-grades of polyethylene that are suited for specific markets and applications.

For example, a premium grade of high-density polyethylene with a relatively high molecular weight (designated as HMW-HDPE) and a bimodal molecular weight distribution can be prepared in two low-pressure reactors operating in series, referred to in the industry as "tandem" reactors. A polymerization catalyst is added to the first reactor under polymerization conditions to produce a relatively very low molecular weight polymer

Table 5.1 Process methods for the manufacture of polyethylene.

Process	Comments
High Pressure	Either a tubular reactor or a stirred autoclave reactor. Operating pressure 15,000–40,000 psi at 150–250°C.
Low-Pressure Slurry	Polyethylene is manufactured as a solid granular material suspended in an organic diluent such as isobutane.
Low-Pressure Solution	Polyethylene is manufactured dissolved in an organic solvent such as cyclohexane. Catalyst may be either homogeneous or heterogeneous.
Low-Pressure Gas Phase	Polyethylene is manufactured as a granular solid without an organic solvent.

Note: Low pressure is usually about 50–500 psi total pressure, which includes ethylene, comonomer, hydrogen and other organic hydrocarbons.

component. Then, the catalyst and polymer produced in the first reactor are transferred to a second polymerization reactor with a different set of polymerization conditions that produces a relatively very high molecular weight polymer component. Therefore, the finished polyethylene material that is transferred out of the second reactor contains a mixture of each of these two polymer components and the finished product exhibits a bimodal molecular weight distribution. This particular grade of HMW-HDPE is marketed as a premium grade of polyethylene that is well suited for the fabrication of merchandise bags that possess improved physical properties. Other premium grades of polyethylene may also be produced in tandem reactors under a different set of polymerization conditions that provide superior products for blow-molding (bottles and large containers) and high-pressure pipe applications.

Multiple reactors may also include the combination of different low-pressure processes. For example, the first reactor may be a gas-phase fluid-bed reactor, while the second reactor may be a slurry process.

Each type of process will be discussed in this chapter with the objective of providing chronological details of important innovations as the polyethylene manufacturing process technology developed from 1939 to 2012, so that future scientists may gain a better understanding of the growth of the polyethylene industry over the past 80 years.

5.3   Global Polyethylene Consumption

Global consumption of polyethylene in 2007 was approximately 66 million metric tons, or 145 billion pounds, with a global capacity utilization rate of 86%. About 106 billion pounds were manufactured using a low-pressure process and 39 billion pounds manufactured with a high-pressure process. Table 5.2 summarizes the consumption by product types [2].

In 2008, the global economy entered a severe recession, with the developed regions of North America and Europe experiencing negative annual gross domestic product (GDP) growth for approximately one to

Table 5.2 Global polyethylene consumption by product type.

Type of Polyethylene (%)	Process High/Low Pressure	2007 Global Consumption	
		Million metric tons	Billion pounds
HDPE (45)	Low	29.7	65.1
LLDPE (28)	Low	18.5	40.7
LDPE (27)	High	17.8	39.2

two years, while emerging markets such as China and Brazil experienced slower economic growth.

However by the end of 2010 consumption continued to expand, with a global polyethylene consumption of 154 billion pounds, which is a 6% increase over the 2007 period summarized above. The consumption data for the individual segments of the polyethylene market for 2010 were HDPE, 69 billion pounds; LLDPE, 46 billion pounds; and LDPE, 39 billion pounds.

5.4   High-Pressure Polyethylene Manufacturing Process

5.4.1   Historical Summary

A high-pressure process was the first method used to manufacture polyethylene.

An excellent summary of the enormous growth of the polyethylene industry from 1935 to 1955 is the text cited above by R. A. V. Raff and J. B. Allison of the Koppers Company, Inc., located in Pittsburgh, Pennsylvania. This classic book (Section 5.1.2) provides a detailed discussion of the polyethylene industry during this important period that established polyethylene as the highest growth thermoplastic material. In addition, the text provides 1,152 references that primarily document the rapid growth period for polyethylene manufactured by the high-pressure process and is a useful source for anyone interested in studying the early history of the polyethylene industry. An early discussion of the low-pressure process, which had just been reported in the early 1950s for the preparation of polyethylene, is also provided in this text.

5.4.2   Details of the Discovery of the High-Pressure Process

It is important to acknowledge a very important development that took place before the systematic investigation by Imperial Chemical Industries (ICI) into the high-pressure chemistry of organic compounds which led to the discovery of high molecular weight polyethylene by ICI scientists in 1933. That development was the research of Dr. A. Michels of the University of Amsterdam, who created a pump capable of providing pressures of 3,000 atmospheres. This engineering accomplishment provided the laboratory equipment needed by Fawcett and Gibson to carry out their research in 1933.

As a result of a high-pressure research program started in 1932 [3] at the Imperial Chemical Industries (ICI) of England, ICI began commercial production of low-density polyethylene (LDPE) in 1939 by a high-pressure, free-radical method [4] to manufacture a branched polyethylene material utilizing a stirred autoclave reactor. The U.S. Patent 2,153,553 was issued to Eric William Fawcett, Reginald Oswald Gibson, and Michael Willcox Perrin of Imperial Chemical Industries on April 11, 1939, and the title page of this historic patent that is responsible for the birth of the global polyethylene industry is shown in Figure 5.1.

The first experiment that provided 8 grams of polyethylene was carried out in 1935 under high ethylene pressure at 180°C. It was later learned that the oxygen in this experiment was added by accident due to a leak in the reactor system. Once scientists discovered that the oxygen was needed as the free-radical source for the initiation of the ethylene polymerization process, the scientists were able to experimentally control the polymerization process. At this time, process conditions were reported as 500–3,000 atm of pressure and a temperature range of 100–300°C. The oxygen content within the polymerization autoclave was controlled by adding varied levels of oxygen to the ethylene supply, thus controlling the ethylene polymerization process. The polyethylene produced in the late1930s had a melting point of about 115°C and a density of 0.91–0.92 g/cc. Based on the fact that this early polyethylene could be drawn into filaments by a technique which was later termed "cold drawing," and that the polyethylene had a relatively high level of crystallinity (60–70%), scientists at the time believed that high-pressure polyethylene was mostly a linear material (i.e., did not contain any short- or long-chain branching) in order to explain these observations. High-pressure polyethylene was later shown by infrared spectroscopy

Patented Apr. 11, 1939   **2.153,553**

UNITED STATES PATENT OFFICE

2,153,553

POLYMERIZATION OF OLEFINS

Eric William Fawcett, Reginald Oswald Gibson,
and Michael Willcox Perrin, Northwich, England, assignors to Imperial Chemical Industries
Limited, a corporation of Great Britain

No Drawing. Application February 2, 1937, Serial
No. 123,722. In Great Britain Feburary 4,
1936

1- Claims.   (Cl. 260—94)

Figure 5.1 Patent for the high-pressure process to manufacture polyethylene.

to contain many more methyl groups than could be accounted for as polymer end-groups [5], suggesting that the initial belief that polyethylene possessed a high degree of linearity was not correct, but did in fact contain pendent side chains. The branching in high-pressure polyethylene was later characterized as a complex mixture of short-chain branches (primarily C_2-C_4) and long-chain branches. At the time, this commercial material was simply referred to as polyethylene (or polythene), as it was the only material available.

5.4.3   Developments during World War II (1940–1945)

Imperial Chemical Industries (ICI) reported polyethylene annual production based in the United Kingdom of 100 tons in 1940, utilizing ethylene derived by dehydrating ethanol. Polyethylene production increased to 1,500 tons by 1945. Importantly, because of the significance of this new material for the war effort, the United States government encouraged both the Union Carbide Corporation and the DuPont Corporation to license the ICI polyethylene process during the early years of WWII. Consequently, by 1943 both companies were producing polyethylene in the United States. DuPont utilized the ICI stirred autoclave reactor for the manufacture of polyethylene, while Union Carbide process engineers developed a new tubular reactor configuration for the manufacture of polyethylene. By the end of the war, the capacity of polyethylene manufactured in the United States exceeded the ICI capacity. During the war, it was realized that the Union Carbide tubular process provided polyethylene with different physical properties than the autoclave-produced material and that the Union Carbide polyethylene was preferred for certain applications. It was later realized that the tubular design provided polyethylene with less long-chain branching.

A summary of the early developments in high-pressure process technology up to the end of World War II are as follows:

- 1933 – Initial discovery of polyethylene by ICI.
- 1939 – First commercial plant (autoclave) built by ICI.
- 1941 – ICI licenses technology to DuPont in the United States.
- 1941 – Union Carbide in the United States builds tubular process pilot plant.
- 1943 – Union Carbide built first commercial tubular plant and DuPont built first autoclave process in the United States.

5.4.4   Post World War II Developments (1945–1956)

Post World War II was an extremely active period in the high-pressure polymerization of ethylene, as a large number of global companies licensed the process to manufacture polyethylene. Key technical findings during this period were disclosed.

High ethylene purity was recognized very early as a necessary part of producing polyethylene resin. The quality of the ethylene gas had to be carefully controlled in order to produce consistent quality resin for commercial applications. Impurities were classified into two kinds, designated as inert impurities and active impurities.

The inert impurities affected the pressure needed in a commercial reactor. As these inert impurities increased in concentration through the recycle loop, the total pressure in the reactor increased. Therefore, steps were taken to remove such impurities in the recycled ethylene feed.

Active impurities were a much more serious concern, as these impurities directly modified the polymerization reaction by acting as:

(a) Crosslinking agents (e.g., acetylene and butadiene);
(b) Copolymerizing monomers (carbon monoxide and propylene);
(c) Chain termination agents (hydrogen, methane, ethane, hydrogen sulfide, carbon dioxide, nitrogen); or
(d) Catalytic agents (oxygen and nitrogen oxides).

Consequently, steps to improve the purity of the ethylene gas were an important aspect in producing higher molecular weight polyethylene while achieving a more controlled process. Early research found that the oxygen content of the ethylene feed stream strongly affected polymerization rate. The allowable oxygen content at about 2000 atmospheres and 165°C was 0.075%. If this limit was passed, then an explosive decomposition reaction occurred with the products being mainly carbon, methane and hydrogen. This type of decomposition episode, which can be viewed as an internal explosive reaction, greatly fouled the polymerization reactor and shut down the polymerization process until the reactor was cleaned. Over the next 30 years or so, one of the primary concerns for the process engineers operating a commercial high-pressure polymerization reactor, was to eliminate such decomposition events (referred to as "decomps" in the industry).

Extensive research was also reported in 1947 by scientists at ICI and DuPont who found that oxygen could be replaced as the free-radical initiator with peroxide compounds [6]. Peroxide compounds are now used exclusively as free-radical initiators for well-controlled high-pressure ethylene polymerization processes in which explosive decompositions have been essentially eliminated from commercial reactors. However, explosive decomposition reactions did continue to plague the high-pressure process from time to time during the 1950s through the 1970s with decreasing frequency.

In 1946, DuPont scientists reported ethylene/vinyl acetate copolymers [7], which were not commercialized until 1961. However, this work provided the wide range of ethylene/vinyl copolymers that have become the driving force behind the continued growth of the high-pressure process today, as these materials are unique to the high-pressure process.

Polyethylene stabilization against both photochemical and thermal degradation was found to be an extremely important part of research programs in support of the early growth in the polyethylene industry. Polyethylene required additives in the finished polyethylene material to greatly improve the useful life of polyethylene in many outdoor applications. The addition of finely divided carbon black was found to be very effective in reducing the effects of sunlight against degradation. Additives such as phenyl-alpha-naphthylamine and diphenyl-para-phenylene diamine were found to be effective antioxidants, especially in electrical applications. Some early patents awarded to Union Carbide and DuPont that addressed the stabilization of polyethylene were issued [8].

Techniques were also developed to determine the molecular weight of polyethylene, as this was an important parameter which determined the physical properties of the polyethylene, and the polymerization process needed to be controlled so that polyethylene with the necessary molecular weight could be produced in a commercial reactor. Polyethylene grades were manufactured with a certain molecular weight and density. The relationship between the number average molecular weight (M_n) and intrinsic viscosity were published and a light-scattering technique was used to find the weight average molecular weight (M_w). The ratio of M_w/M_n was found to directly correlate with the molecular weight distribution of the polyethylene sample [9].

5.4.5   Rapid Growth Period – Demand Exceeded Supply

It is not unusual during the early stage of a product growth cycle for product demand to exceed product supply as new product applications are rapidly introduced. This was the case for the polyethylene industry.

After the end of World War II, polyethylene production in the United States was limited to only Union Carbide Corporation and the DuPont Company which entered the business with the support of the United States government to provide an additional source of polyethylene in support of the war effort. In 1950, the annual production in the United States was 50 million pounds and the product demand was growing extremely rapidly. During this time, polyethylene demand outpaced the supply.

However, the situation changed rapidly in 1952 when an antitrust suit found that ICI must license their polyethylene process to additional companies that wanted to enter into the production of polyethylene. Consequently, within the next year after the court decision, Dow Chemical, Eastman Kodak and National Petrochemical became ICI licensees. Soon, BASF in West Germany became an additional licensor of a high-pressure polyethylene process, so that by 1955 three additional companies in the United States (Spencer Chemical, Monsanto and Koppers) licensed the BASF technology. United States annual production increased rapidly.

Table 5.3 Rapid growth phase of the polyethylene industry in the United States.

Year	United States Production (millions of pounds)
1943	1
1944	3
1945	6
1946	13
1947	16
1948	19
1949	42
1950	55
1951	85
1952	115
1953	144
1954	195
1955	375
1956[a]	530 [1]

[a] Polyethylene production outside the United States in 1956 was 224 million pounds.

The production of polyethylene in the United States from 1943 until 1956 is shown in Table 5.3 [9]. Even with the increased production throughout the late 1950s, polyethylene demand during this period continued to be greater than the supply, which often occurs in the rapid growth period of the product lifecycle.

Polyethylene production outside of the United States in 1956 was 224 million pounds, primarily produced in Great Britain, 86 million pounds; Germany, 69 million pounds; Japan, 20 million pounds; and Canada, 25 million pounds.

5.4.6 Polyethylene Growth (1952–1960)

Examination of the growth of polyethylene in the period between 1952 and 1960 is an especially important time frame. This period begins with the 1952 antitrust suit that ruled that Imperical Chemical Industries (ICI) must license the manufacture of polyethylene by the high-pressure process to additional companies that wanted to enter the polyethylene business, and also includes the beginning of the manufacture of polyethylene by the low-pressure process starting in 1956.

Examination of the data between 1952 through 1956 in Table 5.4 illustrates the enormous growth in the high-pressure process for the manufacture of polyethylene as a result of the court decision. The polyethylene production for 1954 was 195 million pounds, which increased to 375 million pounds in 1955, and is particularly interesting because the 92% annual growth that took place in the manufacture of polyethylene came three years after the antitrust court decision, which is approximately the amount of time necessary to license the ICI technology and construct a new polyethylene plant in order to begin production. Moreover, in 1956,

Table 5.4 Increase in the United States polyethylene production by the high-pressure process.

Year	Production (millions of pounds)	Annual Growth (%)[a]
1952	115	35
1953	144	34
1954	195	35
1955	375	92
1956	519	38

[a]Percent increase in production compared to the previous year.

Source: Ref. [1], page 199.

the annual polyethylene production increased another 144 million pounds (38%) to 519 million pounds, primarily from additional companies entering the polyethylene business.

Table 5.5 shows the United States production of polyethylene from 1956–1960. This period is important because it includes the start of the manufacture of polyethylene utilizing the low-pressure process.

Examination of Table 5.5 clearly shows that after 1956 polyethylene manufactured by both the high-pressure and low-pressure processes experienced very high growth rates.

Total polyethylene production increased about 13% in 1959 and by 18% in 1960, reaching the 1.0 billion pound plateau, almost double the 1956 production.

In 1955, capital costs for a 60 million pound/year high-pressure polyethylene plant was approximately $40 million dollars, or $0.67 per pound of capacity. The cost of polyethylene in 1955 was approximately $0.50/lb. This cost data is summarized below and the 1955 costs were translated into 2010 US dollars. Since 1955 the average annual inflation rate was about 3.9%.

The data in Table 5.6 demonstrates the very high relative costs associated with the manufacture and sale of polyethylene in 1955 using the high-pressure process, and perhaps even more importantly, the relatively high cost of polyethylene in 1955 of $0.50/lb, since in 2010 US dollars the price of polyethylene would be $3.94/lb.

Consequently, it is not readily apparent why the polyethylene industry underwent such enormous growth between 1950–1960 based on the seemingly very high costs associated with the polyethylene business. The growth of the polyethylene business in the 1950s was due primarily to two features:

Table 5.5 Increase in polyethylene production in the United States from 1956–1960.

Year	Production (millions of pounds)		Total Production (millions of pounds)
	High-Pressure	**Low-Pressure**	
1956	519	11 (2%)	530
1957	622	33 (5%)	655
1958	663	90 (12%)	753
1959	714	136 (16%)	850
1960	790	210 (21%)	1,000

1. Polyethylene possessed a very unique set of physical and chemical properties making it the best material for many applications.
2. Polyethylene was cost competitive compared to other materials available at the time.

5.4.6.1 Polyethylene Product Attributes that Resulted in Rapid Growth

The primary markets for polyethylene (mostly LDPE provided by the high-pressure process) during this high growth period are summarized in Table 5.7 for the period 1954–1960 [1].

Table 5.6 Capital costs of a polyethylene plant in 1955 in 2009 US dollars.

Year	1955	2010
Cost of 60 million pound/year in a plant	$40 million	$315 million
Cost per pound of polyethylene capacity	$0.67	$5.27
Price of polyethylene	$0.50	$3.94

Note: Inflation adjustment to convert 1955 dollars into 2009 dollars is an increase by a factor of 7.87. *In 2010 the approximate capital costs for the construction of a state-of-the-art polyethylene manufacturing facility was about $0.2–0.40 US dollars per pound of capacity, with a typical plant annual capacity of 500–1,500 million pounds of polyethylene. The commercial price of polyethylene in 2014 is in the range of $1.00–$1.30 per pound.*

Table 5.7 Markets for polyethylene during the high growth period between 1954 and 1960.

Application[a]	Year			
	1954	1956	1958	1960
Film	70	150	220	330
Coatings	20	35	60	90
Molding	35	80	145	200
Pipe	30	55	75	110
Bottles	7	15	28	45
Electrical	35	60	90	190
Total[b]	197	395	618	965

[a]In millions of pounds.

[b]Total is not the entire production for any particular year as this list consists of only the primary markets.

Source: Ref. [1], page 199.

5.4.6.1.1   Film Properties

Cellophane was a common material used in film for various packaging applications in the 1950s and remains to some extent in use today. Cellophane is produced from raw cellulose, which is a natural product derived from plant matter. The structure of cellophane, which is a biodegradable polymer, is shown in Figure 5.2.

Examination of the structure of cellophane shows that this is a polar, rigid material which accounts for the very different set of chemical and physical properties (see Table 5.8) when compared to polyethylene. Because of these differences, polyethylene became the material of choice for the majority of film applications starting in the 1950s.

However, polyethylene displaced cellophane on both a cost basis and on much improved physical properties required in many film applications.

5.4.6.1.2   Cost Comparison

Cellophane's cost at the time was about $0.60/lb, while the cost of polyethylene was about $0.53/lb. However, polyethylene's density of 0.92 g/cc and cellophane's density of 1.45 g/cc significantly lowered the cost of polyethylene film relative to cellophane fabricated into a film with the same dimensions. For example, the material cost for a 1.0 mil film with an area of 1000 in^2 cost $0.018 for polyethylene while the cost of cellophane was $0.031, thus polyethylene reduced the cost of the film by about 42% [1].

5.4.6.1.3   Property Comparison

Polyethylene also possesses significantly improved chemical and physical properties as compared to cellophane or other materials used in film applications.

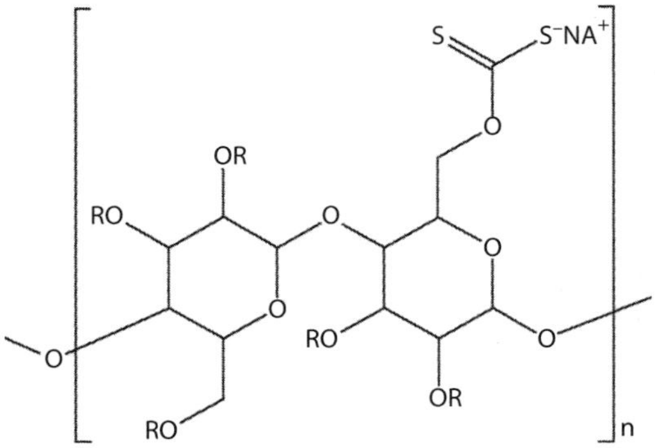

Figure 5.2 Chemical structure of cellophane.

Table 5.8 Properties of polyethylene and cellophane [1].

Property	Cellophane	Polyethylene
Density (g/cc)	1.45	0.92
Elongation (%)	15–25	50–600
Tear Strength (g/mil)	2–10	75–200
Heat Seal (°F)	not sealable	230-300
Folding endurance	Fair	Excellent
Water vapor 90% humidity Permeability (g/h/100 in²)	>100	1.2
Resistance to acids/bases	Poor	Excellent

A comparison of some of these important properties for cellophane and polyethylene are summarized in Table 5.8.

The properties listed in Table 5.8 were responsible for the rapid replacement of cellophane or other cellulose-based packaging materials by polyethylene. Utilizing polyethylene in food packaging applications, such as fresh produce, provided much longer shelf-life and a reduction in waste and spoilage. As can be seen from the physical properties of polyethylene relative to cellulose, polyethylene was more elastic, less likely to tear, heat sealable and much less permeable to water vapor.

5.4.6.1.4 *Pipe Applications*

The chemical inertness of polyethylene to solutions containing harsh chemicals such as inorganic acids, bases and metal salts very quickly made polyethylene a premium pipe material, accounting for the rapid use of polyethylene by the pipe manufacturers. Initial applications were in the chemical industry where the transfer of such solutions is carried out on a routine basis. A representative list of some of the chemicals that are inert to polyethylene for pipe applications is shown below [1].

Examples of Harsh Chemical Solutions Inert to Polyethylene

- Ammonium hydroxide
- Calcium hypochlorite
- Nitric acid
- Sea water
- Sulfuric acid
- Chromic Acid
- Hydrogen peroxide
- Potassium hydroxide

- Phosphoric acid
- Hydrofluoric acid

In many cases polyethylene had a significant cost advantage over other pipe materials such as stainless steel and copper tubing.

Of course, there were organic chemicals that degraded LDPE produced by the high-pressure process in the 1950s, which prevented polyethylene from entering these markets which were dominated by glass- or metal-based containers. For example, gasoline was listed in the 1950s as being an organic solution that attacked polyethylene, but now crosslinked polyethylene is an excellent container for gasoline. In addition, HDPE produced with the Cr-based Phillips catalyst provided an excellent grade of polyethylene by the late 1960s that possessed very good compatibility (measured in self-life and designated in the industry as stress-crack resistance) with lubricating oil used in internal combustion engines. Since that time, polyethylene-based containers have completely replaced other materials such as the metal-based oil or gasoline can, and continue to dominate this market today.

5.4.6.1.5   Electrical Properties

Polyethylene rapidly displaced other materials for electrical applications due to their significantly improved properties for this application. Growth of over 400%, from 35 million lbs/year in1956 to190 million lbs/year in 1960, was achieved. The important product attributes for electrical applications are summarized below.

- Low dielectric constant of 2.3;
- Low power factor of <0.0003;
- High dielectric strength;
- Low water absorption, i.e., low water transmission;
- High tensile strength;
- Chemical inertness;
- Light weight; and
- Resistance to attack by microorganisms such as fungi.

In the early 1950s, scientists at DuPont [10] reported the preparation of polyethylene that was substantially a linear material with a density of 0.95–0.97 g/cc produced by the free-radical polymerization of ethylene at extremely high pressures of 5,000–20,000 atm. This high-density polyethylene had a melting point of 127°C and contained less than one branch per 200 carbon atoms. This very high-pressure process was never commercialized, most likely because of the low-pressure process that was discovered in the 1950s for the manufacture of HDPE by a less costly route.

5.4.6.1.6   *High-Pressure Process (1956–1979)*

By the mid-1950s, modifications were made to the original high-pressure process so that the density range was increased to 0.940 g/cc by ICI, which was designated as the Alkathene brand. Perhaps more importantly, ethylene copolymers were first introduced in 1961 when ethylene/ethyl acrylate copolymers and ethylene/vinyl acetate copolymers were made commercially using the high-pressure process [11].

With the introduction of linear low-density polyethylene (LLDPE) in the 1960s using the solution process and then in the late 1970s by Union Carbide Corporation using the gas-phase fluid-bed process, LLDPE became the preferred material over high-pressure LDPE for many applications. At that time, the future growth of the high-pressure process became somewhat uncertain. In the United States, the last high-pressure expansion was built in 1977 by Mobil Chemical Company in Beaumont, Texas. The United States annual high-pressure capacity has since been unchanged at about 8 billion pounds. However, this has not been the case in the remainder of the world, especially in the Middle East, where low-cost ethylene and low energy prices make the high-pressure process more cost effective. However, it should be noted that in North America, with shale-gas technology that has significantly increased the production of natural gas liquids since about 2010, renewed interest in the construction of additional high-pressure polyethylene capacity in North America seems highly likely over the next several decades.

5.4.7   Worldwide High-Pressure LDPE Capacity Increases (1980–2010)

The manufacture of polyethylene by the high pressure process (tubular and autoclave) continues to add global capacity as shown in Table 5.9.

Table 5.9 Global high-pressure LDPE capacity by year.

Year	Worldwide Capacity (HPPE) (Billion lbs/year)
1980	15
1990	20
2000	30
2010	44

From 1980 to 2000, the high-pressure process exhibited an average annual growth rate of 3.6%, which is higher than expected. In 1980, the gas-phase, fluidized-bed process introduced by Union Carbide Corporation in 1977 under the tradename of UNIPOL, was undergoing rapid growth. At that time, many polyethylene growth forecast reports anticipated a decline in the high-pressure process due to the higher capital costs and operating costs for a high-pressure manufacturing plant compared to the UNIPOL gas-phase process. This clearly has not been the case.

Some recent high-pressure LDPE capacity expansions announced for startup are summarized in Table 5.10.

LyondellBasell's high-pressure tubular reactor (designated as the Lupotech T process) that started production in 2009 is a joint venture partnership with Saudi Ethylene and Polyethylene Company (SEPC) located in AlJubail, Saudi Arabia, and is the largest single-train, high-pressure reactor facility in the world having achieved production rates of 125,000 lb PE/hr. This plant is operating at lower than expected operating costs and the economy of scale achieved with this state-of-the-art design has made the high-pressure process more competitive with the various low-pressure processes available in 2010 [12]. In addition, LyondellBasell has licensed a 330 kt/year plant to a joint venture of Braskem with Mexico's Grupo Idesa for startup in 2015, which is part of a $4.5 billion (USD) petrochemical plant in Coatzacoalcos, Mexico, with 1,115 kt/year of ethylene capacity using ethane as feedstock. The Mexican facility will also include 825 kt/year of low-pressure gas-phase polyethylene capacity using INEOS technology [13].

ExxonMobil reports that their two tubular reactors in Meerhout, Belgium, operate with greater than 98 percent on-line factors, while also maintaining more than 98 percent prime product (polyethylene produced with correct product specifications). It is also important to note that a

Table 5.10 Recent high-pressure LDPE capacity expansions.

Company/location	Capacity[a]	Reactor	Technology	Start-up
Saudi International	440	tubular	ExxonMobil	late 2013
Slovak Republic	440	tubular	LyondellBasell	2012
Borealis	700	tubular	LyondellBasell	2013
Saudi ethylene PE Company (SEPC)	900[b]	tubular	LyondellBasell	2009

[a]Capacity in kt/year.

[b]Operating above the original capacity of 800 million lbs/year. Production rates as high as 125,000 lbs PE/hr reported. Largest HPPE unit worldwide.

state-of-the-art, high-pressure plant undergoes a large number of product transitions as these reactors have a very broad product mix that must be manufactured, which includes different types of polar comonomers and wide variations in the desired comonomer level for individual products.

These high reactor efficiencies reported for the high-pressure process have significantly improved the cost analysis for this process compared to the gas-phase fluidized-bed process, which is often considered the low-cost polyethylene manufacturing process.

The product mix of autoclave and tubular reactors are similar in terms of LDPE homopolymers (0.910–0.935 g/cc) and some specialty grades of polyethylene such as ethylene/vinyl acetate copolymers up to about 30 wt% vinyl acetate (VA). However, the autoclave process provides higher levels of vinyl acetate (40 wt%) in ethylene/VA copolymers and additional specialty grades of polyethylene such as ethylene/methyl acrylate, ethylene/acrylic acid and ethylene/n-butyl acrylate. Polyethylene molecular weight can be varied over a wide range with the high-pressure process, with Melt Index values ($I_{2.16\,kg}$) ranging from 0.15 to 40.

5.4.8   Future of High-Pressure Manufacturing Process

Based on the significant number of high-pressure plants that have either come on stream since 2009, the annual growth rate of the high-pressure process may continue to exceed expectations. The LydondellBasell high-pressure plant in Mexico due for a 2015 start-up [13] supports this forecast.

5.5   Free-Radical Polymerization Mechanism for High-Pressure Polyethylene

The polymerization mechanism for the high-pressure process has been extensively investigated over the past 80 years, but the basic free-radical mechanism accepted for this process was documented in three books published in the 1950s:

- H. Mark and A.V. Tobolsky, *Physical Chemistry of High Polymeric Systems*, Interscience Publishers Inc., New York, 1950.
- P.J. Flory, *Principles of Polymer Chemistry*, Cornell University Press, 1953.

- G.M. Burnett, *Mechanism of Polymer Reactions*, Interscience Publishers Inc., New York, 1954.

The three steps required for the free-radical polymerization mechanism process are: (1) Initiation Step, (2) Propagation Step, and (3) Termination Step.

5.5.1 Initiation Step

Molecular oxygen was utilized by the ICI scientists in 1933 as the free-radical source to initiate the high-pressure ethylene polymerization reaction that provided high molecular weight polyethylene. (A free radical is a highly reactive, unstable, chemical intermediate comprised of an unpaired electron designated by a [•] symbol.) Presently, organic peroxides (R-O-O-R) are utilized as the free-radical source as shown below.

$$R\text{-}O\text{-}O\text{-}R \rightarrow 2\ RO\bullet$$

The free radicals are formed by the thermal decomposition of the initiating organic peroxide. For example, if $R = C_6H_5\text{-}C{=}O$, then this free radical eliminates a CO_2 molecule to provide the $C_6H_5\bullet$ radical that reacts with ethylene as shown below to provide the polymerization initiation free radical.

$$C_6H_5\bullet + H_2C{=}CH_2 \rightarrow C_6H_5\text{-}CH_2\text{-}CH_2\bullet$$

The initiation rate is controlled by the peroxide type, peroxide concentration and reactor temperature.

5.5.2 Propagation Step

The polymerization initiation free radical next reacts with ethylene monomer to form a rapidly growing polymer chain. The very high ethylene pressure (high ethylene concentration) is required in order to increase the rate of chain growth that creates a high molecular weight species before the highly reactive growing free radicals can undergo a free-radical termination step, which terminates the chain growth process.

$$C_6H_5\text{-}CH_2\text{-}CH_2\bullet + n\ H_2C{=}CH_2 \rightarrow C_6H_5\text{-}CH_2\text{-}CH_2\text{-}(CH_2CH_2)_n\text{-}CH_2CH_2\bullet$$

Branching along the polymer backbone is the result of both intramolecular and intermolecular reactions involving the growing free radical illustrated above.

An intramolecular (backbiting) process proposed by M.. J. Roedel in 1953 [14] is believed to be responsible for short-chain branching along the polymer backbone. For example, if the growing free radical bends around along the growing polymer chain, the free radical may abstract a hydrogen atom from a carbon atom five methylene groups from the end of the growing chain. This process forms a new free radical which undergoes addition propagation with ethylene. This is shown below for the abstraction of a hydrogen atom five carbons from the end of the growing chain, which results in the formation of a butyl branch (short-chain branch) along the polymer backbone.

$$
\begin{array}{cc}
\text{H} & \text{H} \\
| & | \\
\text{Polymer} \wedge\wedge\wedge\wedge\wedge\wedge\text{-C-CH}_2\text{CH}_2 \quad \rightarrow & \text{polymer/} \wedge\wedge\wedge\text{-C-CH}_2\text{CH}_2\text{CH}_2\text{CH}_3 \\
\diagup \quad \diagdown & \bullet \;\; \leftarrow \textbf{new chain propagation site} \\
\text{H} \;\; \bullet\text{CH}_2\text{-CH}_2 &
\end{array}
$$

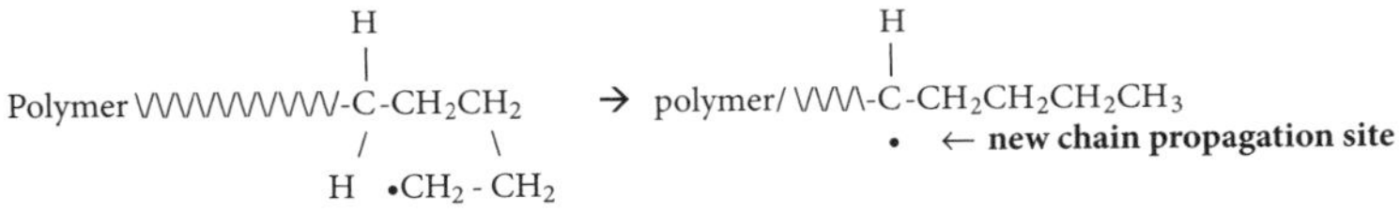

Free radical intramolecular backbiting creates a butyl branch on polymer backbone. (Note: ethylene insertion continues on the newly formed free radical five carbons from the chain end).

An intermolecular reaction is responsible for the formation of long-chain branching where a growing polymer chain abstracts hydrogen from another polymer chain anywhere along the polymer backbone, which initiates polymerization at this newly formed free-radical site. This is illustrated below.

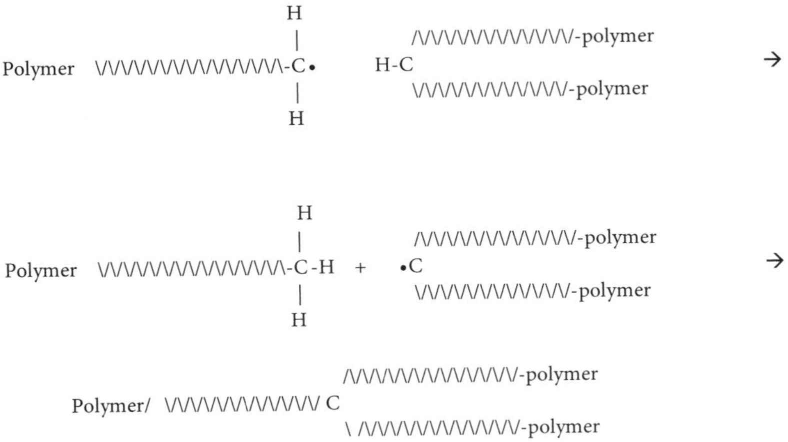

$$
\begin{array}{ccc}
& \text{H} & \\
& | & \wedge\wedge\wedge\wedge\wedge\wedge\wedge\text{-polymer} \\
\text{Polymer} \;\; \wedge\wedge\wedge\wedge\wedge\wedge\wedge\wedge\text{-C}\bullet & \text{H-C} & \rightarrow \\
& | & \wedge\wedge\wedge\wedge\wedge\wedge\wedge\text{-polymer} \\
& \text{H} &
\end{array}
$$

$$
\begin{array}{ccc}
& \text{H} & \\
& | & \wedge\wedge\wedge\wedge\wedge\wedge\wedge\text{-polymer} \\
\text{Polymer} \;\; \wedge\wedge\wedge\wedge\wedge\wedge\wedge\wedge\text{-C-H} \;\; + & \bullet\text{C} & \rightarrow \\
& | & \wedge\wedge\wedge\wedge\wedge\wedge\wedge\text{-polymer} \\
& \text{H} &
\end{array}
$$

$$
\begin{array}{cc}
& \wedge\wedge\wedge\wedge\wedge\wedge\wedge\wedge\text{-polymer} \\
\text{Polymer/} \;\; \wedge\wedge\wedge\wedge\wedge\wedge\wedge \;\; \text{C} & \\
& \diagdown \; \wedge\wedge\wedge\wedge\wedge\wedge\wedge\wedge\text{-polymer}
\end{array}
$$

Addition of chain transfer agents to the polymerization process are used to control the molecular weight of the polyethylene. The chain transfer agent must be capable of donating hydrogen to the growing free radical which terminates the growth of that particular molecule, while the free radical formed from the chain transfer agent will initiate the growth of a new polymer molecule. This reaction is shown below.

Chain Transfer Agent (CT-H):

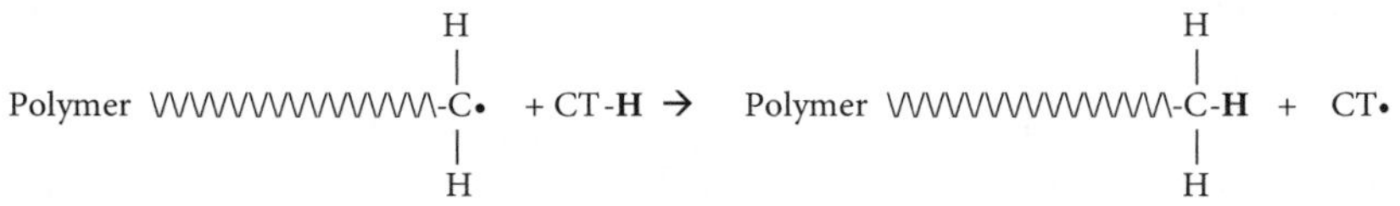

$$
\begin{array}{ccc}
\text{H} & & \text{H}\\
|& & |\\
\text{Polymer } \backslash\!\backslash\!\backslash\!\backslash\!\backslash\!\backslash\text{-C•} \;\; +\,\text{CT-H} \;\rightarrow\; & \text{Polymer } \backslash\!\backslash\!\backslash\!\backslash\text{-C-H} \;+\; \text{CT•}\\
|& & |\\
\text{H} & & \text{H}
\end{array}
$$

where the CT• radical initiates the growth of another polymer chain.

5.5.3   Termination Step

The polymerization reaction may terminate by the interaction of two free radicals as shown below,

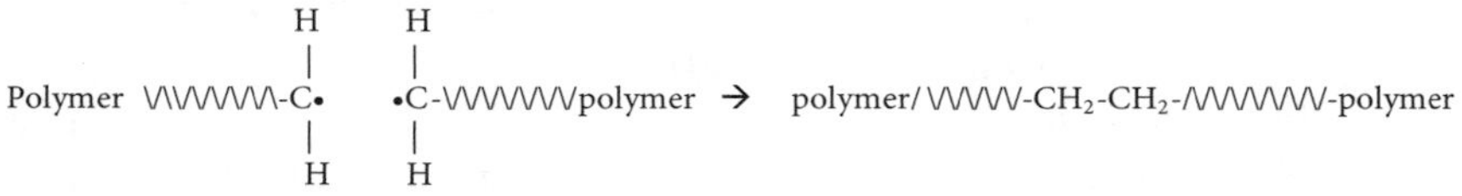

$$
\begin{array}{cc}
\text{H} & \text{H}\\
|& |\\
\text{Polymer } \backslash\backslash\backslash\text{-C•} \quad \text{•C-}\backslash\backslash\backslash\text{polymer} \;\rightarrow\; & \text{polymer/}\backslash\backslash\text{-CH}_2\text{-CH}_2\text{-}\backslash\backslash\text{-polymer}\\
|& |\\
\text{H} & \text{H}
\end{array}
$$

or by a disproportionation reaction which results in a terminal double bond,

$$
\text{Polymer-CH}_2\text{-CH}_2\text{•} \quad \text{•-CH}_2\text{-CH}_2\text{-polymer} \;\rightarrow\; \text{polymer-CH}_2\text{-CH}_3 \;+\; \text{CH}_2\text{=CH-polymer}
$$

Similar to other ethylene polymerization processes, removal of the heat of polymerization from the reactor is a critical feature of the process. In this regard, the tubular high-pressure process has a more efficient heat removal process than the stirred autoclave reactor.

5.6   Organic Peroxides as Free-Radical Source for Initiation Process

The replacement of oxygen as the free-radical source with organic peroxides has led to a more controlled polymerization process which has essentially eliminated highly exothermic decomposition reactions in the reactor (these are referred to as "decomps"), which cause the shutdown of the reactor. Such decomps were caused by the addition of an excess of oxygen or some other impurity to the reactor. These decomps are extremely dangerous as they may lead to the destruction of the reactor and foul the reactor walls with large quantities of carbon. Such shut-downs require that the reactor be cleaned to remove the carbon contamination, resulting in lost production and additional operating costs required to clean and restart the unit.

5.6.1 Types of Organic Peroxides

Some representative examples of commercial sources of organic peroxides as initiators for the manufacture of LDPE are shown in Table 5.11.

Peroxides are available according to the temperature necessary to form the free radicals and are classified accordingly. These include:

(a) Low temperature peroxides (highly active with a one hour half-life below 100°C);
(b) Intermediate temperature peroxides (one hour half-life in the temperature range 100–130°C; or
(c) High temperature peroxides (one hour half-life above 130°C).

These temperature properties determine the polymerization initiation temperature and the length of time the free-radical initiation process is active. For example, low-temperature peroxides possess a relatively short free-radical half-life and offer a low-temperature initiation but also a lower peak temperature, while a high-temperature peroxide with a relatively long

Table 5.11 Some commercial sources of organic peroxides for LDPE initiation.

Company/ Location	Tradename[a]
AkzoNobel/Netherlands	Trigonox
Mitsubishi/Japan	Perbutyl
Arkema Group/France	Luperox

[a]This list does not recommend the use of any of these particular commercial compounds and is only a representative example of tradenames available as of 2012.

Table 5.12   Peroxide family decomposition products.

Peroxide Family	Main decomposition Products
Peroxyesters	CO_2, carboxylic acids, acetone, alkanes, tButyl alcohol
Diacyl peroxides	CO_2, carboxylic acids, alkanes
Peroxydicarbonates	CO_2, alcohols
Monoperoxycarbonates	CO_2, alcohols, acetone, alkanes
Peroxyketals	CO_2, tButyl alcohol, acetone, alkanes, cyclohexanones
Dialkyl peroxides	alcohols, ketones, alkanes
Hydroperoxides	ketones, alcohols, alkanes

free-radical half-life offers higher peak temperatures and a longer lasting initiation process.

The Arkema Group [15] details the type of functional groups present in their Luperox products and provides information into the decomposition products formed by various peroxide types. These decomposition products may be present in the finished polyethylene for commercial applications. This information is necessary to determine if certain decomposition products may interfere with resin color, stability or odor properties. Table 5.12 lists several types of organic peroxides available today and some of the decomposition reaction products.

5.7   Structure of High-Pressure LDPE

5.7.1   General Features

Because the ethylene polymerization mechanism for the manufacture of high-pressure polyethylene involves a growing free-radical active site, the structure of this type of polyethylene is unique in several aspects:

(a) For polyethylene produced from an ethylene homopolymerization process under conditions that result in a density of about 0.92–0.94 g/cc, each polyethylene molecule has a short-chain branching (SCB) content between about 8–40 branches/1000 carbon atoms, while also possessing long-chain branching (LCB). The actual SCB distribution varies with the type of reactor, with a tubular reactor exhibiting a broader branching distribution than a similar polyethylene produced with a stirred-autoclave reactor. The homogeneous SCB distribution accounts for the very high clarity found in LDPE products. The LCB also contributes to the unique set of properties displayed by high-pressure polyethylene. For example, HPPE provides film for shrink-wrap applications. In addition, high melt strength due to the strain hardening of the polymer melt during fabrication contributes to excellent bubble-stability during the blown film-forming process.

(b) The relatively broad MWD of LDPE, compared to LLDPE prepared from a low-pressure process, results in easier fabrication properties providing higher rates (lbs PE processed/hour) of film production. The broader MWD results in a higher level of shear-thinning for the HPPE compared to the narrower MWD of LLDPE resins used in film applications.

(c) The branching characteristics of the LDPE manufactured using the high-pressure process is determined by the type of reactor utilized. The autoclave reactor is a continuous-stirred tank reactor (CSTR) with an agitator to facilitate good mixing. This type of mixing is responsible for a higher level of free-radical interactions, which leads to a more complex molecular structure. The tubular reactor is a plug-flow system, where less mixing leads to fewer free-radical interactions and a less complex molecular structure.

5.7.2 Ethylene at High Pressures

Regardless of the type of reactor utilized for the high-pressure process, it is worthwhile to gain an understanding into the effect on ethylene gas at these extremely high pressures. In an early text published in 1957 by Theodore Kresser on the polyethylene industry [16], Kresser showed the relationship between ethylene gas densities at pressures up to 2,750 atm at 150°C. This relationship is shown in Figure 5.3.

The maximum density of loosely packed undistorted ethylene molecules is at a density value of 0.28 g/cc, which is reached at only 300 atm. Beyond this point, density increases more slowly as the molecules are forced into a quasi-liquid state such that the ethylene density of 0.6 g/cc at 3,000 atm is close to the density of a light organic solvent at room temperature. For example, the density of isopentane at 25°C is 0.616 g/cc. Hence the intermolecular distances approach values found in the liquid state. Consequently, it becomes more apparent how the active free-radical initiators can react with ethylene at extremely high rates to reach high molecular weight values required for a polymer with acceptable physical properties.

Some characteristics of the autoclave reactor and the tubular reactors are summarized in Table 5.13.

The tubular reactor has an inner diameter of < 5 cm but a length of 1–2 km. The tube has multiple injection points and the temperature of polymerization cycles between a wide range depending on the types of peroxides used at the various injection points. The Arkema Group, which markets peroxide compounds for either high-pressure process under the tradename of Luperox [15], provides an example of the temperature profile that takes place in a tubular reactor which is illustrated in Figure 5.4.

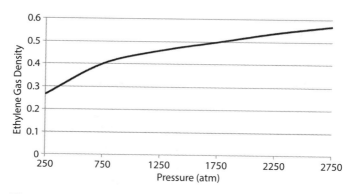

Figure 5.3 Density of ethylene gas at 150°C at pressures up to 2,750 atm [1].

Table 5.13 Characteristics of the autoclave and tubular reactors [15].

Autoclave	Tubular
140–325°C	150–310°C
residence time 20–40 seconds	residence time 2–3 minutes
higher degree of mixing	more plug flow – less mixing
ethylene conversion – 15–30%	ethylene conversion 25–40%
difficult heat removal	easier heat removal with cooling zones
1200–2000 bar operating pressure	2500–3100 bar operating pressure
Single peroxide injection point	Multiple peroxide injection points

The temperature cycles have a significant effect on the structure of the instantaneous polymer produced at a particular time. Increasing temperature reduces polymer molecular weight and increases the relative amount of short-chain branching and long-chain branching. Consequently, the relatively broad molecular weight distribution of LDPE compared to LLDPE is primarily due to the temperature profile within the reactor. It should be noted that the introduction of a chain transfer agent to the reactor also reduces polyethylene molecular weight.

5.7.3   Autoclave Reactor

A typical autoclave reactor consists of a series of two to six stirred reactors in which an initiator (exclusively organic peroxides) is added to each of the reactors. An example of the temperature profile for an autoclave process is illustrated in Figure 5.5.

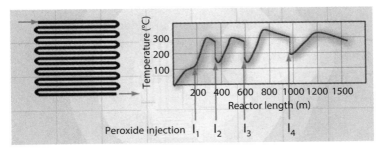

Figure 5.4 Example of the temperature profile that takes place in a tubular reactor.
Source: www.arkema.com.

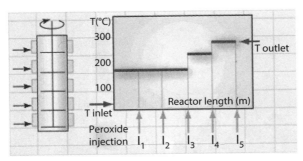

Figure 5.5 Example of temperature profile for an autoclave process.
Source: www.arkema.com.

The reaction temperature is determined by the type and amount of peroxide introduced. The effect of temperature is the same as a tubular reactor with higher temperatures producing lower molecular weight and increased amounts of SCB and LCB. Increase in pressure increases molecular weight and decreases both types of branching.

The details on the type of peroxides used and/or combinations of various peroxides utilized is considered proprietary information, with each company involved in the manufacture of polyethylene with a high pressure process.

The structure of the polyethylene obtained from each process is illustrated in Figure 5.6.

The physical properties of the polyethylene are dependent on the average molecular weight, the molecular weight distribution and the overall branching characteristics, which are the degree of short-chain branching and the degree of long-chain branching.

Union Carbide commercialized the first tubular reactor design in 1943 and it became readily apparent that the LDPE produced with this process was a unique material compared to LDPE produced with the stirred autoclave design employed by ICI. The autoclave-produced material contains a higher level of long-chain branching, as illustrated in Figure 5.6. However, the two high-pressure reactor types also possess different short-chain branching characteristics.

Tubular Process

Autolave Process

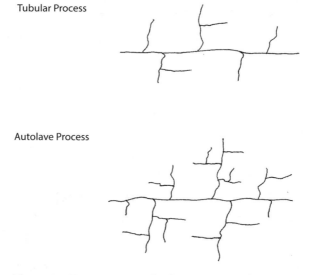

Figure 5.6 Representation of polymer structure for polyethylene produced with high-pressure process.

5.7.4   Characterization of Short-Chain Branching (SCB) in LDPE

In 1982, Wild and coworkers [17] reported the short-chain branching distribution of LDPE produced in each type of reactor design by utilizing the temperature rising elution fractionation (TREF)analytical technique. The autoclave LDPE sample had a 0.924 g/cc density and a 3.0 Melt Index. The tubular LDPE sample had a 0.921 g/cc density and a 2.2 Melt Index. The branching distribution data from each sample is summarized in Figure 5.7.

The tubular reactor provides LDPE with a broader SCB distribution than the LDPE produced in the autoclave reactor. However, the autoclave reactor has a significantly higher SCB content in the region, with 10–15 methyl branches/1000 carbons. Wild also provided data showing the significant differences in the SCB distribution of various LDPE samples prepared by each type of reactor under different polymerization conditions.

It was reported by F. Mirabella, Jr., and E. Ford that the amount of long-chain branching in LDPE samples over a density range of 0.917–0.922 and Melt index values of 0.25–2.5 was about 2.9–3.4 LCB/1000 carbon atoms [18].

The free-radical, backbiting process illustrated above as one aspect of the free-radical polymerization mechanism was originally proposed by

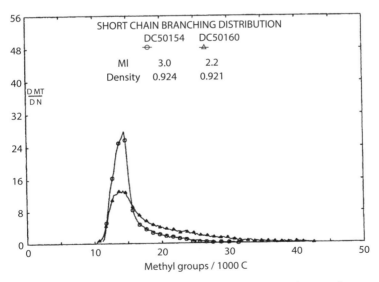

Figure 5.7 Comparison of the short chain branching distribution of LDPE produced from an autoclave reactor (O) and from tubular reactor (▲) using TREF method. Reprinted from [17] with permission from John Wiley and Sons.

M. J. Roedel in 1953 [14] to account for the short-chain branching content in LDPE. More recent characterization of LDPE by ^{13}C-NMR has resulted in a more detailed understanding of the SCB in LDPE, which has shown that LDPE is a complex mixture of various types of branching. Methyl, ethyl, propyl, n-butyl, n-amyl and 2-ethyl hexyl are each present along the polymer backbone. Additional references [19] on the characterization of the short-chain branching content in LDPE are recommended.

5.7.4.1   Structural Differences between LDPE and LLDPE

It is important to understand that the introduction of linear low-density polyethylene (LLDPE) in the 1960s and late 1970s, manufactured by the solution process and the gas-phase process, both operating at relatively low pressure utilizing a transition metal-based catalyst, provided a new type of polyethylene with unique properties that did not result in the elimination of the high-pressure process from commercial use. It opened new markets and applications for the polyethylene industry and resulted in its continued rapid growth.

The key structural differences between high-pressure LDPE and low-pressure LLDPE, each material with a density of ca. 0.918 g/cc, where the LLDPE sample was produced with a high-activity Ti/Mg containing catalyst are summarized below and in Table 5.14.

(a) LLDPE contains only one type of short-chain branch based on the choice of comonomer; ethyl branch from 1-butene; n-butyl branch from 1-hexene and an n-hexyl branch from 1-octene. LDPE has many types of SCB ranging from C_1–C_6 in length.

(b) LLDPE has no detectable long-chain branches, while LDPE has ca. 3–4 LCB/1000 carbon atoms.

(c) LLDPE has a very broad intermolecular SCB distribution ranging from 2–40 SCB/1000 C atoms, with the least branched polymer component accounting for 10–15 wt% of the sample (i.e.,<2 SCB/1000 carbon atoms). This very broad intermolecular SCB distribution provided by a Ti/Mg catalyst indicates that the relative reactivity of various active sites with the comonomer varies over a wide range. LDPE has a relatively uniform branching distribution ranging from 10–40 SCB/1000 carbon atoms, but does not contain a polymer component and has very little SCB.

Table 5.14 Structural differences between LLDPE[a] and LDPE.

Sample Type	SCB Distribution (SCB/1000 C)	MWD (M_w/M_n)	Long-Chain Branching (LCB/1000 C)
LDPE[b]	10–40	5.8–10.2	2.9–3.5
LLDPE[c]	2–40	3.6–3.8	none

[a]The LLDPE sample prepared with a high-activity Ti/Mg Ziegler catalyst.

[b]Density of four samples analyzed 0.917–0.924 g/cc at Melt Index of 0.25–2.5.

[c]Density of foursamples analyzed 0.918 g/cc at Melt Index of 1.0 or 2.0.

(d) LLDPE has a relatively narrow molecular weight distribution with a polydispersity value M_w/M_n of 3–4, while LDPE has a somewhat broader MWD with M_w/M_n values of 5.8–10.2 [18].

5.8   Low-Pressure Process

This section will discuss the evolution of the low-pressure manufacturing process with emphasis on the chronological order in which important technical achievements were reported. For an additional discussion of the status of the low-pressure process as of 2010, the publication by Dennis B. Malpass is recommended [20].

5.8.1   Early History

By the mid 1950s scientists at Phillips Petroleum in the United States lead by J. P. Hogan and R. L. Banks, and scientists in Germany lead by Karl Ziegler, each developed a low-pressure route for the manufacture of polyethylene using transition metal-based catalysts. The low-pressure technology provided a highly linear polyethylene molecular structure which was significantly different than the polyethylene structure provided from the high-pressure manufacturing method, thus leading to the introduction of new nomenclature in order to distinguish these vastly different polyethylene materials. Consequently, the high-pressure-based polyethylene was referred to as low-density polyethylene (LDPE), while the low-pressure-based polyethylene was referred to as high-density polyethylene (HDPE) due to the much higher level of crystallinity and consequently, higher product density, found for the low-pressure material. A review article published by H. R. Sailors and J. P. Hogan of the Phillips Petroleum

Company in Bartlesville, Oklahoma, USA [21], is an extremely important document that summarizes the details on the historical events that took place during this early period (1952–1960). It is highly recommended for anyone interested in understanding the early history of LDPE and HDPE.

The low-pressure process is primarily responsible for the manufacture of the majority of polyethylene produced globally. In 2010, 75% of the polyethylene produced (115 billion pounds) was manufactured with a low-pressure process and 25% (39 billion pounds) was manufactured with a high-pressure process.

5.8.2   Particle-Form Technology for Low-Pressure Process

5.8.2.1   Historical Background

The term "particle-form technology" refers to the process of forming granular polyethylene particles that replicate the shape of the original catalyst particles. Hogan and Banks at Phillips Petroleum Company initially developed particle-form technology in the mid-1950s for chromium-based catalysts. However, the process did not become commercial until about 1961, as the first commercial-scale plant for the manufacture of HDPE which started up in December 31, 1956, used a solution process in which the polyethylene was manufactured as a solution at high temperature (ca. above 135°C). The first commercial plant operating at lower temperature in which the polyethylene particle remained as a solid particle (slurry process) started up in 1961.

Hogan and Banks prepared chromium-based catalysts that were supported on silica by reacting a chromium compound with a porous silica at elevated temperatures. Such silicas have an average particle size of 20–90 μm, a pore volume of ~1.5 cc/g and a surface area of ~300 m²/g. During the polymerization process, one silica-supported catalyst particle generates one polymer particle in which the original silica particle has been sufficiently fragmented to form residual catalyst particles less than 0.1 μm in size. This fragmentation eliminates a visible catalyst residue in the finished polyethylene. These small fragments of catalyst particles are encapsulated into the granular polymer particle. Because of the high activity of the catalyst, catalyst removal steps are not required as part of the polyethylene manufacturing process.

Second generation Ziegler-Natta catalysts of high activity also employ particle-form technology. Therefore, the particle size, the particle size distribution and the porosity of supported second generation Ziegler-Natta catalysts needed to be controlled over a certain range. Such requirements improve the reliability of these catalysts in both the slurry and gas-phase polymerization processes. For example, very small catalyst particles produce very small polymer particles which may cause reactor fouling, while polymerization on very large catalyst particles generates a high level of heat that can melt polymer particles and thus foul the polymerization process.

Based on the requirements outlined above, industrial scientists recognized soon after the discovery of single-site catalysts based on zirconocene/MAO solutions that this type of catalyst needed to be supported on a solid support in order to manufacture polyethylene in commercial reactors based on the slurry process and the gas-phase process. Therefore, suitable supports for the single-site Zr/MAO catalyst systems must meet similar requirements as other catalysts in order to perform satisfactorily as particle-form catalysts.

5.8.2.2 First Commercial Low-Pressure Process by Phillips Petroleum

In April of 1955, Phillips Petroleum began plans to construct a 180 million lb/year high-purity ethylene plant and an adjacent 75 million lb/year HDPE plant. Twenty months later, the Phillips HDPE plant started up on December 31, 1956, in Pasadena, Texas, utilizing the continuous autoclave solution process shown in Figure 5.9. Some important developments that took place before the start-up of this historic plant are summarized in Table 5.15 [21].

The first plant operated in the solution mode for the first four years of operation at a production rate of approximately 9,000 lbs PE/hr. During the first two years of operation, the plant produced only ethylene homopolymers with a Melt Index value of less than one. As outlined by Hogan [22] in 1958, ethylene/1-butene HDPE copolymers were introduced in order to increase the characteristics of the product mix available. Soon after the

Table 5.15 Important developments prior to Phillips plant start-up.

Year	Event
1951	Discovery of the Phillips Cr-based catalyst
1954	Successful pilot plant results allowed for the design of a commercial continuous process for the manufacture of HDPE
1955	1,000 lb/day semi-works plant was on stream
1955/1956	Nine companies in seven countries became licensees
1956	First public disclosure on the Phillips process came in April, 1956, at American Chemical Society National meeting

discovery of the Cr-based Phillips catalyst, product development scientists soon learned that the introduction of short-chain branching into the polymer backbone had a very significant effect on many polymer properties. 1-Butene was the preferred comonomer at the time because it exhibited relatively good reactivity with the catalyst active site and the ethyl branch along the polymer backbone, formed from the incorporation of 1-butene into the polymer structure, had a greater impact on polymer properties than a methyl branch formed from using propylene as the comonomer. Another feature reported by the addition of 1-butene to the process was an increase in catalyst activity, which was unexpected due to the poorer reactivity of 1-butene with the catalyst. This increase in catalyst activity in the presence of other 1-olefins was found to be a common feature of many other ethylene polymerization catalysts such as Ziegler-type catalysts based on titanium.

The addition of short-chain branching into the polyethylene structure greatly improved the environmental stress-cracking resistance (ESCR), lowered the material stiffness (modulus), greatly increased the elasticity (elongation) and improved the load-bearing ability of the polyethylene. Consequently, from 1958–1960, ethylene/1-butene copolymers became the preferred material manufactured in the solution process. Most of the copolymer production produced polyethylene with a density of about 0.950 g/cc with Melt Index values of 0.2 to 12. Copolymers with Melt Index values of 0.1 to 0.6 were used in the fabrication of detergent bottles, sheet, tubing, pipe, wire and cable coating, fibers and extruded film. Copolymers with Melt Index values of 4–12 were used in injection molding applications to fabricate toys and houseware items. Figure 5.8 illustrates the dramatic effect on the load-bearing properties as the polyethylene density is lowered from 0.96 g/cc to 0.95 g/cc by introducing 1-butene into the process.

Examination of Figure 5.8 shows that the time to failure increases from about 50 hours to almost 10,000 hours as the polyethylene density is lowered from 0.96 g/cc to 0.95 g/cc, respectively.

In terms of ESCR, the time to failure improves significantly as density is lowered from 0.96 g/cc to 0.95 g/cc at the same molecular weight, but there is also an additional improvement in ESCR time to failure as the molecular weight is increased, as shown by a decrease in the polyethylene Melt Index. Table 5.16 shows the time to failure (represented as 50% failure in the number of test samples) as Melt Index changes from 6.5 to 0.3 at a density of 0.95 g/cc.

A Melt Index of 0.3 indicates a relatively high molecular weight. Hence, ESCR improves significantly with increasing polyethylene molecular weight. This data is relevant today as the Melt Index of many blow molding grades of polyethylene is approximately 0.3.

The process patent for the polymerization of 1-olefins using the Cr-based Phillips catalyst was issued as U.S. Patent 2,825,721 to J. P. Hogan and R. L. Banks on March 4,1958, and assigned to Phillips Petroleum Company. Claim 1 of this patent was sufficiently broad by claiming a polymerization temperature of up to 500°F, and 1-olefins containing two to eight carbon atoms. However, most claims discussed a lower polymerization tempera-ture (100–450°F) and recovering the polyethylene as a solid particle, thus referring to a slurry polymerization mode. The description of the reactor operation in this patent was sufficiently broad to include a stirred-tank auto-clave as shown in Figures 5.9 and 5.10. However, the stirred tank process

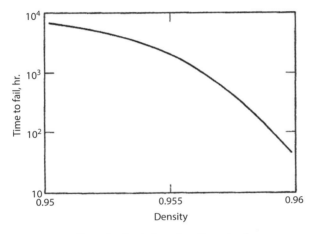

Figure 5.8 Effect of polyethylene density on load-bearing ability with a 20,000 psi static load – Melt Index ca. 0.2–0.4 [22].

Table 5.16 Effect of Melt Index on ESCR F-50 value for ethylene/1-butene copolymers at 0.95 g/cc density [22].

Melt Index 2.16 Kg (g/10 min.)	ESCR (F-50) hours
0.3	350
0.5	175
1.2	55
4.0	20
6.5	10

shown in Figure 5.10, operating at lower temperature and producing polyethylene granular particles, was only investigated in a small pilot plant operation and was not commercialized.

5.8.2.3   Solution Mode Operation

In the solution mode process illustrated in Figure 5.9, the solvent utilized must be an excellent solvent for polyethylene, as the polymer produced is completely dissolved in the solvent. Consequently, the first commercial plant employed cyclohexane as the solvent. Temperature of polymerization was 125–175°C at reactor pressures of 400–500 psig. The cyclohexane solvent is continuously fed to the reactor along with the silica-supported powder catalyst, ethylene and 1-butene. Polymer solution is continuously removed from the stirred autoclave to a holding tank where unreacted ethylene monomer and 1-butene are easily flashed off and recycled to the reactor. The solid catalyst is removed from the solution by filtering and the polyethylene is isolated after the solvent has been removed by steam stripping.

5.8.2.4   Slurry Polymerization Mode

The conversion of the solution polymerization process to a slurry (or suspension) polymerization process (Figure 5.10), required a change to a

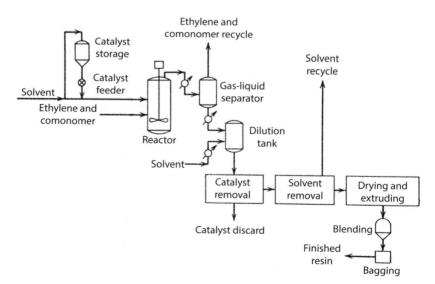

Figure 5.9 Phillips Petroleum Company's first commercial low-pressure solution polymerization plant for the manufacture of high-density polyethylene [23].

solvent that is a relatively poor solvent for polyethylene (such as n-pentane) and a lowering of the polymerization temperature to 90–110°C. For the Phillips catalyst, conversion to the slurry process was especially convenient because the catalyst was supported on a high-surface-area, porous silica particle with a particle size of ca. 5–100 microns. The original silica support possessed a surface area of about 300 meter^2/g and a pore volume of about 1.6 cm^3/g. Consequently, one catalyst particle provided one polymer particle (ca. 500–2,500 microns in diameter) and the original catalyst particle was fragmented into very small aggregates (<0.1 microns) during the polymerization process in which the growth of the polymer within the catalyst particle was responsible for the fragmentation process. This polyethylene particle growth process was designated as "particle-form" technology when the polymerization process involved the slurry method.

The process design illustrated in Figure 5.10 represents the slurry process and depicts a much simpler design. Production steps such as removal of the residual catalyst particle from the polymer solution and the separation of the relatively high-boiling cyclohexane solvent from the polyethylene could be eliminated from the design.

5.8.2.5 Pilot Plant Designs For Particle-Form Reactors – Development History [24]

Phillips initiated a research program in March of 1956 to investigate the commercial manufacture of polyethylene using particle-form technology using a 20- or 30-gallon pilot plant continuous stirred tank reactor (CSTR) design with normal pentane as the slurry solvent. This pilot plant process is illustrated in Figure 5.10.

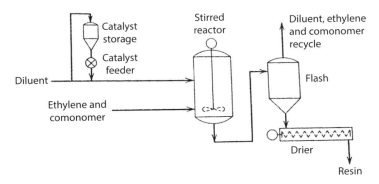

Figure 5.10 Phillips Petroleum Company's low-pressure, slurry polymerization design for the manufacture of high-density polyethylene. This design was only investigated at the pilot plant level [23].

Mixing of the slurry solution to maintain the suspension of the polymer particles in the pentane to prevent fouling of the reactor walls was the primary problem found in operating the CSTR over extended periods.

As discussed above, it was clear that ethylene/1-butene copolymers possessed significantly improved mechanical properties compared to ethylene homopolymer products, so that a commercial particle-form reactor design was needed that could provide polyethylene copolymers over a range of Flow Index values with sustained operability over extended periods of time without reactor shut down. Adding 1-butene to the polymerization process made the autoclave stirred tank reactor even more difficult to operate, as reactor wall fouling problems persisted and in some cases polymer particle morphology was reduced due to some polymer components becoming soluble in the n-pentane.

The operational problems of the stirred autoclave reactor were addressed with a new 20 gallon pilot plant reactor design in which vertical interior baffles were installed parallel to the reactor walls within the stirred tank, which were designed to provide better flow patterns between the reactor wall and the vertical baffles. This new design was designated by Phillips as the "draft-tube" design and this design was proposed to solve the reactor wall fouling problems that prevented sustained operation of the slurry reactor. Process research continued on this draft-tube reactor design until March 1958, when the reactor wall-fouling problems continued to prevent long-term continuous operation of the reactor.

5.8.2.6   *Phillips Pilot Plant Vertical Pipe-Loop Reactor Design*

In June of 1958, a 95-gallon vertical pipe-loop, pilot plant reactor was installed and placed into operation with the objective of solving the reactor wall-fouling problems encountered with the stirred-tank autoclave reactors.

The first United States patent that described the vertical pipe-loop slurry polymerization reactor illustrated in Figure 5.11 was issued to Donald D. Norwood on April 26,1966, as U.S. Patent 3,248,179 and assigned to Phillips Petroleum Company. Earlier applications filed in 1959 were abandoned, thus explaining the relatively late issue date on the Norwood patent. Although Donald Norwood was the only name on the first United States patent issued to Phillips Petroleum, Philips Petroleum credits three process engineers, D. D. Norwood, S. J. Marwil and R. G. Rohling, for the design of the vertical loop reactor [24].

The agitators necessary in the autoclave reactor were no longer required in the vertical pipe-loop system as a new design was disclosed in U.S.

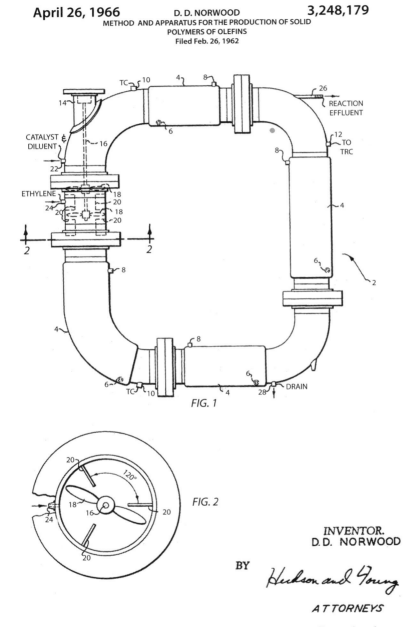

April 26, 1966   D. D. NORWOOD   3,248,179
METHOD AND APPARATUS FOR THE PRODUCTION OF SOLID
POLYMERS OF OLEFINS
Filed Feb. 26, 1962

FIG. 1

FIG. 2

INVENTOR.
D. D. NORWOOD

BY

Hudson and Young

ATTORNEYS

Figure 5.11 Schematic diagram illustrating the Phillips 95-gallon, pilot plant vertical pipe-reactor which was started up in June 1958; U.S. Patent 3,248,179.

Patent 3,226,205 issued to Raymond G. Rohlfing on December 28, 1965, and assigned to Phillips Petroleum Company, which described a reactor impeller with a reagent feed inlet along the shaft.

U.S. Patent 3,152,872 issued to J. S. Scoggin and Harvey S. Kimble on October 13, 1964, and assigned to Phillips Petroleum Company, provides a detailed description of the separation system used to isolate the solid polyethylene granular particles from the liquid diluent used in the vertical pipe-loop reactor. The contributions of Scoggin and Kimble were important in the design and start-up of the first commercial loop reactor in Pasadena, Texas, in 1961. It should be noted that the first vertical loop reactor used n-pentane as the slurry solvent, which was later changed to isopentane and then to isobutane in about 1970.

The 95-gallon, pilot-plant, vertical pipe-loop reactor was designed to eliminate the problem of polymer build-up on reactor surfaces found in the stirred tank auoclave designs by circulation of the reacting slurry around a closed loop at a velocity sufficient to keep the polymer particles suspended. This design was the critical technical development that led to the introduction in 1961 of the Phillips vertical-loop, particle-form reactor on a commercial scale, which is illustrated in Figure 5.12.

5.8.2.7   Operation of the Phillips Pilot Plant Pipe-Loop Reactor

A detailed description of the Phillips pilot plant pipe-loop slurry reactor (Figure 5.11) is found in U.S. Patent 3,248,179. The design consisted of three 4.5 foot, 10 inch, inner-diameter sections and one 4.5 foot, 12 inch inner-diameter pipe which contained the reactor pumping unit as one of the two vertical sections of the design, with either n-pentane or n-hexane as the slurry solvent.

Probably the single most important feature of the vertical-loop slurry-reactor design as described by Donald Norwood is the highly turbulent flow necessary (measured with a Reynolds number of approximately 2,000,000) within the slurry reactor to prevent reactor-wall fouling and allow the process to run for extended periods. Once this critical flow is achieved within the reactor, the entire mass of solids is carried around the loop in a nearly homogeneous flow and there is no concentration gradient from the top to the bottom of the reactor. The velocity of the slurry necessary to maintain this highly turbulent flow varies with the diameter of pipe used in the loop-reactor design and is also a function of other factors such as the solids content of the slurry and the size of the polymer particle. For the pilot plant design, the slurry velocity reported from Example 1 in U.S. Patent 3,248,179 was 6.3 ft/sec, with the reactor maintained at 15.3 wt%

solids at a production rate of 15.0 lbs/hr. However, if catalyst productivity was greater than 2,000 gPE/g catalyst, a flow velocity of 12–14 ft/sec was necessary because of an increase in polymer particle size.

Norwood reported that flow velocities necessary for reactors with an internal loop diameter of up to 40 inches were in the range of 15–45 feet/sec.

In 1961, the slurry vertical loop design [25,26] shown in Figure 5.12 with a single loop design replaced the solution autoclave process, and this process was designated as the Phillips particle-form process for the manufacture of high-density polyethylene. Later designs of the vertical loop reactor by Phillips Petroleum included several folded loops to increase reactor production rate, and this design is illustrated in Figure 5.13.

The introduction of the Phillips loop reactor in 1961 quickly made the solution autoclave reactor obsolete and licensees worldwide adopted the loop design for commercial reactors. By 1962, there were five companies in the United States and six foreign companies producing an annual total of about 450 million pounds of polyethylene. Table 5.17 documents these companies [22].

5.8.3   First Ziegler Catalyst Commercial Process

The Ziegler process for the manufacture of polyethylene followed a route similar to the Phillips process. According to Sailors and Hogan [21], the first

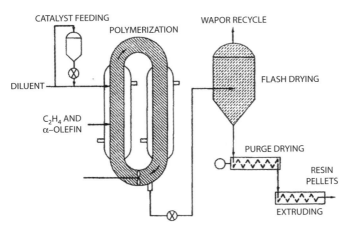

Figure 5.12 1961 Initial schematic illustrating the Phillips particle-form vertical loop process for the manufacture of ethylene homopolymers and copolymers; approximate capacity 75 million pounds polyethylene per year (ca. 9,000 lb PE/hr). Reprinted from [21] with permission from Taylor & Francis Publishing.

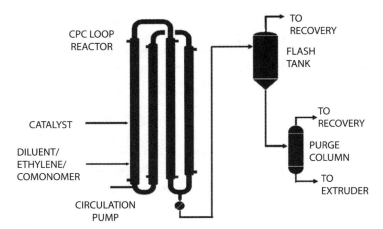

Figure 5.13 Diagram of the modern Chevron-Phillips vertical loop reactor which has an annual capacity of about 1.2 billion lbs polyethylene/year at a production rate of about 120,000–150,000 lb PE/ hr. Reprinted with permission from M. P. McDaniel.

commercial production of polyethylene in a Ziegler-type plant occurred in the Hoechst plant in late 1956, most likely, prior to the start-up of the Phillips low-pressure solution plant on December 31, 1956. In the United States, Hercules began a commercial polyethylene plant using a Ziegler catalyst in 1957. By 1960, annual production of polyethylene with Ziegler catalysts reached about 70 million pounds, while the Phillips process for HDPE was 200 million pounds.

Ziegler licensed the catalyst technology worldwide starting in 1954. The license included use of Ziegler's patent[1] and a 100 page manual describing a laboratory process. Early licensees are summarized in Table 5.18.

However, unlike the Phillips License that included a detailed process design, the early licensees of the Ziegler catalyst developed their own proprietary manufacturing process. Consequently, the early Ziegler commercial plants involved both solution and slurry (suspension) processes utilizing either a batch mode or continuous mode of operation. Polymerization conditions ranged from 50–120°C with ethylene pressures of 150–500 psig. Two concerns with the Ziegler catalyst in a commercial plant were the removal of chloride-containing catalyst residue from the polyethylene and deactivation of the aluminum alkyl needed to activate

---

[1] Some early Ziegler patents were: Belgium 533,362, issued May 5, 1955; Belgium 543,259, issued Dec.17, 1955; and British 799,392, issued Aug. 6, 1958. A United States patent was issued rather late as U.S. Patent 3,546,133, issued Dec. 8, 1970.

Table 5.17 Early licensees of Phillips polyethylene process (1962).

Region	Annual Capacity by Region (millions lbs)
United States	**300–325**
Phillips Chemical – Pasadena, Texas	
Celanese Plastic – Deer Park, Texas	
W.R. Grace – Baton Rouge, Louisiana	
National Petro Chemicals – Deer Park, Texas	
Union Carbide Plastics – Port Lavaca, Texas	
Foreign	**120–130**
Rheinische Olefinwerke – Wesseling, Germany	
British Hydrocarbon Chemicals – Grangemouth, Scotland	
Showa Yuka, K.K. – Kawasaki, Japan	
Solvay et Cie – Rosignano, Italy	
Manufacture Normande de Polyethylenes – Le Havre, France	
Petroclor, Indústrias Quimicas, S.A. – São Paulo, Brazil	
Total	**420–455**

Table 5.18 Early Ziegler licensees.

Company	Location
Hoechst	Germany
Huels	Germany
Ruhrchemie	Germany
Montecatini	Italy
Petrocarbon	Great Britian
Mitsui	Japan
Dow	United States
DuPont	United States
Esso	United States
Goodrich-Gulf	United States
Hercules	United States
Koppers	United States
Monsanto	United States
Union Carbide	United States

a Ziegler catalyst. The use of polar solvents such as methanol were used in this regard.

Commercial resins produced from 1956 to about the mid-1960s with the Ziegler-type catalyst provided a density range from 0.940–0.965 g/cc and a Melt Index range of 0.01–10 [27]. The low Melt Index, relatively high molecular weight polyethylene was produced with a slurry process, while high Melt Index resins useful in injection molding applications were produced in a solution process.

Because licensees of the Ziegler catalyst developed their own proprietary process technology, DuPont Sclair provides an excellent example how a Ziegler licensee developed and expanded both the process, catalyst and product technology, developing new markets and applications for low-pressure polyethylene.

5.8.4 Chevron-Phillips Slurry Loop Process Status as of 2010

The Chevron-Phillips (CP) slurry process is a leading licensor of the slurry loop process for the manufacture of HDPE and LLDPE worldwide. Between 1997 and 2008, the CP process has accounted for the addition of approximately 5.8 billion lbs/year of global polyethylene capacity. Much of the capacity has been added to emerging markets with line capacity exceeding one billion lbs/year. Some capacity additions are summarized in Table 5.19.

5.8.4.1 Reactor Scale

As of 2010, the Chevron-Phillips loop reactor has undergone extensive changes to lower capital costs and increase the capacity of a single line [24]. Reactor capacities have increased from 660–880 million lbs/year in 2005 to 1.2 billion lbs/year as of 2010. The volume of the reactor can be as high as

Table 5.19 Chevron-Phillips (CP) slurry loop licensees from 2003–2008 [28].

Company/Year	Location	Capacity (million lbs/year)
Qatar Chemical Company (2008)	Mesaieed, Qatar	770
Qatar Chemical Company (2003)	Mesaieed, Qatar	1,045
Polietilenos União S.A. (2007)	Santo André, Brazil	440
Maoming Petrochemical (2006)	Maoming, China	770

Note: The additions in Qatar are strategically located due to the low cost of ethane which is separated from the enormous amount of exported natural gas (LNG) by large international oil and gas companies. Source: www.cpchem.com licensees tab.

70,000 gallons (9,350 ft³) with operating rates of 150,000 lb PE/hr depending on the particular resin being manufactured. For example, a general purpose blow molding resin can be produced at such rates with a catalyst productivity of about 2,500 lb PE/lbcatalyst with most of the catalyst residue consisting of silica that has been fragmented to non-visible fragments (i.e.,< 0.1 micron). This improvement in the economies of scale have bought the slurry loop costs in line with the gas-phase process, making these two processes low-cost manufacturing systems. Some of these changes that were made to lower capital costs or operating costs are as follows [24]:

(a) The length of the loop reactor has been increased by making the loop longer and taller. However, it is important to note that the diameter of the loop (24 inches) remained constant over the years, as removing the heat of polymerization to the reactor cooling jacket that surrounds the loop needs to be maintained to avoid higher temperature zones within the slurry. Efficient heat removal is the most critical aspect of maintaining high polymer production rates and reducing reactor fouling.

(b) New higher capacity reactors have maintained a one-pump design.

(c) Engineers have redesigned the method employed to attach the reactor loop to the bulk of the reactor. The steel frame around the loop has been eliminated so that the loop is attached from supports below the loop. Consequently, the loop now rises into the air without visible supports.

(d) Dryers were replaced with simple jacketed flash lines.

(e) Reactors can now run at higher solids level, increasing the content of the loop, and methods to concentrate the slurry have been eliminated.

(f) New reactors now employ high-pressure flash technique so that less energy is used in recompressing the isobutane after flashing.

(g) Direct recycle of isobutane without repurification has been introduced.

5.8.5   Reactor Start-Up

The slurry reactor starts up relatively easily by feeding catalyst to the loop containing isobutane under the preset polymerization conditions. Ethylene partial pressure, polymerization temperature, and comonomer concentration are the key variables. With a Ziegler-type catalyst, which is used for the preparation of injection molding grades of polyethylene, a cocatalyst such as triethylaluminum is required and hydrogen is added as a chain transfer agent to control the molecular weight of the polyethylene. Generally during start-up, the amount of isobutane in the loop is somewhat reduced in order to increase polymer content (solids) more quickly. Full production rates are usually obtained in one to three hours. The amount of granular polymer in the loop for the large units is about 100,000 pounds. Catalyst residence time in the process (polymer in reactor/production rate) is approximately one hour with a production rate of 100,000 lbs/hr.

5.8.6   Product Transitions

One of the most important aspects of any commercial polyethylene reactor is the time it takes to transition to a new product without producing off-grade polyethylene or keeping off-grade material to a minimum.

There are several transition methods used for the loop process.

(a) In the simplest case, when the new product specifications are only slightly different than the resin being manufactured, the reactor conditions may be changed quickly so that no off-grade product is made and no production time is lost. This case may apply when product molecular weight or density is only slightly changed

(b) When the new product specifications are very much different, then the catalyst feed is stopped, the solids content of the slurry is removed, but isobutane feed is continued, or in some cases isobutane feed is increased, then the new polymerization conditions are established, and catalyst feed is restarted. These transitions can also be accomplished in a few hours with no off-grade polyethylene produced.

(c) A rare transition may sometimes be employed where the catalyst in the loop is deactivated by the addition of a chemical poison, e.g., carbon monoxide and then all solids removed while isobutane feed continues. The reactor is restarted after establishing new polymerization conditions. This method is used when one catalyst type is being replaced with a new catalyst type, for example, a Ziegler catalyst followed by a Cr-based catalyst.

5.8.7   Reactor Fouling

Maintaining high annual reactor production requires good reactor continuity with smooth product transitions and low amounts of off-grade material. One aspect of this objective is reducing and/or eliminating reactor fouling.

The prevention of fouling of the reactor within the slurry loop is an important part of the operating procedure. Some process parameters that may change to indicate the possibility of internal fouling are:

(a) Sudden increase/decrease in slurry reactor temperature;
(b) Cooling jacket inlet and outlet temperatures;
(c) Solids velocity in the loop;
(d) Recirculation pump amperage;
(e) Pressure delta when flashing ethylene from discharge chamber.

Fouling may take place on the reactor walls where a coating of polyethylene begins to adhere to the reactor walls, in the recirculation pump, or as part of the granular polyethylene solids that circulate through the loop. If some of the granular polyethylene solids begin to agglomerate, usually due to particles that begin to melt or overheat, larger particles begin to form and circulate as part of the granular polyethylene/isobutane slurry. Granular particle overheating may occur if the ethylene concentration in the isobutane slurry gets too high, creating bubbles in the isobutane slurry. Ethylene solubility in isobutane is a function of reactor temperature. If the solution becomes super saturated then ethylene bubbles can form and reduce the heat transfer from the slurry which may result in polymer particles overheating, initiating polymer particle agglomeration.

Monitoring the cooling jacket inlet temperature and cooling jacket outlet temperature and translating this information into a heat transfer coefficient is also useful in identifying possible fouling. If the reactor walls are beginning to foul with a coating of polyethylene, the heat transfer becomes less efficient, as the polyethylene acts as an insulator, and the heat transfer coefficient reported by the operating system will change. If the slurry solvent is becoming fouled as the solvent/solids recirculate in the loop, this type of fouling may be identified by the operation of the circulating pump. The pump power consumption values may begin to oscillate between higher and lower amperage values to create a phenomenon termed "widebanding." Polyethylene granular particles that begin to overheat and fuse together into larger particles (1–6 inches in diameter for example), will interfere with the flow of the slurry solids in the reactor loop.

Consequently, the power amperage of the recirculation pump will begin to oscillate. If polymer particle overheating continues to a much higher degree, then the formation of extremely large fused polyethylene logs may form, and a reactor shutdown and clean out by opening the reactor is necessary. However, this type of fouling has been greatly diminished in present loop reactors by the introduction of an automatic polymerization kill system that injects a catalyst poison into the reactor to very rapidly terminate the ethylene polymerization process, stopping the exothermic polymerization reaction.

The kill system is automatically used if the computer system, which monitors the key polymerization parameters outlined above, detects serious reactor continuity problems, thus eliminating direct operator action to initiate the kill system. Removing the manual reactor kill decision to the computer model monitoring the polymerization process, eliminates the difficult decision required by the reactor operator to shut down a commercial reactor. However, one downside of the automatic injection of the reactor kill system by the computer system is that it may lead to an occasional unnecessary shut down. However, the benefits of such an automatic kill system have resulted in the almost complete elimination of extreme fouling, which requires the reactor to be opened up and cleaned out by the plant maintenance personnel.

It is important to note that carbon monoxide (which is iso-electronic with an ethylene molecule) is often used as the reactor poison compound because it is a reversible poison that is slowly consumed by the ethylene polymerization catalyst and is incorporated into the growing polymer chain. Carbon monoxide will coordinate with the polymerization active site by displacing the ethylene. Because carbon monoxide is incorporated much more slowly than ethylene into the growing polymer chain, the polymerization process is essentially stopped until the carbon monoxide can be consumed by the catalyst. These reactions are shown in Figure 5.14.

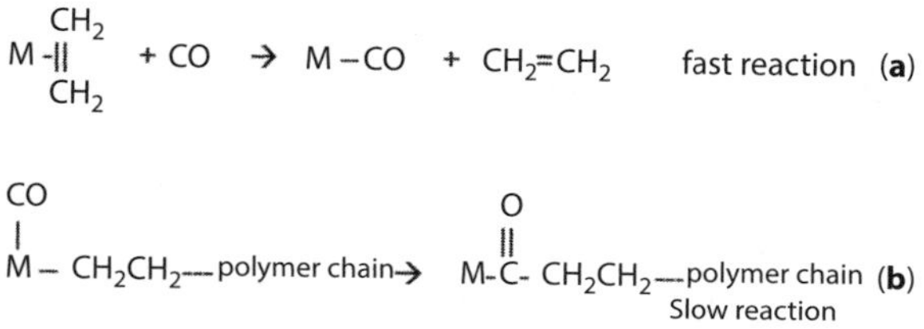

$$M-\|_{CH_2}^{CH_2} + CO \rightarrow M-CO + CH_2=CH_2 \quad \text{fast reaction (a)}$$

$$M-\overset{CO}{\underset{|}{}}CH_2CH_2 - \text{polymer chain} \rightarrow M-\overset{O}{\overset{\|}{C}}-CH_2CH_2 - \text{polymer chain (b)}$$
$$\text{Slow reaction}$$

Figure 5.14 Carbon monoxide incorporation into polyethelene; M is the catalyst active center.

Reaction (A) takes place immediately after the introduction of the carbon monoxide to the polymerization reactor, thus stopping the insertion of ethylene into the growing polymer chain and therefore stopping the heat of polymerization. Reaction (B) will take place very slowly, which results in the incorporation of a carbonyl group into the growing polymer. This slow reaction is responsible for the reversibility of the kill system. Once the added carbon monoxide has been incorporated into the polymer, then the active center will again react with ethylene and the polymerization process will restart.

5.8.8   New Vertical Loop Reactor Design

An improved loop reactor design with cone-shaped zones that transition the loop to a wider diameter section was recently patented by Basell Polyolefine GmbH [29]. This design is illustrated in Figure 5.15.

This new design claims better removal of the heat of polymerization and consequently can operate at higher ethylene pressure with 15 to 17 mol% ethylene concentration possible in the suspension medium, which is isobutane and dissolved monomers. A solids level of >60% is claimed, with 62% solids achieved in the patent example.

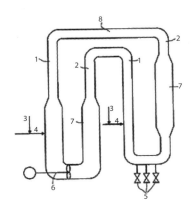

Label	Description
4	Ethylene feed–two loactions
5	Sedimentation legs
6	Isobutane feed into circulation pump
7	Widening region
8	Basic tube diameter

Figure 5.15 New loop reactor design [29].

5.9   Gas-Phase Process

5.9.1   Historical Introduction

One of the earliest reports of a fluidized-bed operation for the production of polyethylene [30] was provided in a British patent issued in 1958. The process was initiated by adding 100 ml of hexane to the polymerization reactor and then adding 3 ml of triethylaluminum and 1 ml of $TiCl_4$ to the hexane, which produces a solid catalyst particle. Ethylene is continuously recycled to the reactor and the hexane is gradually removed by distillation as the heat of polymerization increases the reactor temperature to 65°C, which yields about 50–100 g of polyethylene before the hexane has been completely removed. Polymerization is continued at 85°C by continued ethylene feed through the vessel to maintain a fluidized-bed process so that after 8 hours, 460 g of polyethylene has formed.

Phillips Petroleum also investigated a gas-phase reactor in the late 1950s [31].Lanning and coworkers described a process shown in Figure 5.16 where a chromium-based catalyst is added to a one-liter reactor containing pentane with a stirrer at the bottom of the reactor. Ethylene is fed into the reactor at the bottom.

The experiment was carried out by adding 0.28 g of catalyst with 150 ml of pentane. The reactor was heated to 220°F and the total pressure was set at 450 psig. After 30 minutes, the pentane was flashed off and the polymerization process was continued for 15 hours. Polyethylene was removed at about 1–3 hour intervals and 0.21 g of additional catalyst was added after 8 hours. A total of 849 grams of polyethylene were produced as a dry granular powder. Catalyst productivity was low at about 115 g PE/g cat/hr.

Robert R. Goins received U.S. Patent 2,936,303 on May 10, 1960, which was assigned to Phillips Petroleum Co. It described a three-stage, gas-phase reactor that operated with a countercurrent flow in the presence of inert diluent gases such as ethane or propane. This design is shown in Figure 5.17. The catalyst is fed through line 4 to the top reactor zone 1, where polymerization is initiated on the catalyst particle. Next, the growing polymer particle is transferred through line 9 to zone 2, where the countercurrent gas flow transfers the granular polyethylene to the last polymerization zone 3 through line 10. The finished polymer particle exits the last zone through line 18, which is a product discharge line. Gas flow in Goins' reactor is shown by the arrows, while the heat of polymerization is removed by passing the gases through cooling sections 11, 12 and 13.

In a design that most closely resembles the gas-phase fluidized-bed reactor licensed by INEOS as Innovene G and Univation as UNIPOL

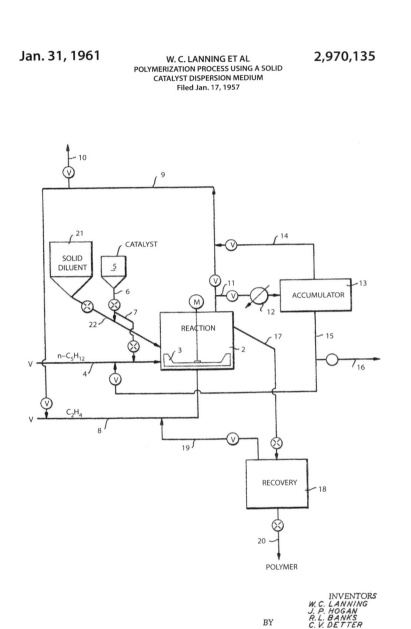

Jan. 31, 1961   W. C. LANNING ET AL   2,970,135
POLYMERIZATION PROCESS USING A SOLID
CATALYST DISPERSION MEDIUM
Filed Jan. 17, 1957

Figure 5.16 Early gas-phase reactor design from Phillips Petroleum [31].

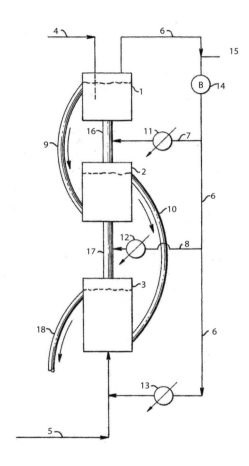

May 10, 1960   R. R. GOINS   2,936,303
                        OLEFIN POLYMERIZATION
                         Filed April 25, 1958

INVENTOR.
R.R. GOINS

BY *Hudson & Young*

ATTORNEYS

Figure 5.17 Early gas-phase fluidized-bed reactor design from Phillips Petroleum Co. [32].

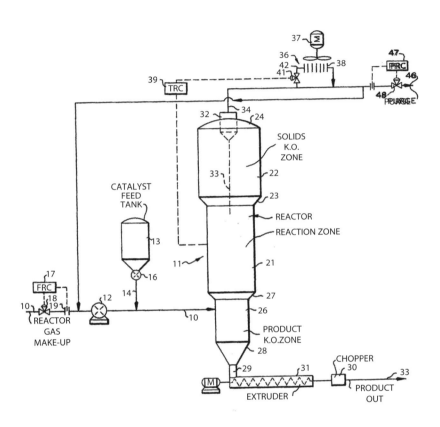

Feb. 27, 1962   R. F. DYE   3,023,203
                          POLYMERIZATION PROCESS
                            Filed Aug. 16, 1958

Figure 5.18 Early gas-phase, fluidized-bed process containing an expanded section [33].

Table 5.20 Dimensions of Dye's reactor.

Section	Diameter (ft)	Height (ft)	Volume (ft³)
Top	18	8	2035
Middle	9	9	572
Lower	4	2	25

technology, was a process developed by Robert F. Dye who received U.S. Patent 3,023,203 assigned to Phillips Petroleum on February 27, 1962, with a filing date of August 16, 1957. This reactor design is the first description of a gas-phase fluidized-bed reactor. The drawing in Figure 5.18 shows two expansion sections in which the diameter increases in order to reduce recycled gas velocity of the ethylene gas as it passes through the reactor. The ethylene gas provides the fluidization in the granular polyethylene bed. The dimensions of Dye's reactor are shown in Table 5.20.

In Figure 5.18, the intermediate zone 21 is the reaction zone which contains granular polyethylene particles. The top zone 22 was designed as a solids knockout (KO) zone which is used to separate granular polyethylene from the fluidization (ethylene) gas. The gas-solids separation is improved further by the addition of a cyclone 32 at the top of the reactor, which is used to return any granular polyethylene that reached this point back to the reaction zone. The bottom zone 26 of the reactor was utilized as the product discharge system. Granular polyethylene particles greater than 20 mesh in diameter will fall into the bottom zone and can be discharged to the extruder used to pelletize the product. The linear velocity of gases in the intermediate (reaction zone) was about 2 feet per second, which was reduced to 0.5 feet per second in the top expansion zone.

Dye provided limited process conditions, but based on a catalyst feed rate of 1.33 lb/hr and an average catalyst residence time of 1.5 hours, the production rate can be estimated from the volume of the reaction zone. Assuming a fluidized bulk density of about 12 lbs/ft³, the reaction zone would contain about 7,000 lbs of polyethylene. A production rate of approximately 4,700 lb/hr would be needed to account for a 1.5 hr catalyst residence time. Polymerization temperature was 210°F and ethylene pressure was 450 psi.

Although Phillips Petroleum did not commercialize a gas-phase process for the manufacture of polyethylene, Sailors and Hogan pointed out that the first commercial gas-phase process came on stream in 1964, with BASF using the licensed Phillips chromium-based catalyst [21]. BASF received U.S. Patent 3,300,457 on January 24, 1967 (filed on May 9, 1963), for a gas-phase process [37].

5.9.2 BASF Early Gas-Phase Reactor

The BASF gas-phase reactor shown in Figuree 5.19 was described as a fluidized-bed reactor with a vertical agitation system to facilitate the fluidization process by the design of the agitation blades. Patent examples were provided for the manufacture of polyethylene and polypropylene. The heat of polymerization was removed by the recycle gas through the reactor walls, and in the case of polypropylene, the addition of the propylene monomer in liquid form. Other key features of this reactor were:

- Monomer recycle through lines 1 and 5
- Catalyst injection – point 7
- Product discharge – point 6
- Ratio of diameter of reaction zone to height of reaction zone was 1:5
- Diameter of expanded section about twice the diameter of reaction zone
- Distance between agitator blade and reactor wall 1–50 mm
- Stirrer speed approximately 120 rpm
- Gas velocity in reaction zone 7–8 cm/sec

Ethylene was polymerized using a Cr-based Phillips catalyst with about 530 psi ethylene and a polymerization temperature of 95°C. The ash content of the polymer was about 0.01% showing very high catalyst productivity of about 10,000 gPE/g catalyst, and the polyethylene molecular weight was 400,000.

5.9.3 Horizontal Gas-Phase Process

Anderson *et al.* received U.S. Patent 3,469,948 on September 30, 1969,which was assigned to Dart Industries, Inc. It described a gas-phase, horizontal reactor system containing a paddle-shaped agitator [34]. The blades on the agitation system were designed to move the granular polymer through the vessel to a discharge valve located at the opposite end of the reactor.

In the 1970s, James L. Jezl and Edwin F. Peters of Standard Oil (Indiana) developed a more advanced cylindrical, horizontal gas-phase reactor for the vapor-phase polymerization of olefins [35] which also utilized a "paddle wheel" agitator to mix the granular polyethylene. This design is shown in Figure 5.20. The unique feature of this design is that the reactor is divided into four polymerization zones that may each be operated under different polymerization conditions so that the reactor operates as

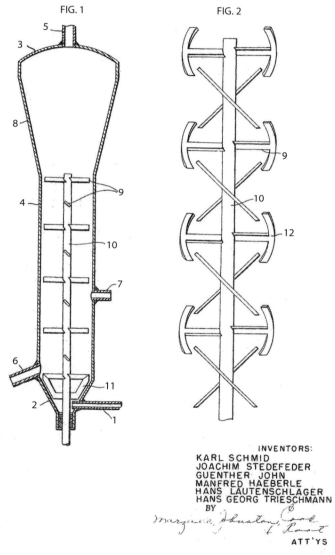

Figure 5.19 BASF early gas-phase reactor.

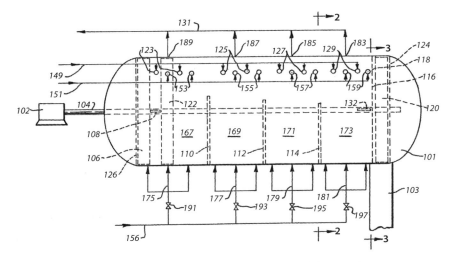

Figure 5.20 Horizontal stirred gas-phase reactor with four different polymerization zones [35].

Table 5.21 Key features of horizontal gas-phase reactor [35].

Marker(s)	Description
167–173	Four individually controllable polymerization sections
110–114	Walls that extend upward to slightly higher than the middle of the reactor; as the solid polymer exceeds the wall height, the polymer falls into the adjacent section moving in the direction of the polymer discharge system.
104	Rotating paddle drive shaft rotating at 5–30 RPM.
156	Vapor recycle line attached to the four vapor recycle inlets 175 to 181. Recycle inlets are designed to prevent fluidization of the granular polymer bed.
191–197	Vapor inlet valves that can be used to control the polymerization conditions in each of the four reactor sections.

four separate reaction vessels in series, thus providing a very wide range of polyethylene molecular structures that may be manufactured with such a design. An important feature of this reactor was the use of an isopentane spray directly into the polymer bed to maintain a certain polymerization temperature and prevent polymer agglomeration.

Key features of the Standard Oil gas-phase reactor are summarized in Table 5.21.

The examples provided in U.S. Patent 4,101,289 demonstrate that polyethylene may be produced with this particular reactor over a wide range of molecular weights and molecular weight distributions, as shown by polydispersity values (Mw/Mn) ranging from about 7.6 to 17.0. Both titanium-based catalysts and Cr-based catalysts were evaluated. Typical ethylene polymerization conditions were 171°F at one end of the reactor and 181°F at the other end of the reactor, with total reactor pressure of 400 psig, and propylene was used as a comonomer to control polyethylene density.

5.9.4   Union Carbide Gas-Phase Reactor

The Union Carbide U.S. Patent 4,003,712 on a gas-phase fluidized-bed reactor was issued to Adam R. Miller on January 18, 1977. This patent was a continuation-in-part of two earlier patent applications; one filed on August 21, 1967 and another filed on July 29, 1970, both of which were abandoned [36]. This design is shown in Figure 5.21.

Union Carbide proposed that their gas-phase process design offered an improved product discharge system relative to the gas-phase process patented by Robert Dye of Phillips Petroleum [33]. The Dye design utilized a product discharge system located at the bottom of the reactor where polyethylene particles would self-collect once the particle became too large (heavy) to remain in the fluidized reaction zone. However, as Union Carbide argued, these large particles continue to generate polymerization heat as they remain active as the particles collect in the discharge system. Consequently, the large granular polymer particles fuse to form larger polymer masses that may plug the product discharge system. Although the BASF gas-phase system [37] did contain an improved product discharge system, similar to the Union Carbide design, the fact that the Union Carbide system did not contain a mechanical stirring component allowed Union Carbide to overcome infringement on the BASF patent.

Key features of the Union Carbide fluidized-bed gas phase reactor are summarized in Table 5.22.

5.9.4.1   Gas Distribution Plate

Because the gas distribution plate was an important aspect of the Union Carbide fluidization process, two U.S. patents, one issued to Montecatini of Milan, Italy, and the other to Mitsui of Tokyo, Japan, that described detailed designs of gas distribution plates at the bottom of a fluidization column that provided improved fluidization of granular solids in a fluidization column, may have proved an obstacle in order for the Union Carbide patent

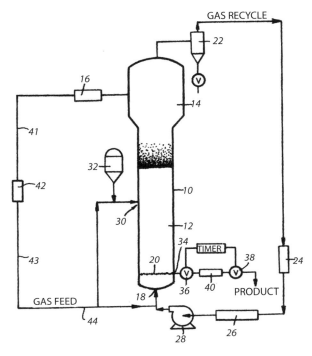

Figure 5.21 Union Carbide fluidized-bed reactor.

Table 5.22 Features of the Union Carbide fluidized bed reactor.

Marker	Description
10	REACTOR
12	Reaction Zone, diameter/height ratio 1/6 to 1/7.5; seed bed required for startup
14	Gas velocity reduction zone; diameter/height ratio 1/1 to ½
16	Gas analyzer for reagent feed control
18	Gas recycle inlet
20	Gas distribution plate
22	Particle collection/return cyclone
24	Particle filter in recycle stream
26	Heat Exchanger
28	Compressor
30	Catalyst injection point; must be placed well above distribution plate
32	Catalyst feed reservoir

to be issue. The Montecatini patent issued to G. Drusco [38] and the Mitsui patent issued to S.Takeuch [39] each patented a porous distribution system that improved the fluidization process by more efficiently passing the fluidization gas into a solid granular material that achieved better mixing of the granular bed confined in the fluidization system.

5.9.5   Gas-Phase Univation Process (2012)

A simplified schematic of the gas-phase process is shown in Figure 5.22. This reactor is in operation worldwide and is now the Univation UNIPOL process, which is a joint venture of ExxonMobil Chemical Company and Dow Chemical Company. Exxon acquired the Mobil assets in late 1999, while Dow acquired the Union Carbide assets in 2000.

5.9.6   Fluidized-Bed Gas-Phase Operation Overview

The gas-phase, fluidized-bed reactor is capable of producing a wide range of polyethylene resins over a density range of about 0.91–0.97 g/cc and a Melt Index range of about 0.01–100. Product limitations are reached in this process if the granular polyethylene becomes brittle (high MI and high density), whereby the granular polyethylene particles begin to fragment,

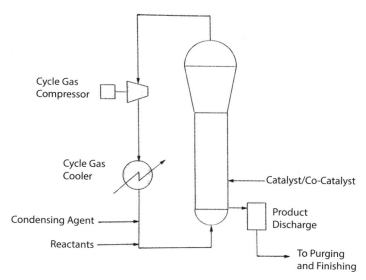

Figure 5.22 UNIPOL gas-phase reactor system. Drawing made from diagram reported at www.univation.com.

leading to a buildup of resin fines that may foul the reactor; or if the polyethylene particles becomes tacky and more difficult to fluidize, leading to possible particle agglomeration and reactor shut down.

The catalysts are supported on porous silica and fed directly to the fluidized reactor as a solid particle, where polymerization usually begins on contact with ethylene and the polymer particle growth process begins. The silica support is fragmented into extremely small fragments (i.e.,< 0.1 microns) which are distributed throughout the polymer particle. Catalyst particle size is usually between about 1–100 microns with a mean particle size in the 30–60 micron range. Depending on catalyst productivity, the polymer particles grow to 500–2,000 microns in diameter and the settled bulk density of the granular polyethylene is about 24–34 lbs/ft³.

Catalyst prepolymerization systems that were once used by some gas-phase reactors in the past, where the catalyst particle is subjected to initial polymerization conditions that promote favorable particle morphology, and then fed to the main reactor, have been eliminated from the process.

Polymerization conditions may vary over a wide range, where polymerization temperatures vary from 60–110°C and total reactor pressure from 100–350 psig. Comonomers are usually limited to 1-butene or 1-hexene to control polymer density, but 1-octene may be used if a particular catalyst exhibits a relatively high reactivity with 1-octene such as metallocene catalyst systems. Resin production rates are controlled by primarily two variables; catalyst feed rate and ethylene partial pressure. In a modern gas-phase process, operating in what Univation designates as "supercondensing mode," production rates for many grades of polyethylene of at least 125,000 lbs/hr are reached.

5.9.6.1   Ethylene Partial Pressure

There are limits as to the ethylene partial pressure that may be used to increase reactor production rates. For example, an increase in ethylene partial pressure above a certain amount may significantly reduce polymer particle morphology, leading to low resin settled bulk density (post reactor) and low fluidized bulk density within the reactor. Low resin bulk density will reduce conveying rates of granular polyethylene as the resin is transferred to another production step. Lower fluidized bulk density will reduce the amount of granular polyethylene in the reactor, which will reduce catalyst residence time (lower catalyst productivity) at a constant production rate. Limits on comonomer feed rates may also limit the ethylene partial pressure obtainable in order to produce polyethylene with a specific density, especially for catalysts with a relatively poor reactivity with higher 1-olefins. Lastly, because hydrogen is used with Ti-containing Ziegler catalysts to control polyethylene molecular weight, ethylene partial pressure values may be limited in order to maintain the necessary hydrogen/ethylene molar ratio to produce polyethylene with a specific high Melt Index value.

5.9.6.2   Catalyst Feed Rate

Increasing catalyst feed rates will increase reactor production rates which will decrease catalyst residence time and consequently, catalyst productivity (lbs PE/lb catalyst). Generally a catalyst productivity of >1,500 lb PE/lb catalyst is used as the minimum acceptable catalyst productivity in a commercial reactor.

Reactor conditions are monitored by a gas chromatographic system that analyzes the vapor content at regular intervals and advanced computer control systems regulate the feed rate of all reagents.

5.9.6.3   Product Discharge

The product discharge system is connected to the side of the reaction zone where product is passed through a valve into a separate chamber. Monomers are removed by degassing and transferred back to the reaction zone, and then the product is transferred to another chamber where additional drying takes place. If an aluminum alkyl was used in the polymerization, then moist nitrogen is added to this vessel to deactivate the aluminum alkyls by hydrolysis. Hydrolysis of the aluminum alkyl forms a volatile alkane and the inorganic $Al(OH)_3$. If the aluminum alkyl is allowed to oxidize first, before the hydrolysis step, then odor-forming byproducts such as alcohols are possible and these decomposition products may remain in the resin as odor-forming sources.

5.9.6.4   Condensing Agent

The early gas-phase reactors operated in the dry mode so that the heat of polymerization was removed only by the recycled gas stream which was passed through a heat exchanging unit and the cooler gas was ejected into the bottom of the reactor. At the time, a large UNIPOL reactor was designated as a G-5000 reactor where the 5000 denoted the volume of the reactor in cubic feet. The G-5000 contained about 100,000 pounds of fluidized granular polyethylene and a production rate of 25,000 to 40,000 lbs per hour was common, resulting in a catalyst residence time of 2.5–4.0 hours.

Production of ethylene/1-hexene copolymers was limited by the reactivity of the catalyst with 1-hexene. Ethylene partial pressure was therefore limited in order to remain in the dry mode and adjust the 1-hexene/ethylene molar ratio at the desired level to produce a LLDPE resin with a density of about 0.916 g/cc.

In 1986, the introduction of the condensing mode of operation increased the production rates by about a factor of 1.6, thereby increasing production rates to 40,000 to 65,000 lbs/hr, which greatly increased the annual capacity of the reactor. The condensing agent is added to the reactor and is a relatively light hydrocarbon such as hexane, which greatly improves the heat removal process and allows much higher operating rates. In1990, the introduction of the super condensing mode increased the production rate by a factor of at least 2.5 over the dry mode capacity. Hence, the G-5000 reactor can now be operated at rates of at least 120,000 lbs/hr for certain grades of polyethylene.

5.9.6.5   Reactor Start-Up/Product Transitions

The gas-phase process is unique in regard to reactor start-up compared to the slurry and solution manufacturing processes. The gas-phase reactor requires the addition of a "seedbed" to the reactor for the start-up operation. In addition, the catalyst type and product specifications of the seed bed should be similar to the product type that will be produced after the start-up is complete. The time required to reach the product specifications depends on the amount of polymer in the seed bed and the production rate during start-up. For example, utilizing a low-bed height during start-up (i.e., reducing the amount of polymer in the seed bed) will significantly shorten the start-up time as the bed is turned over more rapidly. A comparative example is shown in Table 5.23.

The shorter start-up time for the smaller seed bed will also reduce the amount of off-grade polyethylene produced, however, if the startup operation is carried out under the optimum conditions where the process

Table 5.23 Reactor start-up/product transitions.

Seed bed (lbs)	50,000	100,000
Production rate (lbs/hr)	100,000	100,000
Time for four-bed turnovers, hours[a]	2.0	4.0
% seed bed remaining	6.25	6.25

[a]One bed turn over = Bed weight/production rate.

variables are slowly changed during the start-up period, off-grade resin may not be produced. For example, if the Melt Index of the seed bed and the target Melt Index of the production run are only slightly different (i.e., 1.0 MI vs 2.0 MI), then the start-up process may be carried out in such a manner that the product Melt Index may be matched very quickly by producing a Melt Index resin slightly higher than the aim Melt Index, thus producing the target Melt Index in the discharged product. Then the instantaneous resin MI may be lowered as the bed turnover is increased during start-up to bring the entire bed of polyethylene into product specification.

5.9.6.6   Reactor Fouling

Reactor fouling may occur gradually over a long period of time, i.e., weeks to months, or may occur in a few hours. In general terms, a smooth reactor startup and careful reactor monitoring are always important measures to take to prevent fouling. However, serious reactor fouling is relatively rare as the reactor control and monitoring processes have improved over the years.

A gas-phase reactor may undergo resin fouling in several locations. These include: (1) the distributor plate at the bottom of the reaction zone; (2) reactor walls; (3) reactor resin bed; (4) heat exchange unit, or (5) the cycle compressor. However, it is important to note that over the forty years the gas-phase reactor has been in commercial operation the reactor continuity has been greatly improved as newer equipment, better computer operation and reactor monitoring equipment has been developed. In addition, the introduction of condensing mode operation in the 1980s also greatly improved the removal of the heat of polymerization and also reduced reactor static problems. Static charges due to the triboelectric effect as dry granular polyethylene particles contacted other surfaces that created static charge could cause reactor fouling. Reactor operation in the dry mode (non-condensing mode) was more likely to develop static build up in the granular polymer particles.

Reactor walls can foul in primarily two modes. One mode is the formation of "cold skins," where polyethylene particles that are no longer active begin to adhere to the reactor wall, and covering a thermocouple probe attached to the wall that monitors polymerization temperature. This type of fouling results in the polyethylene acting as a heat insulator whereby the thermocouple reads a lower temperature, i.e., "cold skins," than the bulk of the reactor. The second type of wall fouling is similar except in this case the polyethylene particles remain active and generate the heat of polymerization. In this case the polyethylene particles begin to increase in temperature due to poor removal of the heat of polymerization. Therefore, the thermocouple reads a higher temperature, i.e., "high skins," than the remainder of the reactor and these particles begin to melt and fuse into a large mass attached to the reactor wall. This type of fouling is much more likely to create significant operational problems and a partial reactor kill or some other action needs to take place to stop the growth of this fouling.

The fluidized-bed, itself, may also foul as some polymer particle agglomeration takes place usually due to polymer particles over-heating and fusing into larger particles within the fluidized-bed. This type of fouling is more easily controlled by changes in the polymerization conditions. A small amount of polymer-bed fouling is readily detected after the polymer is transferred out of the reactor and undergoes a screening process during resin transfer in the plant. If the screening step, designed to remove relatively large particles (i.e., greater than about 1" in diameter) begins to separate a larger quantity of material, then particle fusing is taking place and steps can be taken to alleviate the problem. This type of fouling is usually not serious and most likely will not lead to lower production rates or reactor shut down for cleaning. However, the buildup of relatively large polymer masses such as a "snowball" size and much larger may lead to fluidization problems, as the large particles are difficult to remain in a fluidized state. This type of fouling is quite rare with well-established catalyst types that have been in commercial operation for many years.

Fouling of the cycle gas compressor and cooling coils is usually limited to a slow buildup of a coating of polymer particles, but is also less likely to occur in recent years

5.9.6.7   Catalyst Requirements for Gas-Phase Fluid-bed Reactor

Excellent reactor continuity: The evaluation of new catalyst types in a fluid-bed reactor over the past 40 years has established that certain types of catalysts are inherently more difficult to run than other types of catalysts. In very general terms, catalysts based on chromium and titanium/magnesium compounds have been less problematic to commercialize. The problems that develop from certain catalyst types can range from high reactor static, unexpected increase in polymerization rate, poor resin morphology during reactor start up, and reactor fouling in the ethylene recirculation system which includes the cycle gas compressor, cooling coils and the distributor plate.

Activity: Catalyst productivity can range significantly between catalyst families and polymerization conditions within a catalyst family from about 2,000–8,000 g polyethylene/g of catalyst. Residual transition metal in the polymer may range from 0.1–10 ppm. For commercial operation a catalyst productivity of > 1500–2000 g PE/g catalyst is usually required.

1-Hexene reactivity: Catalyst reactivity with 1-hexene is important in order to reduce the amount of 1-hexene needed in the reactor. Ziegler-type catalysts based on Ti/Mg compounds exhibit relatively poor reactivity with 1-hexene compared with many other catalysts. Such catalysts can require a 1-hexene/ethylene vapor phase ratio of about 0.12 and higher in a commercial reactor, thus limiting the ethylene partial pressure in the reactor and leading to lower production rates and/or lower catalyst productivity. However, with the super-condensing mode of operation, comonomer fed concerns are rare. Most other catalysts used in a fluid-bed reactor exhibit higher 1-hexene reactivity and, therefore, require lower 1-hexene feed rates and fewer operational problems

Resin morphology/low fines/good feedability: Because most catalysts used in a fluid-bed reactor are supported on silica with a particle size distribution between about 5–100 microns (mean value of 30–60 microns) catalyst feedability to the reactor is not a concern. However, for catalyst types that possess a very high initial rate of polymerization, particle morphology may deteriorate due to a very rapid, uncontrolled fragmentation of the silica during the initial several minutes after the catalyst particle has been introduced into the reactor. Many Ti/Mg-based Ziegler-type catalysts exhibit a high initial rate of polymerization in the presence of high levels of 1-butene or 1-hexene and, therefore, are more likely to produce polyethylene particles with poor morphology. Hence, the manufacture of LLDPE usually produces polyethylene with a lower settled bulk density. Particle morphology problems with LLDPE resins are less likely if ethylene partial pressure and/or polymerization temperature are lowered. In addition, catalyst composition may be altered specifically to improve resin particle morphology. Polymer fines may become a problem if polymer particles begin to fragment in the reactor due to particle brittleness. This may develop for polyethylene grades with a high density and very high melt index (low molecular weight), that produce polymer particles that are more brittle.

5.10   Gas-Phase Process Licensors

The gas-phase process is licensed from several companies. Three sources of gas-phase technology are shown in Table 5.24.

About 55% of the low-pressure polyethylene manufactured in 2009 was prepared utilizing the gas-phase technology of the three companies listed in Table 5.24.

Table 5.24 Three sources of gas-phase technology.

Company	Year	Process	Worldwide 2010 Annual Capacity (Billion lbs)
Union Carbide (Dow)	1968	UNIPOL	36
INEOS	1984	Innovene	12
LyondellBasell	1990	Spherilene	8

5.10.1 Background

Union Carbide Corporation was one of the first companies to license the gas-phase process starting in the late 1960s. Originally, only HDPE was manufactured using one of three chromium-based catalysts. The catalysts were each supported on various grades of silica and fed directly to the polymerization zone as free-flowing powders utilizing Union Carbide-designed catalyst feeding equipment. The catalyst systems used in the late 1960s that were used in the UNIPOL process are shown in Table 5.25.

5.10.2 Gas-Phase Process Company History

1. **Union Carbide/UNIPOL:** Union Carbide introduced the gas-phase, fluid-bed process under the tradename of UNIPOL commercially in the late 1960s using a G-1250 reactor (reactor volume 1250 cubic feet) for the manufacture of HDPE. The LLDPE was added to the product mix in about 1977. Union Carbide formed a joint venture with Exxon in 1996 to pursue the addition of single-site catalysts to the

Table 5.25 Union Carbide chromium-based catalysts for the manufacture of HDPE (1968).

Catalyst Type	Catalyst Components	HDPE Product Types[b]
Cr	Phillips-type catalysts	Blow Molding
UCC type-1	$Bis[(C_6H_5)_3SiO)_2CrO_2/$ DEAEO[a]	HMW-HDPE film Blow Molding
UCC type-2	Chromocene	Injection Molding

[a]Diethyl aluminum ethoxide.

[b]Cr and UCC type-1 catalysts produce resin with a relatively broad molecular weight distribution with UCC type-1 providing a relatively broader MWD than Cr-type resins; UCC type-2 catalysts produce resins with a relatively narrow MWD.

UNIPOL process and formed a joint partnership under the name Univation Technologies LLC in 1997. In 1999, Exxon and Mobil merged. Finally, in 2001, Dow Chemical merged with Union Carbide so that the UNIPOL process evolved from the technical combination of four companies. As of 2010, Univation is considered a joint venture of Dow and ExxonMobil Chemical Companies.

2. **INEOS/Innovene:** British Petroleum (BP) Chemical Company developed the Innovene gas-phase process in 1975 and began licensing it in 1984. In 2005, BP Inc. sold the polyethylene assets to INEOS, which operates the licensing business under the tradename of Innovene. A summary of the licensees for the Innovene gas-phase process, designated as Innovene G, is shown in Table 5.26. Innovene G offers a variety of chromium-based, Ti-based Ziegler-type, and metallocene catalysts under the trademark INcat. The process also allows the use of comonomers 1-octene, 1-hexene and 1-butene for the manufacture of ethylene-based copolymers. Single reactor annual capacities are greater than 500 kt/year, which corresponds to production rates of about 120,000 lbs/hr (60 kt/hr) and provides short transition times between the manufacture of different grades of PE, an important consideration in producing less off-grade product. Usually, off-grade resin may be blended into prime-grade resins in small amounts to consume any such material and improve process economics.

3. **Lyondell/Basell:** Basell was formed over many years from combining the assets of BASF, Hercules, Himont, ICI, Montecatini, Montedison and Royal Dutch Shell. In 2007, Basell merged with United States-based Lyondell Chemical Company, which had acquired the assets of Occidental Chemical Corporation to become LyondellBasell, which is now a leader in the polyethylene industry.

5.10.3   Reactor Size and Configuration

The early UNIPOL reactors licensed by Union Carbide Corporation were 1250 ft^3 in volume and were quickly expanded to 3,000 ft^3 and 5,000 ft^3 and

Table 5.26 Total Innovene G capacity (1975-2005).

Client	Location	Start-up Date	Current Capacity (000 tons per year)
BP Chemicals	Lavera, France	1975	50
BP Chemicals	Lavera, France	1985	240
Quantum	Port Arthur, U.S.A.	1988	100
Ube Industries	Chiba, Japan	1989	50
BP Chemicals	Grangemouth, U.K.	1990	160
TPE	Mab Ta Pud, Thailand	1990	90
CNTIC	Panjin, P.R.C.	1990	125
Chevron	Cedar Bayou, U.S.A.	1990	350
Sinopec / LCIC	Lanzhou, P.R.C.	1991	60
IPCL	Maharashtra, India	1991	220
Erdolchemie	Dormagen, Germany	1991	200
Texas Eastman	Longview, Texas	1993	200
PT Peni	Merak, Indonesia	1993	250
NPC	Arak, Iran	1994	60
Samsung	Seosan, Korea	1994	100
PEMSB	Kerteh, Malaysia	1995	250
CNTIC	Dushanzi, P.R.C.	1995	150
Formosa Plastics	Point Comfort, U.S.A.	1995	200
TPC	Tabriz, Iran	1997	100
PTPeni Third Train	Merak, Indonesia	1998	200
Westlake	Sulphur, Louisiana	1998	220
Chevron	Orange, Texas	1998	250
Bataan PE	Philippines	2000	200
Formosa Plastics	Taiwan	2000	240
Sidpec	Alexandria, Egypt	2000	225
BP	Grangemouth, U.K.	2001	320
SECO	Shanghai, P.R.C.	2005	600

(Continued)

Table 5.26   *(Continued)*

Client	Location	Start-up Date	Current Capacity (000 tons per year)
Amir Kabir	Bandar Imam, Iran	2005	300
Total Capacity			5,510

INEOS began licensing Innovene G in November 1984. Since then a total of 28 licenses have been granted worldwide. The total production capacity of *Innovene* now exceeds five million tons per annum.

Source: www.ineos.com/polyethylene

were designated G-1250, G-3000 and G-5000, respectively. As of 2010, the 5,000 ft^3 reactor is the standard size.

The annual capacity of the gas-phase process is up to 1.2 billion pounds, which translates to production rates of up to 150,000 lbs/hr assuming an 8,000 hour production year. Reactors are available as either single- or dual-reactor configuration. However, the trend is to produce complex molecular structures, such as polyethylene with a bimodal molecular weight distribution, in a single reactor with more complex catalyst systems to eliminate the need for two reactors in series (tandem configuration). All three licensors of the gas-phase process offer a dual reactor configuration with the first dual gas-phase process going into operation in 1996 by Union Carbide Corporation and designated as UNIPOL II.

5.10.4   Gas Phase and Slurry Loop in Series

Borealis introduced a dual reactor configuration in 1995 in Finland combining a slurry-loop reactor with a gas-phase reactor under the tradename of Borstar Polyethylene that is available for license. The 280 million pound per year reactor is unique in that the first reactor is a slurry-loop process utilizing propane as diluent under supercritical conditions rather than the usual slurry diluent isobutane. Supercritical propane offers lower polymer solubility than isobutane so that the risk of reactor fouling is lower with supercritical propane.

5.11   Solution Process

5.11.1   Historical Introduction

Phillips Petroleum's first commercial plant for the manufacture of HDPE with their Cr-based catalyst used a solution process from the plant start-up on December 31, 1956. This solution process was in use for four years until Phillips converted to an n-pentane vertical-loop process, which was introduced in 1961 [25,26]. The operating conditions for the Phillips solution process are listed in Table 5.27.

Two unique features of a solution polymerization process is the need to operate at relatively higher ethylene pressures in order to maintain a sufficient ethylene concentration due to the higher temperatures used, and the need to utilize an organic solvent such as cyclohexane that readily dissolves polyethylene. For the Phillips catalyst, polymerization temperature was used to control polymer molecular weight.

In 1955, one of the first examples of high-pressure, solution polymerization using a Ziegler-type catalyst was reported by Dow Chemical [40] using a high-pressure tubular polymerization process in which ethylene gas is added to an oil containing a heterogeneous Ziegler catalyst prepared by adding $TiCl_4$ and triisobutylaluminum to the oil. Polymerization conditions were 16,000–23,000 psig at 200–250°C and a catalyst residence time of about 8 minutes. The polyethylene molecular weight and density were in a commercial range as indicated by a Melt Index ($I_{2.16kg}$) value of 0.3–1.4 and density of 0.934–0.949 g/cc.

DuPont reported similar results using a stirred autoclave reactor operating at about 3,000 psig and 222°C using cyclohexane as the solvent [41]. The DuPont catalyst was prepared with $TiCl_4$ and $LiAl(C_{11}H_{22})_4$.

5.12   DuPont Sclair Process

DuPont introduced a commercial solution process in the 1960s with the tradename of Sclair that reportedly used a Ziegler-type catalyst based on both a vanadium compound (VCl_4 or $VOCl_3$) and a titanium compound ($TiCl_4$) in the presence of an aluminum alkyl cocatalyst. Operating conditions were above 200°C and 1,000 psi [42].

Table 5.27 Operating conditions for the Phillips solution process.

Solvent:	Cyclohexane
Temperature:	125–175°C
Pressure:	400–500 psig
Dates of Operation:	1957–1960*

*Prepared only ethylene homopolymers for the first two years, then utilized 1-butene as a comonomer to prepare polyethylene with lower density.

5.12.1   Background

DuPont's Sclair process was extremely important technology which was introduced in the 1960s, as it offered commercial ethylene copolymers at densities as low as 0.916 g/cc with improved physical properties over comparable polyethylene resins produced with the high-pressure process. In addition, DuPont scientists also developed improved catalysts for the Sclair process based on vanadium compounds that produced homogeneous branching distribution of the short-chain branches introduced along the polymer backbone by incorporation of a comonomer such as 1-butene into the polymer structure.

Although some catalyst aspects of the Sclair technology were presented in Chapter 4 of this book, additional details are presented below.

In a classic Canadian Patent 849081 awarded to Clayton T. Elston and assigned to DuPont on August 11, 1970, data is presented that show that the catalysts utilized in the Sclair process provide ethylene/1-olefin copolymers and terpolymers with a homogeneous branching distribution, which Elston defined as the random incorporation of the comonomer into the polymer backbone and a constant ethylene/comonomer ratio independent of molecular weight. For comparison, Elston defined heterogeneous branching distribution as an ethylene/comonomer ratio that varies with polyethylene molecular weight. According to Elston, the preferred molecular structure for ethylene/1-olefin copolymers is a homogeneous branching distribution due to improved physical properties of such materials when compared to ethylene copolymers with a heterogeneous branching distribution. Elston demonstrated that physical properties are dependent on polyethylene structural features such as:

- Narrow molecular weight distribution
- Comonomer type
- Homogeneous comonomer distribution

Copolymers with the structural features outlined above produce polyethylene films with reduced haze, higher gloss, higher dart impact strength, lower heat seal temperature, lower solvent extractables, and less film orientation, as characterized by less change between machine direction (MD) properties and transverse direction (TD) properties.

Some typical polymerization data provided by Elston is summarized in Table 5.28.

The data show that polyethylene with a homogeneous branching distribution is provided by a vanadium catalyst based on $VO(OBu)_3$ with

Table 5.28 Polymerization data from Canadian Patent 849081.*

Example	1	2	3	4
Branching Type[a]	hetero	homo	hetero	homo
Comonomer	1-butene	1-butene	1-octene	1-octene
Melting point (°C)	119.6	105.6	119.7	110.8
Density @ 1.0 MI	0.9187	0.9148	0.9207	0.9203
Catalyst	VOCl$_3$	VO(OBu)$_3$	VOCl$_3$/TiCl$_4$	VO(OBu)$_3$
Cocatalyst	isoprenyl Al[b]	EtAlCl$_2$	isoprenylAl	EtAlCl$_2$/Et$_2$AlCl

Conditions: Polymerization temperature 100–176°C; Residence time 2–6 minutes.

[a]hetero = heterogeneous; homo = homogeneous

[b]See J.J. Ligi and D.B. Malpass, *Encyclopedia of Chemical Processing and Design*, Marcel Dekker, NY, Vol. 3, p. 32, 1977for a detailed discussion of isoprenylaluminum (IPRA).

*Issued August 11, 1970 to C. T. Elston; assigned to DuPont.

EtAlCl$_2$ or EtAlCl$_2$/Et$_2$AlCl as cocatalyst. The melting point data of the four examples shown above suggest that the polyethylene with a homogeneous branching distribution is consistent with a single-site polymerization catalyst similar to catalysts discovered by Kaminsky in the late 1970s based on a zirconocene compound and methylalumoxane as cocatalyst.

Elston examined other vanadium-based catalysts based on VO(OR)$_2$Cl (where R was n-decyl) in which the polymerization was carried out in the presence of a diene monomer that reacted with the catalyst to produce long-chain branching in the polymer structure and chlorine containing organic ester that increased catalyst activity [43].

On February 28, 1978, A. W. Anderson and G. S. Stamatoff received U.S. Patent 4,076,698. It was assigned to DuPont de Nemours and Company and provided DuPont the composition of matter on ethylene/1-olefin copolymers in which the 1-olefin contained 5 to 18 carbons. The original patent was filed on January 4, 1957. After a series of court proceedings that took place in the early 1980s, the court awarded DuPont the composition of matter claim on these copolymers. Hence, any polyethylene producer that offered ethylene/1-hexene or ethylene 1-octene copolymers from approximately 1978–1995 was obligated to pay DuPont a royalty. The copolymers produced by Anderson and coworkers were prepared with the solution process using $TiCl_4$ + $VOCl_3$ (Ti/V) molar ratio of 3 and tri-isobutyl aluminum as cocatalyst. Polymerization conditions were 220°C at 1500 psi total pressure in cyclohexane as solvent and catalyst residence time of 5 minutes.

Also, in the late 1960s, Dutch State Mines (DSM) introduced another commercial solution process (Stamicarbon process) which also used a Ziegler-type catalyst with a Grignard reagent as an important part of the catalyst synthesis [42].

5.13   Solution Process (2012)

5.13.1   Overview

The solution process produces polyethylene dissolved in an organic solvent such as cyclohexane at 130–200°C and 500–1,500 psi ethylene and is consequently the preferred process for many specialty grades of polyethylene that are difficult to prepare in either the slurry- or gas-phase process. The solution process readily adapts to a wide range of comonomers (1-hexene, 1-octene, styrene and cyclic olefins) over a very broad range of comonomer content. The only limitation is polyethylene grades with a relatively very high molecular weight that are either insoluble in cyclohexane or form a solution with a high viscosity.

Based on a reactor configuration involving two or more reactors in series, the solution process is probably the most flexible low-pressure polyethylene manufacturing process due to the molecular structure control of each of the polymer components prepared in each of the reactors and the wide range of resins that can be produced with this process. Although manufacturing costs are somewhat higher with the solution process, compared to slurry- and gas-phase processes, the fast transition times and the capability of preparing value-added premium grades of polyethylene can overcome manufacturing cost disadvantages.

Nova Chemicals (Alberta, Canada), Dow Chemical (Freeport, Texas) and DSM (Odessa, Texas)(note: in 2002 DSM sold its petrochemical business to SABIC) are three companies with solution polymerization capacity in North America.

Nova Chemicals presently operates more than a 1.5 billion lb/yr solution phase polymerization process in Alberta, Canada and a second plant in Ontario. Nova obtained the Sclair technology from DuPont in 1994.

As of 2009, Nova had a worldwide annual capacity of approximately 5.4 billion pounds manufactured with the Sclairtech process. Key features of this process are relatively small, 2800–9000 liter, reactors that result in shorter reactor residence times and faster transitions to other polyethylene grades. The reactors possess intensive mixing capability and may operate as single or dual reactors in series for the manufacture of polyethylene with a more complex molecular structure in which the molecular weight and comonomer content of the polymer component produced in each of the two reactors can be controlled over a wide range. Polyethylene with a density of 0.905–0.965 g/cc and a Melt Index value ($I_{2.16\,kg}$) of 0.4 to 150 may be produced. The advanced Sclairtech process utilizes Ziegler and single-site catalysts with 1-octene or 1-butene as the comonomers.

5.13.2   Bimodal MWD in Solution Reactors

Nova also reports a single-site catalyst based on an organometallic complex with a metal center (Ti, Hf or Zr), and containing a cyclopentadienyl-type ligand and a phosphinimine ligand with methylalumoxane as activator, that produces bimodal MWD resins in the advanced Sclairtech reactor system [44]. The resins provide premium properties in pipe applications.

The Sclairtech technology also offers lower cost ethylene as an ethylene feed stream of 85% ethylene and 15% ethane may be used in the process. The cost savings comes from lower ethylene purification requirements in the ethylene plant.

5.13.3   Dowlex Solution Process

The Dow Chemical Company developed the solution polymerization process with the Dowlex tradename in the 1970s to manufacture ethylene/1-octene linear low-density polyethylene (LLDPE) copolymers. The process is proprietary and not available for license. The ethylene/1-octene copolymers exhibit many superior physical properties relative to other ethylene copolymers based on 1-butene or 1-hexene.

However, since the 1990s, Dow scientists discovered a single-site designated as a constrained geometry catalyst system which has been used to significantly expand the product mix available from the Dow solution process. Dow presently produces copolymers based on 1-butene, 1-hexene and 1-octene over a wide density and Melt Index range. Note: the reader is referred to Chapter 4 of this book, which discusses the Dow catalyst in more detail.

Chemical Market Resources [45] has reported that the Dowlex process consists of two stirred tank reactors in series with an isoparaffin (mixture of C_8 and C_9) solvent operating at 160°C and approximately 320 psig with a total residence time from both reactors of about 30 minutes. Such reactors allow for polyethylene with complex molecular structures to be prepared as the process conditions of each reactor may be varied to control the polymer composition, molecular weight and weight fraction of each of the two polymer components that make up the finished polyethylene resin.

References

1. T.O.J. Kresser, *Polyethylene*, Reinhold Publishing Corp., New York, 1957.
2. SRI Consulting data from www.sriconsulting.com.
3. J.C. Swallow, *Polythene, 2nd Ed.*, A. Renfrew, and P. Morgan, Eds., Interscience Publishers, New York, pp. 1-10, 1960.
4. E.W. Fawcett, R.O. Gibson, M.W. Perrin, J.G. Patton, and E.G. Williams, British Patent 471,590, assigned to Imperial Chemical Industries, Sept. 6, 1937.
5. J.J. Fox, and A.E. Martin, *Trans. Faraday Soc.*, Vol. 36, p. 897, 1940.
6. British Patent 587,378, issued April 23, 1947 to E. Hunter et al. and assigned to Imperial Chemical Industries (ICI); and British Patent 584,794 issued January 27, 1947 andassigned to DuPont. See also U.S. Patent 2,409,996 issued October 22, 1946 to M.J. Roedel and assigned to DuPont.
7. L. Squires,U.S. Patent 2,395,381, Feb. 19, 1946; W.E. Hanford, et al., U.S. Patent 2,473,996, June 21, 1949;M.J. Roedel, et al., U.S. Patent 2,703,794, March 8, 1955; all assigned to DuPont.
8. L. Seed, U.S. Patent 2,542,783, assigned to Union Carbide, Feb. 20, 1951; M.F. Gribbins, U.S. Patent 2,519,755, assigned to E.I. DuPont de Nemours & Co., Aug. 22, 1950; Y. Conwell, et al., U.S. Patent 2,698,463, assigned to DuPont, Jan. 4, 1955.
9. R.A.V.Raff, and J.B. Allison,*Polyethylene*;High Polymers Series, Vol. XI, Interscience Publishers, Inc., New York, 1956.
10. A.W. Larcher, and D.C. Pease, U.S. Patent 2816883; filed Aug. 2, 1951; assignedto DuPont, Dec. 17, 1957.
11. R.A.V. Raff, and K.W. Doak, *Crystalline Olefin Copolymers, Part I*;High Polymers Series, Vol. XX, Interscience Publishers, p.7, 1965.
12. Press Release from LyondellBasell, Rotterdam, Netherlands, Aug. 6, 2009.
13. *Plastics News*,Dec. 9, 2013, p. 13.
14. M.J. Roedel, *J. Am. Chem. Soc.*, Vol. 75, p. 6110, 1953.
15. www.arkema-inc.com,accessed Dec. 21, 2009.
16. T.O.J. Kresser, *Polyethylene*, Reinhold Publishing Corp., New York, pp. 67-68,1957.
17. L. Wild, T.R. Ryle, D.C. Knobeloch, and I.R. Peat, *J. Polymer Science: Polymer Physics Ed.*, Vol. 20, p. 441, 1982.
18. F.M. Mirabella, Jr., and E.A. Ford, *J. Polymer Science, Part B: Polymer Physics*, Vol. 25, pp. 777-790, 1987.

19. T. Usami, *Macromolecules*, Vol. 17, p. 1756, 1984; [17a]J.C. Randall, *J. Applied Polymer Sci.*, Vol. 22, p. 585, 1978, [17a]G. Levy, *Macromolecules*, Vol. 12, p. 41, 1978,

20. D.B. Malpass, *Introduction to Industrial Polyethylene*,Scrivener Publishing, Salem, MA,and John Wiley & Sons, Hoboken, New Jersey, Chap. 7, pp. 85-96, 2010.

21. H.R. Sailors, and J.P. Hogan, *J. Macromol. Science: Part A: Chemistry*,Vol. A15, Iss. 7,pp. 1377-1402, 1981.

22. J.P. Hogan, in: *Copolymerization*;High Polymers Series, Vol. XVIII, Interscience Publishing, John Wiley and Sons, G.E. Ham, Ed., Chap. 3, pp. 89-113. 1964.

23. J.P. Hogan, in: *Copolymerization*; High Polymers Series, Vol. XVIII, Interscience Publishing, John Wiley & Sons, G.E. Ham, Ed., Chap. 3, p. 89, 1964.

24. M.P. McDaniel, Chevron Phillips Chemical Co., personal communication, 2009.

25. D.D. Norwood,1996, Phillips vertical loop reactor, U.S. Patent 3,248,179, issued April 26, 1966, and assigned to Phillips Petroleum Co.

26. R.G. Rohlfing, 1965, U.S. Patent 3,226,205, issued Dec. 28, 1965, and assigned to Phillips Petroluem Co.

27. R.A.V. Raff, and K.W. Doak, *Crystalline Olefin Copolymers, Part I*;High Polymers Series,Vol. XX, Interscience Publishers, p.365, 1965.

28. www.cpchem.com, accessed Dec. 5, 2009.

29. S. Mihan, et al., U.S. Patent 7,553,916,assigned to Basell Polyolefine GmbH, June 30, 2009.

30. A.G. Gelsenberg Benzin, British Patent 791,889, March 12, 1958. Reference taken from R.A.V. Raff and K.W. Doak, *Crystalline Olefin Copolymers, Part I*; High Polymers Series, Vol. XX, Interscience Publishers, p.386 (1965).

31. W.C. Lanning, et al. U.S. Patent 2,970,135, assigned to Phillips Petroleum Co., Jan. 31, 1961; and R.F. Dye, U.S. Patent 3,023,203, assigned to Phillips Petroleum Co., Feb. 27, 1962.

32. R.R. Goins, U.S. Patent 2,936,303, assigned to Phillips Petroleum Co., May 10, 1960;

33. R.F. Dye, U.S. Patent 3,023,203, assigned to Phillips Petroleum, Feb. 27, 1962.

34. J.E. Anderson, et al., U.S. Patent 3,469,948, assigned to Dart Industries, Sept. 30, 1969.

35. J.L. Jezl, and E.F. Peters, U.S. Patent 4,101,289, assigned to Standard Oil (Indiana), July 18, 1978.See also other related patents: J.W. Shepard et al., U.S. Patent 3,957,448, assigned to Standard Oil Co., May 18, 1976; Jezl, et al., U.S. Patent 3,965,083, assigned to Standard Oil Co., June 22, 1976; Peters, et al., U.S. Patent 3,971,768, assigned to Standard Oil Co., July 27, 1976; and Jezl, et al., U.S. Patent 3,970,611, assigned to Standard Oil Co., July 20, 1976.

36. A.R. Miller, U.S. Patent 4,003,712,assigned to Union Cabide Corp., Jan. 18, 1977.

37. K. Schmid, etal., U.S. Patent 3,300,457,assigned toBASF, Ludwigshafen, Germany,Jan. 24, 1967.

38. G. Drusco, U.S. Patent 3,298,792,assigned to Montecatini, Milan, Italy, Jan. 17, 1967.

39. S. Takeuchi, U.S. Patent 3,463,617,assigned to Mitsui, Tokyo, Japan, Aug. 26, 1969.

40. E.B. Barnes, J.E. Thomson and G.A. Klumb, Jr., U.S. Patent 2,882,264, filed Oct.10, 1955, assigned to Dow Chemical Company, April 14, 1959.

41. British Patent 783,487, assigned to E.I. DuPont de Nemours & Company; Sept. 25, 1957.

42. J. Boor, Jr., Ziegler-Natta Catalysts and Polymerizations, Academic Press, p. 172, 1979.

43. C.T. Elston, U.S. Patent 3,984,610,assigned to DuPont of Canada Ltd., Oct. 5, 1976.

44. A. Kazakov, etal., U.S. Patent 6,858,677,assigned to Nova Chemicals International, Feb. 22, 2005.

45. Chemical Market Resources, Inc., New Generation Polyolefins, June/July Vol. 7, No. 6, p. 21, 2002.

6

Fabrication of Polyethylene

6.1 Introduction

The fabrication of polyethylene into commercial products is the business objective of the polyethylene industry. The applications of polyethylene may be very broadly defined in the following categories:

- Packaging
- Containers
- Transportation
- Medical
- Coatings

Each product category consists of thousands of individual products that have applications that touch essentially every aspect of human existence. The net result of these products is to improve the standard of living of society across the globe.

This chapter will discuss the primary methods that are used to shape polyethylene in the molten state into the end-use products. These fabrication methods are generally divided into extrusion and molding methods.

- **Extrusion methods:** Extrusion is used for polyethylene films, sheets, wire and cable, coatings and pipe. Film is manufactured by either a blown-film or casting process.
- **Molding methods:** Molding is divided into primarily injection molding, blow molding, rotational molding and thermoforming.

Other subjects that will also be discussed that impact the polyethylene fabrication business are the wide array of additives that are blended into polyethylene to improve the performance of the end-use products and the rheological properties of polyethylene which explain the flow of molten polyethylene that takes place during the fabrication process.

6.1.1   Fabrication Business

Unlike the companies that manufacture polyethylene as a raw material, which are approximately 30 global petrochemical corporations with annual revenue of about 5–60 billion US dollars, there are thousands of companies across the globe involved in the fabrication of polyethylene. These companies range in size from companies with annual revenue of about 0.5 million USD employing a few people, to relatively very large companies with annual revenue in excess of 1–10 billion USD and several thousand employees. These differences are primarily due to the amount of capital required to engage in these businesses. The manufacture of polyethylene is a capital intensive business requiring 1–5 billion USD to construct a petrochemical plant, while a fabrication business, or a business that supports the fabrication process, requires much less capital, perhaps only a 5–10 thousand USD investment to start a very small business.

6.1.2   Terms and Definitions Important in Polyethylene Fabrication

Thermoplastic   Polyethylene is a thermoplastic which can be softened or melted into a molten, viscous phase in which the structure of the material is unchanged during the heating process. On cooling, the molten phase returns to a solid state and the cycle may be repeated, making polyethylene an excellent material to be recycled.

Molecular weight

The molecular weight of a polyethylene molecule may be expressed as the degree of polymerization (n) of any particular molecule $(CH_2CH_2)_n$. Commercial grades of polyethylene have an average degree of polymerization from about 1,000 to about 10,000. The molecular weight of commercially useful polyethylene is expressed as a Melt Index (MI) value and is the mass of polyethylene that flows through a small orifice in ten minutes when the polymer sample is heated to 190°C with a mass of 2.16 Kg placed on the molten polyethylene sample. These measurements are carried out according to the American Society for Testing Materials (ASTM) standards. The majority of commercial grades of polyethylene have MI values of about 0.1 to about 50, where 0.1 represents a polyethylene sample with a relatively high molecular weight and high viscosity and a value of 50 represents a polyethylene sample with a relatively very low molecular weight and low viscosity.

Molecular weight distribution

The molecular weight distribution of a particular grade of polyethylene is an important parameter that has a significant effect on the rate at which the polyethylene may be fabricated into end-use products and the mechanical properties exhibited by the polyethylene. The ratio of the weight average molecular weight (M_w) and the number average molecular weight (M_n) is defined as the polydispersity of a polyethylene sample. Commercial grades of polyethylene exhibit a polydispersity value of from 2.0 (very narrow MWD) to about 20 (very broad MWD). Commercially, the molecular weight distribution is often measured by a Melt Flow Ratio value which is obtained by determining the amount of polymer, heated to 190°C, that flows through an orifice with a high load (HLMI) weight applied to the molten polyethylene sample (21.6 Kg) compared to the amount of polyethylene that flows through the orifice with a lighter weight (2.16Kg) termed Melt Index (MI). This Melt Flow Ratio (MFR) is HLMI/MI. MFR values of about 15–20 indicate a polymer sample with a very narrow MWD (polydispersity value of 2.0), while an MFR value of 120 represents a polymer sample with a very broad MWD or polydispersity value of about 20. In addition, the MFR value is also a relative measure of the degree of "shear-thinning," or decrease in melt viscosity demonstrated by a particular sample.

Viscoelasticity

Viscoelasticity is a property characteristic of all plastics. Plastics have both solid state properties such as elasticity, strength and shape retention and liquid properties such as flow, which is dependent on variables such as temperature and load. Materials that demonstrate both viscous and elastic behavior under an applied stress are designated as viscoelastic.

Viscosity	Viscosity is the resistance to steady flow shown within the body of the material and is an internal friction or a measure of a molten polymer's resistance to flow. Viscosity is measured with a viscometer over a wide range of shear rates at a constant temperature. Liquids with a constant viscosity over a range of shear rates are designated as Newtonian liquids.
Non-Newtonian flow	A decrease in apparent viscosity of a liquid with an increase in shearing rate applied to that liquid is characteristic of a material with non-Newtonian flow. Molten polyethylene exhibits a decrease in viscosity with increasing shear stress and is, therefore, a non-Newtonian liquid.
Shear-thinning	The change in viscosity over a range of shear rates is designated as shear-thinning. The amount of shear-thinning in any particular polyethylene sample over a range of shear rates is an important parameter for fabricating polyethylene into end-use items. Higher levels of shear-thinning improve the processability of the polyethylene and provide faster processing rates and lower production costs.
Melt strength	Once a polyethylene sample is melted to create a viscous liquid, the resistance of the melted polyethylene to stretching is termed the melt strength. Melt strength is measured as the maximum amount of tension that may be applied to a molten sample without the sample breaking. From a physical point of view, polymer molecule entanglements, increasing polymer molecular weight and molecular weight distribution can be viewed as parameters that enhance melt strength.
Strain-hardening	Strain-hardening can be viewed as an increase in the stiffness of a molten polyethylene sample as the sample is elongated (stretched), or an increase in elongational viscosity. The LDPE, which contains long-chain branching in its molecular structure, is the only type of polyethylene that exhibits strain-hardening.

6.1.3   Development of Melt Index Instrument

One of the most common instruments associated with the polyethylene industry is the Melt Index instrument that is found in research, product and manufacturing facilities around the globe. For commercial applications, polyethylene is sold according to Melt Index values summarized in material specification data sheets, usually by the Melt Index value determined with a 2.16 kg weight ($MI_{2.16}$).

The Melt Index instrument was developed by Imperial Chemical Industries (ICI) scientists [1] in the late 1930s in order to test polyethylene flow at 190°C. The instrument consists of a heated vertical barrel with a small die with an orifice of about 2 mm placed at the bottom of the barrel. The barrel holds about 4–6 grams of polyethylene. A weight is applied to the molten polyethylene by placing a piston holding a weight into the top of the barrel, and the rate that the molten polyethylene flows through the orifice is measured in g/10 minutes.

The ICI scientists used this instrument to provide information on the rate that polyethylene samples flowed through the instrument. Relatively higher flows indicated polyethylene with a relatively lower molecular weight. However, the instrument was also used by ICI scientists to measure the elastic properties of the molten polyethylene. The elasticity was expressed as a % recovery once the polyethylene sample (strand) that flowed through the die cooled to room temperature. The % recovery value was calculated from the diameter of the polymer strand (D_s) and the diameter of the orifice (D_o):

$$\% \text{ recovery} = 100[(D_s\text{-}D_o)/D_o]$$

6.1.4   Polyethyene Product Space for the Fabrication of Finished Products

Figure 6.1 shows the density and Melt Index values of various grades of polyethylene that are fabricated into finished polyethylene products. The

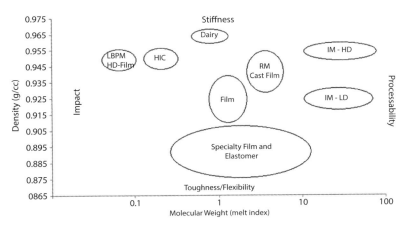

Figure 6.1 Polyethylene product space identifing various grades of polyethylene used in various fabrication techniques.

type of polyethylene used for various fabrication techniques are outlined below.

LBPM:	Large part blow molding
HIC:	Household industrial chemical (bottles)
HD-Film:	HMW-HDPE film
Dairy:	Milk bottle resin, ethylene homopolymer
IM-HD:	Injection molding, HDPE.
RM/CastFilm:	Rotational molding or cast film process
Film:	Blown film process
IM-LD:	Injection molding, LLDPE

6.2   Early History of Polyethylene Fabrication (1940–1953)

From the late 1930s through the end of World War II, polyethylene was primarily used for the manufacture of wire and cable in support of the war effort. Scientists and engineers were involved in the extrusion-coating process required for wire and cable applications. Figure 6.2 shows the extrusion equipment used to coat wire with a polyethylene jacket. Note that the application date in Britain was November 3, 1941, long before the patent was issued in the United States as U.S. Patent 2,384,224 dated September 4, 1945.

The equipment used an extrusion barrel containing a metal screw that transferred the polyethylene through a heating zone surrounding the extrusion barrel to melt the polyethylene. Next, the molten polyethylene entered a die (the die is positioned at a right angle to the extrusion screw) that was modified to permit a wire to pass through the center of the die before the polyethylene-coated wire emerged from the die. As discussed earlier, Imperial Chemical Industries (ICI) was the only source of polyethylene for the war effort until 1943, when Union Carbide Corporation and the DuPont Corporation were each granted a license from ICI for the production of polyethylene in the United States in support of the war effort. Before the end of the war, polyethylene production in the United States exceeded the ICI production in the United Kingdom. Annual polyethylene production in the United States was 1, 3 and 6 million pounds for 1943, 1944 and 1945, respectively. Consequently, from 1933 until about 1943,

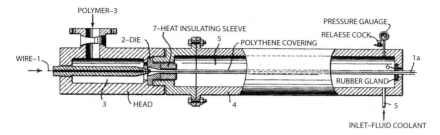

Figure 6.2 Extrusion of thermoplastic (polyethylene) materials for production of polyethylene-coated wire and cable key parts include: 1. Wire; 2. Die; 3. Polyethylene; 4. Cooling chamber; 5. Cooling water. Example 1; wire, was 1.2 mm in diameter; polyethylene coating overall diameter was 7 mm; extrusion rate 10 feet/minute, Temperature 130°C; the covering on the wire is set solid and free of voids on exiting the cooling chamber.

polyethylene research and development were primarily carried out by ICI scientists, primarily for extrusion coating of wire.

An important patent issued to Edmond G. Williams (assigned to ICI) describing the extrusion of thermoplastic (polyethylene) materials for production of polyethylene-coated wire and cable is shown in Figure 6.2. The drawing illustrates the process.

The rate of the wire-coating process was limited to such a rate that a smooth covering of the wire was produced without voids in the polyethylene coating. The development of improved methods for forming smooth, void-free sheets or coatings on wire were patented by Daniel E. Strain and Walter V. Osgood in U.S. Patent 2,480,615, issued to DuPont on August 30, 1949, in which a specific die geometry was used to subject the molten polyethylene to very high shearing stresses with sufficient frictional heat-developing forces so that smooth sheets or wire-coatings of polyethylene could be produced at higher rates.

6.2.1   Post World War II

The introduction of polyethylene for consumer applications began immediately at the conclusion of World War II, when polyethylene capacity increases at ICI, DuPont and Union Carbide accelerated the rapid growth of the polyethylene industry. Fabrication methods were developed for applications in film, pipe, injection molding and blow molding. Polyethylene consumption in 1953 in the United States was about 144 million pounds, which was a 24-fold increase over the 1945 production. For the year 1953, the applications for polyethylene by the fabrication method are summarized in Table 6.1.

Table 6.1 Consumption of polyethylene in the
United States in 1953.

Method	(%)	pounds (million)
Film	32	46
Wire and Cable	22	32
Pipe	17	25
Injection Molding	14	20
Paper/Film Coating	10	14
Blow Molding	5	7

6.3 Stabilization of Polyethylene

6.3.1 Introduction

Polyethylene is subjected to various environments during the fabrication process and during the service life of a wide variety of products. These environments include:

- Heat
- Oxidation
- Hydrolysis
- Attack by host materials (liquids and gases)
- Ultraviolet light and other forms of radiation
- Electricity

Various inorganic and organic materials are added to polyethylene to prevent degradation and/or improve performance. The amount and type of additive depend on the end-use application. Additives are incorporated into polyethylene at various stages of the fabrication process depending on the application. However, it is necessary to blend many additives into the granular polyethylene at the manufacturing facility, where the polyethylene is converted to pellets prior to being packaged into individual containers and classified as a specific "Grade" of polyethylene designed for certain end-use applications. The initial additive package also stabilizes the polyethylene before the material is subjected to high temperatures and shear during processing into finished polyethylene products. Developing additives that prevent the oxidative degradation of polyethylene was one of the first objectives of scientists at ICI, DuPont, and Union Carbide, which were the first companies to manufacture polyethylene in the 1940s.

6.3.2   Thermal Oxidation Mechanism (1920–1960)

The autocatalytic thermal oxidation of hydrocarbons at slightly elevated temperatures has been known for over 90 years [3]. As an example of this process, the autoxidation of the saturated hydrocarbon octadecane at 105°C is shown in Figure 6.3, where oxygen uptake was measured over 100 hours.

Several historic publications that provided early evidence into the chemistry involved in this process were published between the 1920s and 1940s. Christiansen [4] and Backstrom [5] provided data in 1924 and 1927, respectively, that a chain reaction was involved. Next, in the early 1930s, Backstrom [6] and Ziegler and Ewald [7] demonstrated that the oxidation of hydrocarbons was initiated by free radicals. In 1942, Farmer and coworkers reported that an "autocatalysis" process involving free radicals and the formation of hydroperoxides was involved [8]. This 1942 paper identified the three steps responsible for the auto-oxidation of polyethylene that result in the degradation of polyethylene properties. These three steps were free-radical initiation, propagation and termination.

Moureu and Dufraisse classified compounds that slowed the oxidation process on hydrocarbons as "antioxygens" [9].

Finally, between 1949 and 1954, Bateman and Bolland consolidated these early studies into a mechanism responsible for the autocatalytic oxidation process [10,11]. The three-step reaction mechanism developed by

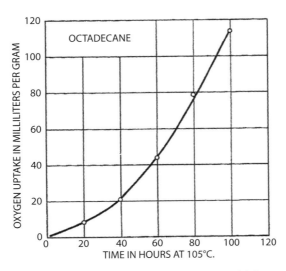

Figure 6.3 Autocatalytic oxidation of octadecane; Ref. [3], page 833.

them is recognized today as the process involved in the oxidation of polyethylene. The mechanism is divided into three stages:

(a) Initiation	Polymer $\rightarrow R\bullet$	(6.1)
(b) Propagation	$R\bullet + O_2 \rightarrow R\text{-}O\text{-}O\bullet$	(6.2)
	$R\text{-}O\text{-}O\bullet + polymer \rightarrow ROOH + R'\bullet$	(6.3)
	$ROOH \rightarrow RO\bullet + \bullet OH$	(6.4)
(c) Termination	$2R\bullet \rightarrow R\text{-}R$	(6.5)
	$R\bullet + R\text{-}O\text{-}O\bullet \rightarrow R\text{-}O\text{-}O\text{-}R$	(6.6)
	$2R\text{-}O\text{-}O\bullet \rightarrow R\text{-}O_4 - R$	(6.7)

The rate of Equation 6.2 is much greater than the rate of Equation 6.3; hence, deactivation of the R-O-O• species constitutes a key step in oxidation inhibition, which is the role of the primary antioxidant. The homolytic cleavage of the hydroperoxide species, ROOH, Equation 6.4, leads to the autocatalytic nature of the oxidation. Thus, both the rates of Equations 6.3 and 6.4 account for the rate of oxidation. The role of a secondary antioxidant such as a phosphite, $P(OR)_3$, is to act as a peroxide decomposer by oxidation of the phosphite to a phosphate $(OR)_3P=O$.

In 1961, the infrared spectrum of an oxidized sample of polyethylene showing the carbonyl and hydroxyl groups was reported. This spectrum is shown in Figure 6.4.

More recent work into the complex nature of polyethylene oxidation has been reported with the objective of identifying oxidation products more completely and also gaining a better understanding of chain scission and crosslinking reactions that take place during oxidation of polyethylene.

The formation of the OH-containing species and the carbonyl-containing species in the IR spectrum shown in Figure 6.4 were investigated in more detail [12,13].

$$ROOH \rightarrow RO\bullet + \bullet OH \rightarrow R\bullet + H2O/ROH \quad \text{(OH containing species)} \qquad (6.8)$$

$$2RO_2\bullet \rightarrow ROH + R'(C=O)R'' \quad \text{(Alcohol and ketone)} \qquad (6.9)$$

However, as pointed out by Lacoste and coworkers [14,15], the reactions expressed above and as Equations 6.1–6.9 do not explain polyethylene backbone scission or the formation of more complex decomposition products such as esters, lactones, carboxylic acids and unsaturation [16–18].

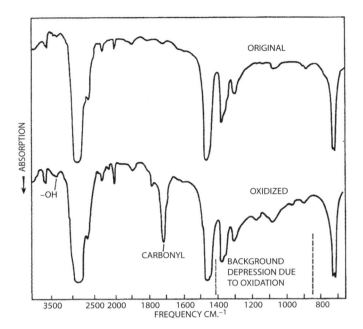

Figure 6.4 Infrared spectra reported in 1961 of a branched polyethylene before and after thermal oxidation; the carbonyl group (C=O) repesenting the primary functional group. Ref. [3], page 832; Reprinted with permission from *Modern Plastics*.

Backbone scission is the essential factor necessary for the embrittlement of polyethylene and leading to the performance failure of polyethylene.

One mechanism proposed [14,15] for chain scission and the formation of a terminal aldehyde group, that could later be further oxidized to form a carboxylic acid, is shown below from the homolytic cleavage of the hydroperoxide group along the polymer backbone.

(chain scission)

---CH2-C(H)-CH2--- → ---CH2-C(H)=O   +   •CH2-----polymer
      | terminal aldehyde   terminal free radical
      O•

To explain the formation of vinyl and trans-vinylene products, Gugumus [16,17] proposed chain-scission reactions involving a secondary hydroperoxide group.

Lacoste and coworkers [14] investigated the hydroperoxide decomposition products in polyethylene samples that were first oxidized with 10 Mrad of gamma irradiation in air at 40°C, and then underwent a second treatment of either heating, UV irradiation or gamma irradiation. A LLDPE sample in which all additives were removed prior to the first oxidation step was used for this investigation. The second treatment was carried out in the absence of additional oxygen. In this manner, the decomposition of the hydroperoxide groups formed during the first oxidation step could be monitored by IR spectroscopy. Table 6.2 shows the amount of the wide variety of oxygen-containing functional groups identified by Lacoste and coworkers in the first oxidized step. The concentration of various functional groups present was given as mmol/kg.

In a second treatment step, the oxidized LLDPE sample was heated in vacuum at 85°C for 150 hours with the concentration of each sample measured at 25 hour intervals. These results showed that the sec-hydroperoxide groups were eliminated after about 50 hours and the $-CH=CH2$ concentration was reduced 50%. All other functional groups increased in concentration except for the vinylidene group, which remained unchanged.

6.3.3 Polyethylene Melt Processing

Scientists at Dow Chemical Company [19] examined thermal scission and crosslinking of various types of polyethylene during melt processing. Polyethylene undergoes melt processing in an extruder where the molten polymer is exposed to rather severe conditions such as high temperatures (ca. 150–350°C), often in the presence of oxygen, and shear forces at high

Table 6.2 Oxidation products and unsaturation in the firstoxidized LLDPE samples.

Functional Group	Concentration (mmol/kg)
Sec-OOH	40
Tert-OOH	0
$-CH_2-(C=O)OH$	18
$-CH_2-OH$	13
$-CH_2(C=O)CH_2-$	45
$-C=O-O-CH_2-$	0.6
-Lactone	2
$-CH=CH_2$	19
Trans-CH=CH-	20
$>C=CH_2$	2

This oxidation step was carried out in air at 40°C using 10 Mrad gamma irradiation [14,15].

temperature. During this time, the polymer can undergo degradation in which chain scission and/or crosslinking occurs. In addition, oxygen-containing functional groups can be introduced into the structure (ketones, aldehydes, carboxylic acids) that could affect polymer color or odor.

6.3.3.1 Scission

There are two primary mechanisms for chain scission. The first was designated as "oxidative scission," which involves beta-cleavage of oxygen-centered alkoxy radicals with a relatively low activation energy of ca. 59 kJ/mol. This type of scission dominates at moderate temperatures or under conditions of high oxygen concentration. The second scission mechanism was designated "thermal scission," and involves beta-cleavage of alkyl radicals. This scission mechanism produces vinyl and vinylidene groups depending on the type of alkyl radical involved. Activation energy of 84–117 kJ/mol was determined.

6.3.3.2 Crosslinking

An investigation into crosslinking under thermo-oxidation conditions with HDPE prepared with a Ziegler catalyst and a Phillips catalyst demonstrated the importance of vinyl concentration in the polyethylene for crosslinking to occur. The HDPE produced with a Ziegler catalyst contains a relatively low vinyl concentration because hydrogen is used with a Ziegler catalyst to control polyethylene molecular weight that produces saturated end groups. The HDPE prepared with a Cr-based Phillips-type catalyst produces polyethylene with a high vinyl concentration (each molecule has a vinyl-end group). The Ziegler-prepared HDPE does not crosslink under extrusion conditions, while the HDPE prepared with the Cr-based catalyst does undergo crosslinking. Crosslinking or chain branching is attributed to the addition of alkyl radicals to vinyl groups as shown below. This is an energetically favored reaction resulting in the formation of a sigma bond at the expense of a π bond.

Crosslinking or chain-branching reaction: $R \bullet + \text{-C=CH}_2 \rightarrow \bullet\text{C-CH2-R}$

The LLDPE resin that was hydrogenated to eliminate unsaturation did not crosslink under extrusion conditions, while the LLDPE that contained some vinyl unsaturation did undergo some crosslinking as indicated by a decrease in the vinyl concentration. A ^{13}C-NMR study showed that the decrease in vinyl concentration was accompanied by the formation of long-chain branching (LCB).

Table 6.3 Temperature dependence of crosslinking and scission.

Temperature (°C)	vinyl's/1000 carbon (a)		
	500 seconds	**1500 seconds**	**3000 seconds**
225	0.90	0.70	0.55
275	0.75	0.55	0.35
350	0.72	0.62	0.50

Data from Figure 13 in reference [19].

6.3.3.3   Temperature Dependence of Crosslinking and Scission

An example of the degree of crosslinking taking place in a HDPE resin with a high vinyl content (prepared with a Phillips catalyst) at various extrusion temperatures was investigated by Johnston and Morrison [19] by measuring the decrease in the vinyl content of the resin at various intervals of time and temperature. The data is summarized in Table 6.3.

The data show that the highest degree of crosslinking in the HDPE prepared with a Phillips catalyst takes place at 275°C after 50 minutes of treatment. It is important to note that HDPE prepared with a Phillips-type catalyst with a broad molecular weight distribution and used in applications such as blow molding, HMW-HDPE film and premium pipe are often "tailored" in an extruder, where a certain amount of crosslinking is introduced into the resin. The resin tailoring is carried out to modify melt properties such as die swell, melt strength or melt elasticity needed to provide better properties in the fabricated item.

An important conclusion by Johnston and Morrison was that the antioxidant package greatly inhibits the oxidation rate and reduces both crosslinking and chain scission. However, most antioxidants are better peroxy radical scavengers than alkyl radical scavengers and, therefore, inhibit oxidative chain scission more than the alkyl-radical-dominated crosslinking reaction. Consequently, at high extrusion temperature, antioxidants shift the crosslinking/scission process towards crosslinking.

6.4   Historical Overview of Some Common Polyethylene Additives

6.4.1   Polyethylene Additives (1935–1955)

In the early stage of the polyethylene industry, research carried out from about 1935 to the mid-1950s will be reviewed in order to identify the scientists that made important contributions in identifying various types of compounds which were added to polyethylene to improve product performance. Many additives that were developed in the 1920s and 1930s for the stabilization of natural rubber for the manufacture of tires were readily available for scientists working in the polyethylene industry. For example, carbon black and antioxidants that were developed for the rubber industry were used as additives for polyethylene.

6.4.1.1   Carbon Black

In 1857, Thomas Hancock first showed that black-colored rubber exhibited less deterioration when exposed to sunlight than white rubber. This finding has since been used, most extensively, in the use of carbon black in the manufacture of tires. However, in the late 1930s, scientists at ICI found that the addition of carbon black to polyethylene had even greater benefits. The addition of 25–50 wt% of dispersed carbon black to polyethylene was first described by B. J. Habgood in British Patent 532,665, filed in Great Britain on May 30, 1939, and issued as U.S. Patent 2,316,418 on April 13, 1943. The addition of carbon black to polyethylene provided a composition with increased hardness, increased tensile strength and an increased electrical conductivity, making these materials especially well suited as dielectric-protective coatings for the manufacture of wire and cable. A wide variety of carbon blacks were studied with particle sizes of 40–120 millimicrons.

6.4.1.2   Antioxidants

A new class of antioxidants obtained from the condensation reaction of diarylamines and formaldehyde was reported by scientists at B.F. Goodrich in U.S. Patent 1,890,916 issued on December 13, 1932, for the stabilization of rubber. In one example, four moles of diphenylamine was reacted with 0.5 moles of formaldehyde to yield tetra phenyl methylene diamine with a melting point of 103–106°C. This class of antioxidant was effectively used by scientists at ICI for improving the resistance of polyethylene to oxidation.

6.4.1.3   Flame Retardants

An invention which provided an additive for polyethylene that produced polyethylene with self-extinguishing properties under an ASTM flammability test and flame-retarding properties according to an Underwriters Laboratories flame retardant test, was disclosed in U.S. Patent 2,480,298 issued to W. B. Happoldt on August 30, 1948, and assigned to DuPont Company. The combination of antimony trioxide (20–35 wt%), at least

6 wt% of a solid chlorinated hydrocarbon with a high chlorine content (55–80 wt%Cl) and 59–74 wt% polyethylene, provided a polyethylene combination with the flame-retarding properties discussed above.

6.4.1.4   Lubricants

The Bakelite Division of Union Carbide received U.S. Patent 2,462,331 on February 22, 1949, for the use of carboxylic acid esters or metal salts of carboxylic acids as additives to polyethylene, that improved the processing of polyethylene into film and sheets. The addition of 0.2–0.5 wt% of these compounds to polyethylene was reported to cause the ready release of the polyethylene film or sheet from hot milling and calendaring rolls at elevated temperatures (115–150°C) to provide films and sheets with good surface appearance. Examples of these releasing agents were calcium stearate, aluminum laurate and aluminum tristearate.

6.4.1.5   Anti-Static Agents

The addition of water-insoluble aliphatic alcohols or ethylene-oxide condensation products to polyethylene were in U.S. Patent 2,525,691, issued on October 10, 1950, to ICI scientists. Examples of alcohols used as anti-static agents were a saturated aliphatic alcohol(s) such as myristic, cetyl and stearyl alcohol in combination with an unsaturated alcohol such as oleyl alcohol. The second type of anti-static agent was obtained from the interaction of ethylene oxide with a mixture of a saturated and an unsaturated alcohol.

6.4.1.6   Calcium Carbonate as Filler

The addition of a mineral such as calcium carbonate to polyethylene was found to increase the toughness of the polyethylene and increase the softening point temperature. The ICI scientists obtained U.S. Patent 2,466,038 on April 5, 1949, describing a composition that was produced by the addition of calcium carbonate (30 wt%) with a particle size of less than 0.1 micron and polyethylene (70 wt%).

6.5   Examples of Additives Presently Used in the Polyethylene Industry (2012)

Presently, there are a wide variety of compounds that may be added to polyethylene to improve the performance of polyethylene in end-use applications. Because of the complex nature of this subject, this section will review some of the more important additives used in the polyethylene industry. For additional information on this topic the reader is referred to a recent text published by Michael Tolinski entitled *Additives for Polyolefins*[20].

The types of additives used in the polyethylene industry are:

- Antioxidants – primary and secondary
- UV Stabilizers
- Catalyst component scavengers
- Processing aid/lubricants, release agents
- Anti-block agents – silica/talc
- Flame retardant
- Color – other pigments; liquid/solid color agents
- Anti-static agents
- Foaming agents
- Slip agents/wetting agents/coupling agents
- Antifog
- Antibacterial
- Mineral fillers

6.5.1   Antioxidants

Two representative examples of primary antioxidants that are commonly used in the polyethylene industry that act as free-radical scavengers to terminate free radicals, and thus interrupt the oxidation process, are shown in Figures 6.5 and 6.6. Both compounds are sold by BASF Resins under the Irganox tradename. Note that BASF acquired this brand from Ciba-Geigy. It should also be noted that the composition of matter patents that were issued on many commonly used compounds up to about the mid-1990s have expired and, therefore, other companies may market the same material under a generic brand name. The structure of two common antioxidants in use today are given in Figures 6.5 and 6.6.

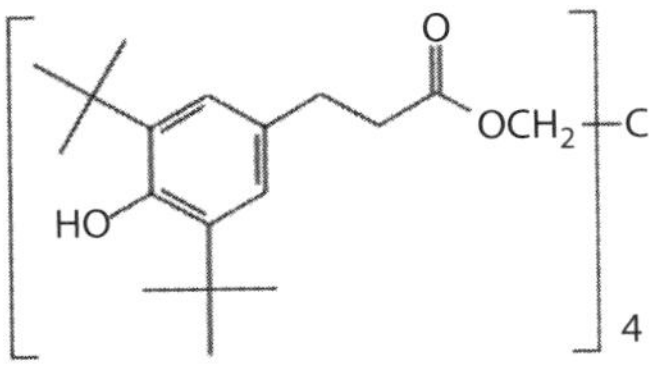

Figure 6.5 Structure of IRGANOX 1010.

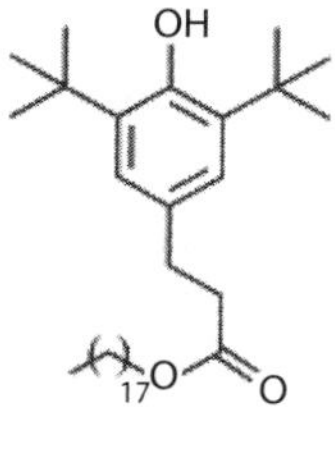

Irganox 1076

Figure 6.6 Structure of IRGANOX 1076.

Figure 6.7 Free radical termination.

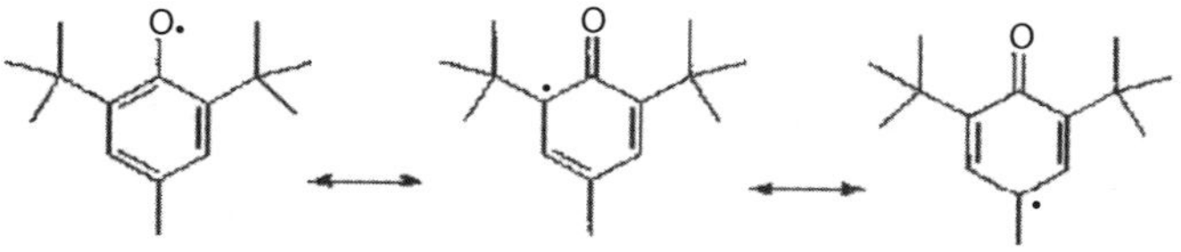

Figure 6.8 Resonance stabilization of hindered phenol.

Both antioxidants contain hindered phenolic groups, common in similar products manufactured by other suppliers of primary antioxidants, that eliminate free radicals (R•, RO•, or ROO•) by donating the hydrogen atom attached to the phenolic group and forming a relatively stable freeradical that isomerizes between different resonance structures. This process is illustrated in Figures 6.7 and 6.8.

6.5.2 Secondary Antioxidants

A secondary antioxidant is used in conjunction with a primary antioxidant to improve the thermal oxidative stability of polyethylene. The secondary antioxidant is a hydroperoxide decomposer that reduces the peroxy group to an alcohol or ether, consequently preventing the homolytic cleavage of the ROOH group to RO• and •OH.

This reaction is shown below and an example of a commercial product available from Chemtura (Alkanox 240) is shown in Figure 6.9.

$$P(OR)_3 \ + \ R'OOH \ \rightarrow \ O{=}P(OR)_3 \ + \ R'OH$$

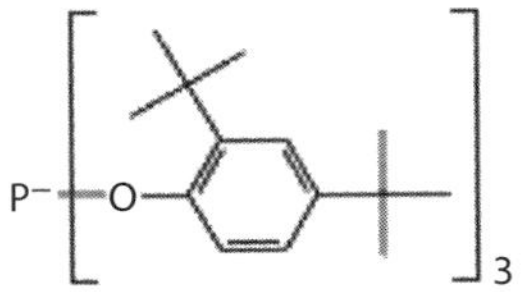

Figure 6.9 ALKANOX 240; Chemtura sold its antioxidant and UV stabilizer business to SK Capital Partners in May, 2013.

6.5.3   UV-Light Stabilizers

Ultraviolet (UV)stabilizers protect the polyethylene from the effects of sunlight, which increases the life of the final product by maintaining the physical properties necessary for a particular application. There are two types of additives that interfere with the degradation of polyethylene due to interaction with UV radiation. The first type is compounds that absorb UV light and therefore protect polyethylene by creating a barrier to the UV light. Carbon black was an early UV stabilizer in wide use today primarily in pipe and wire and cable applications. Compounds such as benzophenone, benzotriazole or hydroxyphenyl triazine fall into this category and typical structures are shown in Figure 6.10.

The second type of UV stabilizer is hindered amine light stabilizers (HALS) that are an extremely efficient stabilizer against the photodegradation process. They are compounds that do not absorb UV radiation, but act to inhibit degradation. Relatively high levels of stabilization are achieved with relatively low levels of concentration because the HALS are not consumed during the stabilization process (unlike primary and secondary antioxidants).

An example of this type of stabilizer is shown in Figure 6.11 as Lowilite 19 from Chemtura Corporation.

6.5.4   Mineral Fillers/Reinforcing Agents

The addition of inorganic compounds to polyethylene is carried out for a variety of reasons. Fillers are sometimes only low-cost particulate materials used to extend the polyethylene and, therefore, lower the cost of the fabricated item. However, fillers usually affect some other aspect of the finished product such as the stiffness (modulus) or the processability of the polyethylene to increase rates of fabrication. Fillers may also reduce mold shrinkage and the thermal expansion coefficient. Filler such as mica may increase heat resistance. Some examples of these compounds are discussed below.

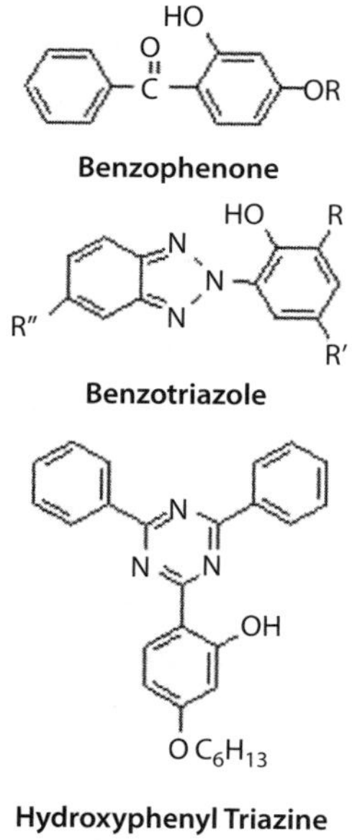

Benzophenone

Benzotriazole

Hydroxyphenyl Triazine

Figure 6.10 Examples of UV-light absorbing compounds.

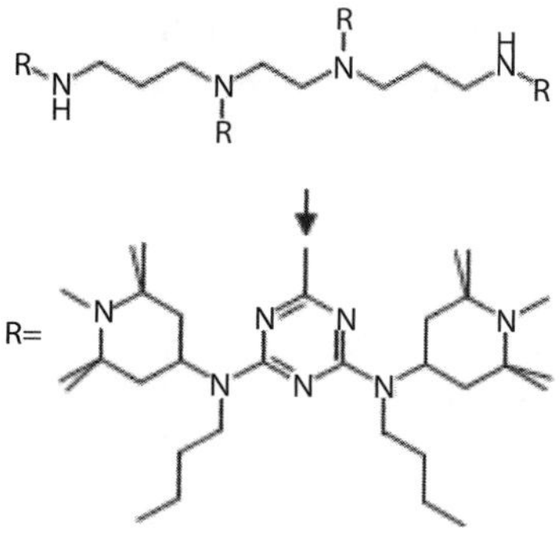

Figure 6.11 Lowilite 19; molecular weight 2286 and melting point 115–150°C; Source: www. Chemtura; Now owned by SK Capital Partners.

a. Ziegler-type catalysts contain acidic chloride atoms (i.e., Ti-Cl and Al-Cl bonds) that hydrolyze to form hydrogen chloride that may cause corrosion or color problems during processing or in end-use applications. A chloride scavenger such as calcium stearate can be used to neutralize the hydrolyzable chloride atoms. In addition, calcium stearate also acts as a processing aid providing higher fabrication rates and lower power requirements.

b. Calcium carbonate added to HMW-HDPE resin at levels of up to 15 wt% serves several functions: (a) increase film-fabrication rates of 5–10% (depending on loading) by decreasing melt pressure and motor load; (b) decrease amount of heat necessary to melt the resin; (c) increase thermal conductivity of the film; (d) modify the surface energy (increase in film coefficient of friction), improving the stacking of the finished merchandise bag or grocery sack; and (e) improve film toughness by increasing the dart drop strength under certain fabrication conditions [21]. Note that the particle size and particle size distribution of fillers such as calcium carbonate is an important consideration and must be evaluated on a case by case basis.

c. Titanium dioxide/carbon black, TiO_2, is a common pigment used when a brilliant, opaque white is desired. Carbon black is both a pigment and a stabilizer that prevents degradation by absorbing and preventing the penetration of ultraviolet light beyond the surface of the item.

d. Talc is the softest of all minerals and is a hydrated magnesium silicate mineral with the composition $Mg_3Si_4O_{10}(OH)_2$. The addition of talc to polyethylene used in food packaging reduces oxygen and water vapor transmission. The addition of talc to polyethylene pipe improves the creep resistance of the pipe over extended periods. Creep is the deformation of the pipe under a load such as soil. Other benefits of talc as an additive are an increase in stiffness, improved thermal conductivity which can improve fabrication rates, and talc with a particle size of about 2 microns can act as a nucleation agent, improving the rates of crystallization in polyethylene.

e. Coupling agents: The interfacial adhesion of the filler material such as an inorganic mineral to the polyethylene is often poor due to the difference in surface energy of the inorganic material and the polyethylene. Inorganic compounds are often hydrophilic while polyethylene is hydrophobic, thus creating a stress at the boundary of the two materials. The most common coupling agents used are silane compounds that contain one type of functional group that can react with the inorganic material such as a Si-OH group and a second type of functionality such as an alkyl group (sometimes containing a $C=C$ double bond) that is compatible with the polyethylene.

6.5.5   Lubricants

An *external* lubricant is a low molecular weight organic material that is poorly soluble in polyethylene and will migrate to the surface of the molten polymer to form a slippery coating during the processing operation, thereby producing a smooth surface for the molded article and acting as a release agent. Salts of stearic acid such as calcium stearate or zinc stearate are examples of external lubricants. An *internal* lubricant is soluble in the molten polyethylene and operates by lowering the viscosity of the polyethylene, thereby decreasing energy consumption during fabrication and increasing production rates. Relatively low molecular weight polyethylene (such as a polyethylene wax) is an internal lubricant. The addition of ca. 5–15% LDPE that has a broader molecular weight distribution than LLDPE manufactured with a Ziegler catalyst will also improve the processability of LLDPE in blown-film applications. However, a small decrease in film toughness properties such as dart impact or tear strength may occur, but additive levels are chosen to provide a balance of acceptable film properties to improved processability.

6.5.6   Blowing Agents

Chemical blowing agents are compounds that decompose to liberate a gas *in situ* such as carbon dioxide, while physical blowing agents are volatile compounds that are soluble in molten polyethylene but simply vaporize after heating and when exposed to reduced pressure, as molten polyethylene is exposed to the atmosphere as it exits the fabrication process.

6.5.7   Flame Retardants

Because organic polymers such as polyethylene contain only carbon and hydrogen and are therefore flammable, flame retardants are added during the processing of polyethylene to reduce flammability. Flame retardants contain large amounts of chlorine or bromine that are thought to interrupt the formation of organic free radicals that are part of the flame propagation process.

6.5.8   Antiblock Agents

The term "blocking" refers to the tendency of two layers of polyethylene film to adhere to each other making the separation of the two layers more difficult. Consequently, a relatively high degree of blocking will interfere with the unwinding of the film from the take-up roll during a film fabrication step or the separation of the two films while opening a finished merchandise bag. The blocking of two films is due to dipole-dipole interactions between the carbon and hydrogen atoms of the surface molecules that make up polyethylene. This attractive force decreases exponentially with increasing distance between the atoms (layers), so that making the polyethylene surface more irregular (rough) will dramatically decrease the blocking forces by increasing the distance between the surface atoms. Another cause of blocking between two layers of polyethylene film is the molecular structure of the polyethylene. Relatively lower molecular weight, highly-branched polyethylene molecules tend to migrate to the polymer surface through the amorphous phase of the polymer film, creating an adhesive layer at the boundary of the two films. This is more common in the case for LLDPE resins prepared with titanium-based Ziegler catalysts that produce polyethylene with a heterogeneous (non-uniform) branching distribution.

Natural and synthetic silica, specialty grade talcs[1] and calcium carbonate are the primary inorganic materials used as antiblocking agents. Natural silica mined and classified as diatomaceous earth is available under the tradename Celite. Inorganic antiblocking agents function to introduce a small amount of surface roughness onto the surface of the film, therefore increasing the distance between the two layers of film. The LLDPE and LDPE formulated with 0.1–0.5 wt% of synthetic silica (i.e., Cab-O-Sil from Cabot Corporation) or specialty grade talc (i.e.,Specialty Minerals Inc.'s Optibloc®) will eliminate blocking problems in most films.

Commercial antiblock agents may be coated in order to modify the surface energy of the antiblock agent prior to adding it to the polyethylene. For example, polyethylene glycol (Carbowax 4600) was added to natural silica (Celite 219) to modify the surface of the antiblocking agent for improving film clarity[22]. The LLDPE and LDPE formulated with calcium carbonate at 0.5–2.0 wt% will also eliminate film blocking problems. Because of the softness of talc compared to silica, master batches formulated with talc are less likely to cause machine wear problems over extended periods and, therefore, talc is the preferred material as an antiblock agent. The particle size of antiblock agents is normally in the range of 0.5–5 microns. The elimination of any particles greater than 5 microns in diameter is critical to prevent film gel problems or film failure problems.

---

[1] For additional information on talcs as antiblocking agents, contact: Specialty Minerals Inc., 35 Highland Avenue, Bethlehem, PA, 18017.

6.6   Rheological Properties of Polyethylene

In terms of the polyethylene industry, Rheology is the science of investigating polyethylene in the molten state and more specifically the behavior of the molten polyethylene as it flows with an applied shear stress.

Almost all polyethylene is conveyed during the fabrication process through an extruder, which is a machine that melts the polyethylene and conveys the polyethylene through the machine using a rotating screw.

Because polyethylene is a non-Newtonian liquid, the viscosity of polyethylene decreases with increasing shear, which is beneficial in the fabricating process because this property makes polyethylene easier to fabricate. Figures 6.12 and 6.13 show the relationship between the viscosity of

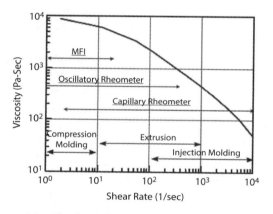

Figure 6.12 Rheology of polymer processing demonstrating the shear-thinning behavior of molten polyethylene.

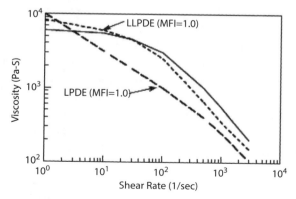

Figure 6.13 Rheological comparisons (190°C) for LDPE with a relatively broad MWD and LLDPE with a relatively narrow MWD.

(a)   (b)   (c)

Figure 6.14 Picture of molecular orientation during shear-thinning process. Reprinted from B. J. Edwards *et al.*, J. of Non-Newtonian Fluid Mechanics, Vol. 152, p. 168 (2008) with permission from Elsevier Publishing.

the molten polyethylene at various shear rates for extrusion and injection molding.

Figure 6.14 shows the molecular rearrangement that takes place during increasing shear rates from lower levels of shear to higher levels of shear, (a), (b) and (c), respectively. Although Professor Edwards has used low molecular weight polyethylene waxes in this modeling work, this research demonstrates the type of molecular rearrangement that takes place in the molten state as polyethylene undergoes shear forces during the fabrication process. This type of molecular rearrangement is responsible for the decrease in viscosity exhibited by non-Newtonian liquids under shear forces. It should be noted, however, that commercial grades of polyethylene with a much higher molecular weight than that used by Professor Edwards would not reach a condition where the polyethylene molecules become completely aligned, as shown in Figure 6.14; Step (c).

6.7   Fabrication of Film

6.7.1   Introduction

Polyethylene films are fabricated primarily by two methods: Blown Film and Cast Film.

Blown film is usually produced at a film thickness of < 20 mils (0.020 inches, 0.5 mm), but primarily processed at 2 mils or less with thinner films of 0.5–1.0 mil, the most common thickness. Thin films are much more cost effective as long as the film has sufficient toughness for the particular application. The blown film process may produce a monolayer film or through coextrusion techniques, where multiple layers of different plastics are used to produce specialty films for more demanding applications where barrier properties are needed.

Cast films are produced by extruding the molten polyethylene through a horizontal slit (designated as a die) directly onto a chill roll, which is usually water cooled and chrome coated to produce a very smooth film. Cast film can be produced at higher rates due to the simple nature of the process.

Some of the applications for polyethylene film are summarized below:

- Agricultural film
- Bread bags
- Deli wrap
- Food packaging
- Freezer film
- Garment film
- General purpose packaging
- Geomembrane films
- Heavy-duty shipping sacks

- Liners
- Retail bags
- Multilayer packaging film
- Pouch bags
- Produce bags
- Trash bags
- Stretch wrapping films
- Pallet stretch wrapping

The molecular structure of polyethylene used in film applications can vary over a very wide range as compared to the molecular structure of polyethylene used in other fabrication methods. For film applications the density and Melt Index (MI) of the polyethylene may vary from about 0.900–0.960 g/cc with MI ($I_{2.16 kg}$) values of about 0.05–2.0. In addition, the molecular weight distribution of polyethylene used in film applications may vary over a wide range from polydispersity values (M_w/M_n) of 2–20.

6.8   Blown Film Extrusion

6.8.1   Description of Blown Film Extrusion

Blown film extrusion is the most common method to fabricate polyethylene film. The process involves extruding molten polyethylene through a circular die and then inflating the interior of the molten polyethylene as it exits the die to form a bubble. The die thickness is about 20–100 times thicker than the final film thickness, so that the molten polyethylene is stretched during the time that the molten polyethylene exits the die and before the time the molten polyethylene cools to a solid. Important aspects of this process are a uniform film thickness, the time it takes to cool the molten, amorphous polyethylene to a semicrystalline solid and the molecular structure of the polyethylene used to fabricate the film. During this cooling process the volume of the polyethylene shrinks as the solid material has a higher density than the molten polyethylene.

6.8.2   History of Polyethylene Rapid Growth in Film Applications

In the 1950s, polyethylene film for food packaging became an enormously important growth area for the polyethylene industry. Polyethylene was able to replace other packaging materials such as paper and cellophane and open up new markets for food packaging such as fresh produce. There were several product attributes of polyethylene over other packaging materials that accounted for this rapid growth.

6.8.2.1   Cost

Although polyethylene cost about $0.55/lb in the mid1950s, a seemingly very high price which translates to about $3.50/lb in 2010 dollars, polyethylene was able to displace cellophane on a cost basis. On a volume basis, a 1000 in², one mil film of polyethylene cost about $0.018 while the same volume of cellophane cost $0.031, providing polyethylene with a 42% lower cost than cellophane. In addition, an improvement in fabrication techniques in the 1950s and the availability of higher density polyethylene that provided higher film stiffness with improved handling characteristics, allowed polyethylene film producers to reduce film thickness to 0.5 mil, offering additional packaging cost savings [24].

6.8.2.2   Shelf-Life

The advantages of packaging food in polyethylene film are a reduction in waste and spoilage due to a significant increase in shelf-life, convenience for the retailer in improved rates of handling and stocking food items, convenience for the consumer in easier handing and the opportunity for the food company to promote their brand with packaging labeling. Other advantages of prepackaged food are an increase in self-service while shopping, protection of food from damage while transported and, in the case of fruit, an acceleration of the ripening process.

6.8.3   Blown Film Apparatus

A schematic diagram of the blown-film-forming process is illustrated in Figure 6.15 and outlined in Table 6.4.

A modern day blown film process is shown in Figure 6.16. Based on the shape of the bubble as the molten polymer exits the die, this line is processing LLDPE. Advanced designs are utilized to cool and stabilize the bubble (referred to as a Bubble Cage) at high line rates to form films of a uniform thickness. Single to multilayer films are possible.

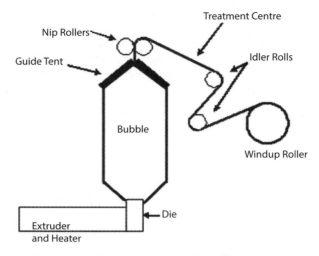

Figure 6.15 Simple diagram of a blow film apparatus.

6.8.4 Multilayer Films

Multilayer films are used as high-barrier structures to supply more advanced films for packaging perishable food such as cheese and meat. Three-, five-, seven- and nine-layer films are now possible. General Films, Inc. (Covington, OH) announced the start-up of a nine-layer film system in 2007 using a Battenfield Gloucester Engineering Co. Inc.(Gloucester, MA) film system. Multilayer films contain several different types of polymer material that each serve a specific function. A five-layer film contains three different plastics. A tie layer material such as nylon, an oxygen barrier layer (ethylene vinyl alcohol) and polyethylene (LLDPE) providing five layers, as shown below.

LLDPE
Tie Layer
Oxygen Barrier
Tie Layer
LLDPE

The LLDPE layers provide the bulk of the film body, while the tie layers adhere to both the LLDPE and the oxygen barrier polymer allowing the film to maintain a strong structural foundation. The oxygen barrier layer exhibits low permeability to oxygen to increase shelf-life of the package.

Table 6.4 Major components of the blow film apparatus.

Component	Description
Extruder	An extruder consists of a heated barrel containing a screw that turns to mix the molten polyethylene to form a homogeneous, amorphous material that passes through to an exit die.
Die/Air Source	The die is a circular apparatus with a slit around the circumference of the die where the molten polyethylene exits the extruder to begin the blown-film-forming step.
Bubble	An air source in the center of the die is used to form the bubble using internal air pressure and expand the molten polyethylene to a diameter much larger than the die. This process determines the width of the polyethylene film. Typically, the bubble expands to 3 to 4 times the diameter of the die. This expansion is referred to as the "blow-up" ratio and is the first step in reducing the thickness of the molten polyethylene as it exits the die. For example, molten polyethylene exiting a 40 mil die gap (0.040 inches) is reduced to a thickness of 10.0 mil for a blow-up ratio of 4. This reduction of the molten polyethylene thickness in the direction perpendicular to the direction the bubble is moving is referred to as the "Transverse Direction" (TD). The reduction of the molten polyethylene thickness is also achieved in the direction the bubble is moving, which is referred to as the "Machine Direction" (MD). This is carried out with the nip rollers discussed below. Bubble cooling may be carried out with a cooling collar around the exterior of the bubble and may also be cooled from the interior of the bubble. The "frost line" height is the point on the bubble where the molten polyethylene changes phase to a solid material, as indicated by a haze around the circumference of the bubble.
Guide Tent	As the polyethylene bubble passes under the guide tent the bubble is collapsed to a two-layer film.
Nip Rollers	The rotation rate of the nip rollers determines the final thickness of the polyethylene film. For example, if the nip rollers rotate at a rate ten times the rate the molten polyethylene exits the die, and then the molten polyethylene is reduced by an additional factor of ten. In our example above, the 10.0 mil thick polyethylene would be reduced to a final thickness of 1.0 mil. Reducing the thickness of the molten polyethylene in the direction parallel to the movement of the bubble is referred to as the "Machine Direction" (MD).
Treatment Center	The surface of polyethylene is often oxidized by an electrical discharge or plasma so that the film will accept dyes and be printable.
Finishing Rollers/ Windup Roller	These rollers are used to collect the finished "Lay Flat" film for storage or transfer to another process.

Figure 6.16 Modern day blow film apparatus.

6.8.5 Low-Density Polyethylene Films

The three types of low-density polyethylene used for blown film applications are high-pressure LDPE and LLDPE prepared with either a titanium-based Ziegler-type catalyst or a single-site catalyst with each type having a unique structure.

The LDPE has an intermediate MWD, uniform branching distribution and long-chain branching. The LLDPE prepared with a Ziegler-type catalyst has a narrow MWD and non-uniform branching distribution. The LLDPE prepared with a single-site catalyst has a uniform branching distribution and a very narrow MWD.

Some typical properties of these resins for film applications are summarized in Table 6.5.

Dart drop values, tear strength, clarity, heat seal and processability are important properties for film applications. Depending on the application, the film processor bases the selection of a particular type of LDPE/LLDPE resin on the properties required. For example, a garment bag cover for dry cleaning may require good film clarity where dart strength and

Table 6.5 Typical resins and properties for blown film applications (2.0 mil film).

Resin Type	LDPE	LLDPE	LLDPE
Catalyst Type	free radical	Ti/Ziegler	Ti/Single-site
Density (g/cc)	0.924	0.920	0.920
Melt Index ($I_{2.16}$)	0.75	1.0	0.85
Elongation to break (%)			
MD/TD	390/570	830/890	600/650
Dart Drop (g)	170	290	780
Elmendorf Tear (g)			
MD/TD	350/260	900/1200	710/1000
Relative Film Clarity	High	Low	High
Relative Heat Seal Temp.	Low	Higher	Low
Processability[a]	High	Intermediate	Low
Extrusion Melt Temp. (°C)	160–190	232	216

[a]Processability is proportional to MWD; LDPE has a FI/MI value of about 60, LLDPE Ziegler based FI/MI value of about 28 and single-site LLDPE has a FI/MI value of about 16. These are estimates by the author. Data not reported in Dow data sheets.

Source: Dow Chemical Company website.

tear strength values are less important. Therefore, LDPE may be the better resin for this application based on relatively high film fabrication rates and thin film gauge. On the other hand, a packaging application where film strength and excellent film elasticity are required would most likely require a LLDPE-type resin. Note the very high dart impact value of 780 g for the LLDPE produced with Dow's single-site catalyst. In addition, the resins with a relatively narrow MWD exhibit less film orientation.

6.8.6   LLDPE with a Broad MWD

Most LLDPE resins manufactured have a relatively narrow molecular weight distribution as indicated by Melt Flow Ratio values (MFR = FI/MI) of 16–30 or polydispersity values of 2–4. Films produced from this type of polyethylene have higher dart impact and tear properties as shown in the Table 6.5. However, LLDPE resin with a MWD broader than LDPE is used in geomembrane applications such as landfill covers and caps and other heavy-gauge film applications.

The LLDPE with a broad MWD can be produced in a single reactor with a Cr-based catalyst or in tandem reactors such as two gas-phase reactors operated in series such as the UNIPOL II process introduced in the early 1990s.

Chevron Phillips manufactures several broad MWD LLDPE resins in a slurry loop reactor using a Cr-based catalyst that generates 1-hexene, *in situ*, used as comonomer. Table 6.6 summarizes some property data for this type of polyethene film.

In landfill applications, this type of film needs to have a very high environmental stress crack resistance to withstand degradation by various waste components found in a landfill.

Using UNIPOL II process technology, Union Carbide introduced "easy flow" LLDPE products that had better processability than LLDPE with a narrow MWD and LDPE with an intermediate MWD. Union Carbide reported processability and product physical property data for these new resins [25]. Some of the data is summarized in Table 6.7 which compares a UNIPOL II "easyflow" resin with a high-pressure LDPE.

Examination of the data in Table 6.7 shows that the UNIPOL II resin processes with excellent bubble stability and similar output rate and head pressure as a LDPE resin with a lower Melt Index value. In addition, the UNIPOL II resin exhibited a higher modulus, dart-drop strength and higher puncture energy than the LDPE resin.

6.8.7   Blown Film Process for HMW-HDPE Film

Fabrication of high molecular weight, high-density polyethylene (HMW-HDPE) into 0.5 mil film, is primarily used for producing merchandise bags and grocery bags (designated as T-shirt bags) with high stiffness and durability. Premium bags in this market are made with polyethylene with a bimodal molecular weight distribution, although polyethylene with a unimodal, broad MWD can also be used but provide lower dart drop strength. Polyethylene for HMW-HDPE film applications has a HLMI value of about 6–9, an MI of about 0.05 to 0.08 and a density of about 0.950–0.955 g/cc.

Table 6.6 Typical specifications and properties for LLDPE with a broad MWD.

Property	Value
Density (g/cc)	0.922
Flow Index (21.6 kg)	15
Elongation to break (%)	750
ESCR (F-50) hours	>1,500

Source: Chevron-Phillips data sheet.

Table 6.7 Comparison of easy flow LLDPE with high-pressure LDPE [25].

Resin		LDPE (HP)	LLDPE (UNIPOL II)
Resin Properties			
Melt Index (I2.16)		0.9	1.4
Flow Index (I21.6)		58	125
FI/MI (MWD)		64	98
Density (g/cc)		0.924	0.924
Processability			
Melt Temp (°F)		408	400
Head Pressure (psi)		3400	3470
Resin Output Rate (lb/hr/rpm)		4.8	4.7
Bubble Stability (qualitative)		Very Good	Excellent
Physical Properties			
Tensile Strength (psi)	MD	3230	4880
	TD	2550	3430
Secant Modulus (psi)	MD	30,300	39,600
	TD	33,700	49,900
Tensile Impact (lb/ft3)	MD	445	1520
	TD	580	760
Elmendorf Tear (g/mil)	MD	184	136
	TD	262	852
Dart Drop (g/mil)		87	122
Puncture Energy (in-lb/mil)		8.2	21.1

Polyethylene with a bimodal MWD may be manufactured with two or more polymerization reactors in series, with two reactors the most common configuration. In the two reactor process, one reactor produces a polymer component with a relatively very high molecular weight (HLMI of about 0.2–0.3) and containing a small amount of branching by utilizing a comonomer such as 1-butene or 1-hexene. The second reactor produces a polymer component with a relatively very low molecular weight (MI of about 1,000 or higher) with this polymer component an ethylene homopolymer. Tandem reactors may be either both gas phase (UNIPOL II) or both slurry (Mitsui) or one gas phase and one slurry (Borstar).

However, in the early 1990s, Mobil Chemical Company scientists developed a bimetallic catalyst based on combining a Ziegler-type Ti-containing catalyst with a zirconocene/methylalumoxane-based single-site catalyst that produced HMW-HDPE with a bimodal MWD in a single gas-phase reactor. Recently, Chevron Phillips reported a dual single-site catalyst that also produces a bimodal MWD, HMW-HDPE resin in a single slurry-loop reactor. Details of these catalysts are discussed in Chapter 4 of this book.

6.8.8   High-Stalk Extrusion

An unusual feature of fabricating HMW-HDPE into film is the length the molten polyethylene travels during the film-forming process before the bubble is inflated. The bubble shape is illustrated in Figure 6.17 and is referred to as "high-stalk" extrusion. This extrusion shape is used in order to produce film with more balanced orientation (more balanced properties in the MD and TD direction). After the melt exits the die, the melt is drawn down by about a factor of 20 in the machine direction before the bubble is inflated with a blow-up ratio of 4, i.e., the diameter of the bubble is four times the diameter of the die. Hence, the melt is drawn down by a factor of 80. Consequently, a final film thickness of 0.5 mils is obtained with a die

Figure 6.17 Illustration of high-stalk extrusion of HMW-HDPE polyethylene.

gap of 40 mils. Note that the broad MWD and high molecular weight of the polyethylene resin used for HDPE film are molecular features that provide sufficient melt strength for this type of extrusion to be used.

Table 6.8 summarizes the properties of the polyethylene used for the manufacture of high-density polyethylene film, the processing conditions and the film toughness properties of the finished film, which include the dart impact strength and tear strength.

The machine direction tear strength of the film is relatively low, so that a HMW-HDPE grocery bag made from a resin that increases the MD tear value would provide a significant advantage in the marketplace.

6.8.9 Cast Film Line

The cast film process was developed in the late 1940s primarily by scientists at ICI, DuPont and Union Carbide. An early description of the cast film process was published in British Patents 474,426 (ICI) and 561,373 (DuPont). The cast film concept filed on June 26, 1948, and published as U.S. Patent 2,586,820, issued on February 26, 1952, to W.F. Hemperly *et al.* and assigned to Union Carbide is shown in Figure 6.18.

Table 6.8 Typical properties of premium HMW-HDPE film prepared with two reactor slurry polymerization process.

Flow Index	**8.0**
Melt Index	0.05
Flow Index/Melt Index	160
Density (g/cc)	0.950
Melt Pressure, psi	6450
Bubble Stability	Very Good
Film Quality	High – low gels
Die Gap	40 mil
Extrusion Rate (lb /hr)	120
Blow Up Ratio	4
Stalk Height; (inches)	28
Film Thickness (mil)	0.5
Dart Drop (F 50%) g	365
Elmendorf Tear (g/mil)MD	25

Source: U.S. Patent 5,539,076; Mitsui Resin; 50 mm Alpine Film Line.

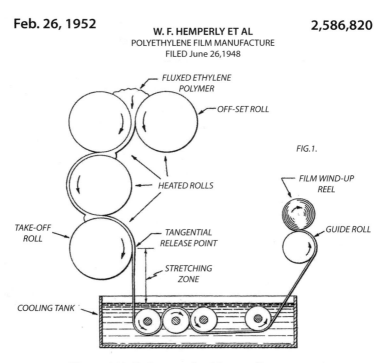

Figure 6.18 Early example of the cast film process.

The drawing illustrates the important features necessary for the cast film process which include:(a) the melting of the polyethylene onto a heated roll, (b) a "stretching zone" where the molten polyethylene film thickness is reduced to a thinner gauge by moving the take-up roll at a faster rate than the take-off roll, and (c) the cooling tank where the polyethylene material solidifies to a thin film.

An early description of the cast film process was published in *Modern Plastics* in 1952 [26] which was very similar to the cast-film scheme shown in Figure 6.19 which represents the process used today. The cast film process involves the extrusion of polyethylene through a die to form a thin molten layer of material that is drawn down to a thinner gauge based on the difference between the rate the molten polymer exits the extruder and the rate that the molten polymer is cooled onto a chill roll. The film cools very rapidly on the chill roll and film orientation is only in the machine direction. Optical properties are usually better with cast film as compared to blown film and line rates may be higher with the cast film process. A cast film line may also involve coextrusion of several layers of different types of thermoplastics to fabricate specialty films with improved properties. A

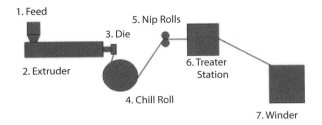

Figure 6.19 Schematic of cast film process.

Table 6.9 Polyethylene used in the cast film process.

Polyethylene Type[a]	Melt Index $(I_{2 \cdot 16}\,kg)$	MWD[b]	Density (g/cc)
LLDPE – single site	0.5	Very Narrow	0.903
LLDPE – single site	4.0	Very Narrow	0.904
LLDPE	1–6	Narrow	0.917–0.941
LDPE	2–6.4	Intermediate	0.922–0.928
HDPE	4.4	Narrow	0.952

[a]Data from Dow Chemical Company.

[b]Narrow MWD with polydispersity value of approximately 3.5–4.0; intermediate MWD from high pressure process.

schematic of the cast film process is illustrated in Figure 6.19. Some resin specifications for polethylene used in the cast film process are summarized in Table 6.9.

Markets for cast film are: (a) stretch film that is used in pallet wrapping and other types of shipping to secure the load, (b) high clarity films, and (c) more recently cast film is used in fabricating premium-grade trash bags.

6.8.10 Pipe Applications

Polyethylene for pipe applications has a relatively high molecular weight and a broad molecular weight distribution. Some premium grades have a bimodal MWD. Table 6.10 summarizes the product specifications for several suppliers of pipe-grade polyethylene for various applications.

Polyethylene pipe systems are an environmentally responsible material with significant cost and life-expectancy advantages over other materials such as metal and concrete. These advantages are due to the superior properties of polyethylene that include physical properties such as flexibility,

Table 6.10 Product specifications for several suppliers of pipe-grade polyethylene.

Source[a]	Application	MI[b]	HLMI[c]	Density (g/cc)
Chevron Phillips	pressure pipe	---	8.0	0.945
Chevron Phillips	corrugated pipe	---	21.0	0.953
Dow Chemical	sewage	0.08	8.5	0.945
Dow Chemical	natural gas	0.16	9.5	0.941
ExxonMobil	drainage	0.25	26	0.953

[a]Data from each company website to represent various Grades available for certain applications.

[b]MI is $I_{2.16kg}$.

[c]HLMI, High Load Melt Index from $I_{21.6kg}$.

chemical inertness, light weight and toughness. In addition, installation advantages such as longer length pipe and leak-proof joints lower installation costs, making polyethylene a preferred material for a wide variety of pipe applications. Pipe applications provide products for aboveground, surface, buried and floating or submerged marine applications.

Pipe applications include the transportation of drinking water, irrigation water, sewer, and drainage, industrial chemicals as liquids or gases, and natural gas. It is important to note that polyethylene pipe has a history of conveying natural gas and liquid oil-based materials with an improved safety record. For example, natural gas pipelines in earthquake-prone regions such as Japan have experienced much fewer failures after earthquakes, providing an important safety incentive for future growth.

Pipe diameters vary over a very wide range, from a few millimeters diameter for tubing to diameters of five feet or greater for liquid transportation applications.

A pipe application that has shown strong growth in recent years is crosslinked polyethylene for residential plumbing systems. These systems offer lower installation cost compared with copper piping, and some unique advantages such as less wait-time for hot water to reach the point of use, which also lowers energy usage for the hot water system.

Borealis is a major supplier of polyethylene resins for crosslinking polyethylene pipe applications under the tradename BorPEX, offering resins with a HLMI (MI $_{21.6\,kg}$) between about 3–10 over densities of 0.944–0.955 g/cc. AkzoNobel supplies several organic peroxides for polyethylene crosslinking reaction under the tradename Trigonox. With organic peroxides, the crosslinking reaction takes place while the polyethylene is in the molten state during the pipe extrusion step, which may be carried out using twin-screw extruders for better melt-mixing characteristics. Although other methods are available for crosslinking the pipe after the extrusion step such as using infrared heaters, and this technology is available from Crosslink Finland Oy under the tradename Unipex [27].

6.9   Fabrication of Polyethylene with Molding Methods

Several molding methods are used to fabricate polyethylene products and these methods are blow molding, injection molding, rotational molding and thermoforming.

6.9.1   Blow Molding

Details into the blow molding process have been recently summarized by Norman C. Lee [28] in which the engineering aspects of the process are primarily discussed. This section will briefly summarize the key components of the blow molding process, but will focus more on the polymer properties necessary for the blow molding process to provide high performance products.

6.9.1.1   Brief History of Blow Molding (ca. 1850–1960)

The process of blow molding goes back thousands of years to the skilled craftsmen that molded glass into a wide variety of shapes where the "blowing" was provided by the craftsperson's lungs as he molded the molten glass.

However, the invention of a machine for performing the blow molding process, as pointed out by Lee, was disclosed in a United States patent for blow molding natural latex, and was issued in the 1850s to Samuel Armstrong.(See ref. [29] for a brief history of blow molding ca. 1850–1960).

An improved design was disclosed in U.S. Patent 237,168 issued to W. B. Carpenter on February 1, 1881, and assigned to the Celluloid Manufacturing Company, and was the first patent for the processing of an extruded molten plastic material into a parison for blow molding. A parison is a hollow molten tube of plastic that is formed by allowing the molten plastic to pass through a circular die. Carpenter described the "molding of hollow forms with celluloid or like plastic material," where the parison was referred to as a "blank tube." The blank tube was enclosed in the mold prior to the blowing process in which the molten plastic material is pushed

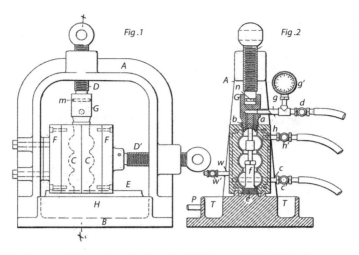

Figure 6.20 Blow molding design disclosed in U.S. Patent 237,168. Drawing provides two views of the apparatus.

against the surface of the mold during the blowing process. In addition, the mold contained a jacket for the circulation of cooling fluids and also allowed for venting after the blow molding step was carried out.

The diagram of this blow molding design patented by Carpenter is shown in Figure 6.20. Note that the drawing on the left more clearly shows the two halves of the mold labeled "C" in a closed orientation.

Blow molding applications for plastics increased rapidly in the 1940s with the development of low-density polyethylene manufactured by a high-pressure process. The LDPE offered greatly improved product attributes over glass, metal and other plastic material used in the blow molding process. Polyethylene offered a non-breakable container that was a significant advantage over glass and was an inert container for liquids such as inorganic acids and bases that were corrosive to metal or handled much more safely in a plastic container because of the non-breakable attribute of polyethylene. In many cases polyethylene also had a significant cost advantage over other materials.

Some corrosive chemical solutions which are inert to polyethylene that were packaged in blow-molded containers using LDPE are listed below and were some of the first applications for high-pressure LDPE.

- Ammonium hydroxide
- Chromic acid
- Calcium hypochlorite

- Hydrogen peroxide
- Nitric acid
- Potassium hydroxide
- Sea water
- Phosphoric acid
- Sulfuric acid
- Hydrofluoric acid

Of course, there were organic chemicals that attacked LDPE produced by the high-pressure process in the 1950s that prevented blow-molded containers from entering many applications. For example, gasoline, lubrication oil, acetone, ethyl acetate and ethyl ether were listed in the 1950s as being organic liquids that attack polyethylene [30], but various grades of polyethylene are now available that are very inert to such organic liquids. The molecular structure of polyethylene resins available today, as specialty grades of polyethylene, have been specifically formulated to pass a variety of ASTM testing to improve shelf-life. For example, improved polyethylene resins are available today which are suitable containers for motor oil, gasoline, soaps and many other organic-based liquids, which were not available in the 1950s. For blow molding applications these markets are referred to as HIC (household industrial chemicals) applications.

Plax Corporation (later purchased by Monsanto Chemical Company) developed the first blow molding machine in 1931 and in the 1940s was also the first company to manufacture the first LDPE squeeze-bottle (containing an underarm deodorant named "Stoppette") using blow molding [29]. In this application, the polyethylene bottle served two roles, as a container and as a dispenser [31].

In 1950, Continental Can Company received the first patent for a continuous extrusion blow molding machine. The term continuous applies to the constant extrusion of the parison as compared to an intermittent parison forming process. Plax Company and Continental Can Company owned a strong patent position on the blow molding process until about 1960. After the patents expired, Zarn Company pioneered the first HDPE milk bottle for Borden Dairy in North Carolina using Uniloy machines [29].

6.9.1.2   Environmental Stress Crack Resistance

Environmental stress crack resistance (ESCR) is an important property for many applications because it estimates the service life of polyethylene in any particular application. As pointed out by A. Lustiger [32], the term "Environmental Stress Cracking" was defined in 1959 by J. B. Howard to describe the failure of polyethylene bottles containing hydrofluoric acid and polyethylene cables that were exposed to methanol. Howard's definition of environmental stress cracking is "failure in surface-initiated brittle fraction of a polyethylene specimen or part under polyaxial stress in contact with a medium in the absence of which fracture does not occur under the same conditions of stress."

However, tests for determining the cracking sensitivity of polyethylene that was exposed to liquids such as detergents were reported earlier in 1951 by J. B. DeCoste [33] in which a razorcut was made on molded samples, which were then bent in the direction of the razor cut, after which the specimen sample was placed into a test tube containing a "cracking agent" which was a detergent or some other liquid under consideration. Tests were usually carried out at 50°C. This early work by DeCoste was the basis for the ASTM "bent strip" test to determine the stress-crack resistance of polyethylene samples, where the time for one-half of the samples to fail was noted as the stress-crack resistance. The ASTM test that was under development in 1955 used IGEPAL CO-630 obtained from General Dyestuff Corp., where IGEPAL was described as an alkyl, aryl polyethylene glycol which was the stress-cracking reagent.

The addition of crosslinking agents such as organic peroxides to polyethylene has also provided a method to alter the polyethylene structure during the fabrication process in order to provide a material that improves ESCR for many applications. Crosslinked polyethylene is an excellent container for gasoline and has completely replaced the metal-based gasoline container that was common many years ago.

6.9.1.3   Types of Blow Molding Machines

There are three basic types of blow molding machines which are shown in Table 6.11.

The details of each type of machine are very clearly described in Lee's text [28] and the reader is directed to this book for additional information.

Table 6.11 Types of blow molding machines.

Machine Type	Comments
Injection blow molding	Smaller bottles, injection molded neck, no scrap
Extrusion blow molding	Most common method; bottles from 0.2 liter > 300 gallon
	Can fabricate bottles with handles
Stretch blow molding	Intermediate size bottles; 0.5–2 liter

However, a diagram of a typical extrusion blow molding machine is illustrated in Figure 6.21, with the important features of the parison as the molten polyethylene exits the die also shown.

Important aspects of the blow molding process are as follows:

a. As the molten polyethylene (noted as the dark area in both drawings in Figure 6.21) exits the extrusion process, the polymer is redirected 90° from a horizontal to a vertical flow. It is important that the molten polymer flows to form a parison with uniform wall thickness. The parison is a hollow, pipe-shaped mass of molten polyethylene. The parison must flow slightly further than the bottom of the water-cooled mold. The mold closes and "blow air" is directed into the center of the parison, and the molten polymer is expanded to fill the surface of the mold. The polymer cools rapidly and the mold opens and the molded article is ejected from the mold.

b. The proper formation of the parison is the critical feature of the blow molding process. The molten polymer that forms the parison requires melt strength to avoid "sag," which is a parison that is thinner at the top than the bottom. In addition, parison swell needs to be within certain limits as the amount of swell correlates with the amount of resin required to form the fabricated article (i.e., bottle weight). Parison swell or die swell is due to an increase in the thickness of the molten polymer as the polymer exits the die. The parison wall exceeds the thickness

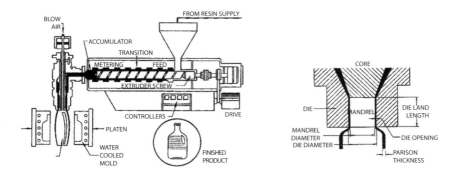

Extrusion Blow Molding Macine   **Parison Formation**

Figure 6.21 Extrusion blow molding machine and parison formation. Reprinted from [28] with permission from Carl Hanser Verlag Munich.

of the die. In the laboratory, a Melt Index machine is used to measure relative die swell in a particular sample of HDPE. This measurement may be designated as "annular die swell" (ADS).

The HDPE prepared with the Cr-based Phillips catalyst is the primary resin used in blow molding to manufacture various types of bottles. The two primary HDPE bottle applications are the milk bottle resin and the household industrial chemical (HIC) resin, with typical resin specifications summarized in Table 6.12.

6.9.1.4   Method to Decrease Die Swell

Modifying the molecular structure of HDPE in the extruder using free-radical initiators such as peroxides can reduce die swell (bottle weight) and broaden the MWD of the base resin to provide faster bottle fabrication rates [34,35]. The addition of 100 ppm of an organic peroxide to HDPE prepared with a Phillips catalyst lowered bottle weight 4.5%, as shown in the data in Table 6.13.

The addition of air (40–160 ppm based on resin weight) to the extruder has also been reported [36] to modify the molecular structure of HDPE resin with either a unimodal or bimodal molecular weight distribution. The

Table 6.12 Primary blow molding bottle applications.

Resin	MI	FI	FI/MI	Density	Key Product Requirements
Milk Bottle	1.0	80.0	80	0.962 g/cc	stiffness/no odor[a]
HIC Bottle	0.5	40	80	0.954 g/cc	ESCR/shelf life[b]

[a] Ethylene homopolymer to increase stiffness and reduce odor.

[b] Environmental stress crack resistance (ESCR) is a measure of the shelf life of the container and contents. Liquids such as oil, laundry detergent and soaps require high ESCR.

Table 6.13 Peroxide-modified HDPE [35].

Sample	PPM Peroxide	MI ($I_{2.16}$ kg)	FI ($I_{21.6}$ kg)	FI/MI[a]	ADS[b]	Bottle Weight (grams)
Control	none	0.56	44	80	0.875	31.1
Experimental	100 ppm	0.27	33	123	0.845	29.7

[a] FI/MI is directly proportional to the molecular weight distribution of the polyethylene sample. Higher FI/MI values indicate the polyethylene will experience a greater degree of shear thinning (lower apparent melt viscosity) as the polymer is extruded under shear.

[b] Annular die swell is equal to the (thickness of melt index strand – die diameter)/die diameter.

Table 6.14 Addition of air to the extruder to modify the molecular structure of HDPE [36].

	Control Sample	*Air Modified Sample*
Air	none	160 ppm
MI	0.42	0.21
FI	39.4	32
FI/MI	93	156
Parison Drop Time (sec)	1.25	1.10
Extruder Pressure (psi)	2227	2,180

air-modified resin has improved processability as indicated by an increase in Melt Flow Ratio (FI/MI), better parison melt strength and lower die swell. This data is summarized in Table 6.14.

The weight and diameter of the parison is controlled by the geometry of the annular die opening and the resin being extruded. As the molten polymer is extruded through the die, it swells to a thickness and diameter that is greater than the dimensions of the die opening. The amount of swell is dictated by a complex relationship between the internal shape of the die and the viscoelastic properties of the polymer.

In order to sell a product in many of the segments of the blow molding market, it must match the swell characteristics of the established products in that market. Small-scale swell tests have been used by some HDPE resin producers to develop new blow molding grades as well as for quality control of their existing grades. In order to be effective predictors of performance, the small-scale tests must operate at a temperature and shear rate similar to that normally found in the customer's equipment. An annular die can give information on weight and diameter swell; a capillary die will not be able to distinguish between the two.

Small-scale tests are generally best for process quality control. When developing a product using a new catalyst or a new process, the small-scale tests can only provide general guidelines that must be periodically field tested to ensure the desired performance.

6.9.1.5   Milk Bottle Resin

Blow-molded bottles using LDPE were partly responsible for the rapid growth of polyethylene in the late 1940s and 1950s. However, with the introduction of high-density polyethylene in the late 1950s, a high stiffness resin suitable for the plastic milk bottle was introduced.

The blow-molded milk bottle resin produced with the Phillips Cr-based catalyst is an ethylene homopolymer with a density of 0.964 g/cc and Melt Index (MI) of 0.7 and Flow Index of approximately 70–80 (Chevron Phillips Marlex Grade EHM 6007). The ethylene homopolymer provides a relatively high modulus and reduces the possibility of odor problems due to ethylene/1-olefin oligomers in the resin. The high density of this resin is made possible for packaging water-based liquids such as milk, fruit juices and drinking water because these liquids are extremely inert towards degrading polyethylene. A container with such a high density would be unsuitable for packaging soaps and oils where high ESCR is needed.

Table 6.15 provides the amount of HDPE used in various types of fabrication applications. Although this data was limited to the United States, the trend would qualitatively apply to the global market.

6.9.2   Injection Molding

6.9.2.1   Introduction

Injection molding is unique compared to the other fabrication techniques used in the polyethylene industry because of the relatively low molecular

Table 6.15 Usage of HDPE by application in the United States.

Market	Billion Pounds Annually	Percent Market Share
Extrusion		
Film (up to 12 mils)	2.33	15.6%
Sheet (over 12 mils)	0.77	5.1%
Pipe and conduit, corrugated	0.66	4.3%
Pipe and conduit, noncurrogated	1.28	8.5%
Other extruded products	0.47	3.1%
Rotomolding	0.15	1.0%
Injection Molding	2.37	15.7%
Blow Molding	4.90	32.5%
Resellers and Compounders	1.87	12.4%
All other uses	0.27	1.8%
Total	15.06	100.0%

weight and relatively low viscosity of the molten polyethylene used in the process. The low viscosity is required because the molten polyethylene needs to flow into all regions of the mold to completely fill the mold to acquire the necessary shape. The time to fill the mold decreases with viscosity. In addition, polyethylene with a narrow molecular weight distribution reduces polymer stress in the molten state and decreases melt relaxation times to reduce residual stress in the finished molded part, reducing shrinkage and warpage problems.

Although a wide variety of other thermoplastics are used in the injection molding process, polyethylene manufacturers provide the process engineer with a large number of different grades over a very wide range of molecular weight and density, so that the physical properties of the finished article such as flexibility, modulus (stiffness) and tensile strength may be easily modified. Several types of polyethylene are used today in injection molding applications, including the standard grades of LDPE, LLDPE and HDPE, and more recently developed products that were identified from single-site ethylene polymerization catalysts that include lower density, highly elastic, flexible grades of polyethylene. Some typical product data for injection molding applications are summarized in Table 6.16.

6.9.2.2   Polyethylene Shrinkage

Polyethylene undergoes considerable shrinkage during the solidification process. The density of molten polyethylene is about 0.7 to 0.8 g/cc, depending on melt temperature. In injection molding, the fabricated part is cooled as rapidly as possible in order to obtain fast production rates, so the polyethylene undergoes a large amount of shrinkage in a short time

Table 6.16 Various grades of polyethylene used in injection molding applications.

Polyethylene Type[a]	Melt Index ($I_{2.16}$ kg)	Density Range (g/cc)[c]	Molecular Weight Distribution
HDPE	2–70[b]	0.940–0.965	Narrow
LDPE	2–60	0.918–0.923	Intermediate
LLDPE	20–160	0.917–0.933	Narrow

[a] Molecular weight distributions as indicated by a polydispersity value of 3–5 for HDPE and LLDPE.

[b] Most applications require Melt Index in the 5–20 range.

[c] Density is a function of polyethylene molecular weight. A high Melt Index polyethylene at a particular density will contain less comonomer than a one Melt Index polyethylene at the same density due to faster crystallization rates.

during the solidification process. Consequently, fast melt relaxation times are required in order to prevent warpage in the finished part. Therefore, polyethylene with a relatively low molecular weight and a narrow molecular weight distribution provides relatively faster melt relaxation times to prevent part warpage in the injection molding process, and are therefore the preferred type of polyethylene in injection molding applications.

In addition, the amount of melt shrinkage varies with the density of the polyethylene used in the product application. Injection molding with HDPE is more difficult than with LDPE or LLDPE because the amount of shrinkage that occurs during the solidification process is higher in the case of HDPE. For example, the difference in the density of the molten polyethylene and the density in the finished product is larger if HDPE is used in injection molding, i.e., 0.96 g/cc in the solid part and about 0.75 g/cc in the molten phase or a difference of 0.21 g/cc. For LLDPE, the shrinkage during the solidification process is less, i.e., 0.92 g/cc in the solid part and 0.75 g/cc in the molten phase or a difference of 0.17 g/cc.

6.9.2.3   New Product Applications

As of 2010, injection molding applications are being developed for polyethylene prepared with single-site catalysts. Product development scientists at Borealis have reported [37] a LLDPE with a Melt Index of 30–50 and density of 0.921–0.926 g/cc with improved resistance to cooking oils compared to a similar product based on a polyethylene prepared with a multi-site Ziegler catalyst. The molecular structure of the polyethylene prepared with the single-site catalyst has a narrower molecular weight distribution and a more homogeneous branching distribution relative to the polyethylene prepared with the Ziegler catalyst.

Blends of two samples of polyethylene in which each blend component was prepared with a single-site catalyst has also been reported [38] to provide an injection molding grade of HDPE with greatly improved environmental stress crack resistance (ESCR). A commercial grade of HDPE suitable for injection molding and prepared with a Ziegler catalyst was used as a control material. This data is shown in Table 6.17.

6.9.2.4   History of Injection Molding Process

Several books published on injection molding are an excellent source into the growth of the injection molding process over the years. Some references covering different periods of the injection molding process are summarized in Table 6.18.

Table 6.17 Blends of two samples of polyethylene.

Sample	Wt%	Melt Index ($I_{2.16}$ kg)	Density	Mw/Mn	ESCR
Blend-component-1	30	0.45	0.919	2.59	
Blend-component-2	70	56.6	0.970		
Final Sample	100	4.8	0.955	3.8	>605 hours
Commercial Sample	100	6.5	0.952	3.6	4.5

Table 6.18 Textbooks published on the injection molding process at three different periods.

Publication Date	Authors/Title/Publisher
1972	Irvin I. Rubin, *Injection Molding Theory and Practice*, John Wiley and Sons
1987	Joseph B. Dym, Injection Molds and Molding: A Practical Manual, Van Nostrand Reinhold
1994	Herbert Rees, *Injection Molding Technology*, Hanser Publishers
2007	Tim A. Osswald, Lih-Sheng Turng and Paul Gramann, *Injection Molding Handbook*, Hanser Publishers

The first thermoplastic injection molding application is credited to John Hyatt in 1868, who received U.S. Patent 133,229 on November 19, 1872. Hyatt molded celluloid into billiard balls.

British Patents 471,590 and 499,333 issued January 23, 1939, to E. W. Fawcett, R. O. Gibson and coworkers, and assigned to Imperial Chemical Industries (ICI) described early methods to injection mold polyethylene. A more detailed description of an injection molding machine was described in British Patent 567,375, issued to E. Shipton and W.N. Hill on February 12, 1945, in which the importance of the mold temperature was investigated and remains a critical parameter today in the injection molding process. A mold temperature was below the softening point of the polyethylene to permit the article to harden in a reasonable time, but a mold that is too cold will result in strains developing (e.g., warpage and shrinkage) in the finished molded part.

A publication by A. D. Ferguson [39] describes the combination of the screw feed used in the extrusion of molten polyethylene with the injection molding process, which was the basis of the present day injection molding process.

A diagram of the injection molding process is shown in Figure 6.22. The extruder is similar to other fabrication techniques where the granular resin enters a hopper, where the material is transferred to a heated barrel and a reciprocating screw moves the molten plastic to the molding cavity. The molding cavity is unique to this process. The molten polyethylene is forced under pressure into a closed mold that is continually cooled. After the molded part cools, the mold opens and the fabricated part is ejected from the mold. An injection molded article can usually be identified by the ejector pin marks that are usually present on the molded part.

Similar to most other fabrication techniques, the introduction of LDPE just prior to World War II created rapid growth in the process. The design shown in Figure 6.22 utilizing a screw injection process was built in 1946 by James Watson Hendry and is essentially the basic design in use today.

6.9.2.5   Some Aspects of the Machine Design

The melting of the polyethylene is primarily accomplished by the turning of the screw which converts mechanical (frictional) energy into heat, but the melting process is also aided by heating bands around the extruder barrel.

Although the extrusion process that transports the molten polyethylene through the extruder barrel exerts some pressure on the plastic, this pressure is too low for the injection process, hence an injection process is required to fill the mold.

The primary difference between injection molding machines is in the design of the injection process. In a plunger-type design, a predetermined

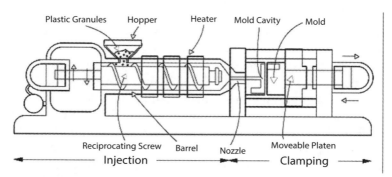

Figure 6.22 Injection molding machine. Source: www.en.wikipedia.org/wiki/injection_molding.

amount (shot) of molten polyethylene is injected into the mold with a piston.

Another injection process is used with reciprocating screw injection molding machines which were introduced in the mid-1950s in Europe and in the early 1960s in North America [40]. In this design, the screw rotates but also moves backward because of low pressure exerted by the hydraulic cylinder, allowing the molten polyethylene to accumulate in the end of the barrel near the mold. The amount of backward movement of the screw determines the amount of plastic that will be injected into the mold. The screw is forced forward, acting like a piston, by increasing the pressure exerted by the hydraulic fluid in the injection cylinder at the rear of the machine. A diagram of this type of machine is shown in Figure 6.23. Note the injection cylinder at the rear of the machine which povides the force to fill the mold with molten polyethylene.

6.9.2.6   Mold Design

Proper mold design is the most important aspect of a successful injection molding application to produce a high quality product. Elimination of shrinkage and warpage in the molded article is the primary quality objective. As shown in Figure 6.24, the injection molding process may produce very complex parts.

It is important that a mold can be easily opened and the manufactured part easily ejected from the mold. Ejection pins are built into the mold to

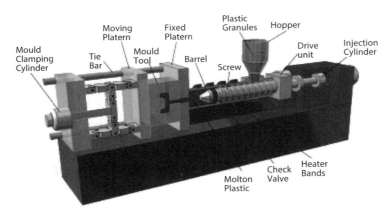

Figure 6.23 Diagram of an injection molding machine. Source: www.rutlandplastics. co.uk/ moulding_machine.shtml.

Figure 6.24 Examples of complex parts fabricated with injection molding. Source: www. engineeringindustries.com. Company: Engineering Industries Inc., 407 South Nine Mound Road, Verona, Wisconsin, USA.

facilitate the removal of the molded item. Some important features of the mold design are listed below.

a. Provide a route for the molten polyethylene to flow from the screw barrel to the mold cavity.

b. Allow air to escape from the mold as the mold is filled.

c. Minimize wall thickness and when possible design molds with uniform wall thickness. If changes in wall thickness are necessary, the thicker walls need to be introduced gradually to eliminate intersecting walls of non-uniform thickness. Wall thickness for polyethylene is usually in the range of 0.03–0.20 inches.

d. Corners need to be round to eliminate sharp corners. Sharp corners interfere with the flow of the plastic as the mold fills and introduce stress concentration in the molded part.

e. Avoid designs with large flat surfaces. Use ribs to increase the bending stiffness. A rib is a perpendicular support wall added to a flat surface. The perpendicular wall intersects the flat surface with a rounded corner.

f. Deep shapes need to be tapered to facilitate part removal during the ejection step.

g. Mold temperature control is necessary for high quality parts. Mold temperature control requires a circulating cooling fluid and cooling channels in the mold in order to remove heat from the molten polyethylene at a controlled but rapid rate. The temperature difference between the molten polyethylene and the cooling fluid determines the rate of heat removal.

6.10   Rotational Molding

6.10.1   Background History

The use of biaxial rotation and heat for materials fabrication dates to R. Peters of Britain in 1855 for the manufacture of hollow artillery shells [41,42]. Rotational molding of polyethylene started in the early 1950s.

Rotational molding is a low-cost fabrication method for the manufacture of hollow parts carried out by simultaneously heating and rotating a mold containing polyethylene powder [41,42]. The heat melts the polyethylene powder to a molten state allowing the polyethylene to coat the interior of the mold without applying internal pressure to the mold as in injection and blow molding. Rotational molding can produce extremely large parts such as storage containers ranging in size from about 5 to greater than 20,000 gallons, and wall thickness up to about one inch. Rotation around two perpendicular axes (biaxial rotation) is required to achieve a uniform thickness within the entire mold. The rotational molding process is unique compared to other molding methods because:

a. A polyethylene powder is added to the mold to facilitate rapid molding;
b. The powder is melted within the mold;
c. Usually no scrap material to recycle;
d. Lower cost mold because of less severe molding conditions;
e. Finished parts have essentially no internal stress because polyethylene melt is not subjected to shear during molding process.

Rotational molding is especially useful for the manufacture of very large parts and containers. For example, container volume is limited in the blow-molding process by the size and weight of the parison which is formed during the process. A very heavy parison is subject to sagging before the application of the blowing step, which will cause a non-uniform part thickness.

Some of the important variables that need to be carefully monitored during the rotational molding process are:

a. *Mold Design*

Molds need to provide a smooth, nonporous surface with good heat conductivity to allow for controlled heating and cooling. Two-piece molds manufactured from cast aluminum or sheet-metal are common.

b. *Resin Particle Size and Particle Size Distribution*

Polyethylene is usually ground to a particle size of 35 mesh (<0.0197 inch or 0.50 mm) to provide a particle size suitable for rotational molding.

c. *Oven Temperature/Heating Time*

The molding may take place either in an oven that contains the entire mold or the mold may be directly heated with hot air from an external flame as the mold rotates during the molding process. Oven temperature controls the rate at which the polyethylene powder is completely melted and the flow ability (viscosity) of the molten polyethylene contained in the mold to create a uniform distribution of the polyethylene. Excessive heating during the molding step must be avoided to avoid polyethylene degradation and loss of mechanical properties such as impact strength of the finished part.

d. *Cooling Rate*

Although the rotational molding process is carried out with very little shear on the polymer melt, the cooling rate of the polyethylene after the fabrication step remains important in order to avoid excessive shrinkage and warpage. This is due to the crystallization process that takes place as the polymer melt cools from an amorphous, less dense state to a solid, semicrystalline state, higher density solid.

e. *Resin Properties (Zero-Shear Viscosity)*

The viscosity of the polyethylene material at very low shear-rate is an important parameter for the type of polyethylene used in the rotational molding process because of the very low shear rate that is present during the molding process. This property is referred to as zero-shear viscosity of the resin.

Table 6.19 shows some typical resin specifications for polyethylene suitable for the rotational molding process. The relatively higher Melt Index

Table 6.19 Commercial grades of polyethylene for rotational molding applications.

Company/Grade	MI ($I_{2.16}$ kg)	Density (g/cc)
ChevronPhillips	2.0	0.943
	3.0	0.939
	6.0	0.936/0.945
LyondellBasell/Lupolen	4.0	0.9395
	7.5	0.9355

Note: Resins in the table above represent a few examples of rotational molding grades of polyethylene available in 2012. Data obtained from the company website accessed in March, 2012.

values provide lower viscosity of the molten polyethylene to aid in the fabrication step.

6.11   Thermoforming

Thermoforming is a fabrication technique that first requires the polyethylene sample to be converted into a flat film or sheet of variable thickness. Once this intermediate film/sheet is formed, the thermoforming step takes place by heating the film/sheet to a softening point within a mold that replicates the shape of the final product. A vacuum within the molding step is usually utilized to evacuate trapped air from the mold and to force the softened polyethylene sheet into the contours of the mold.

Thermoforming is commonly used to fabricate disposable cups, containers, lids, trays and other items used in the food, medical and retail packaging applications. Thick gauge polyethylene sheets may be used to thermoform very large parts such as refrigerator panels or liners.

6.11.1   Thin- and Thick-Gauge Thermoforming

A film/sheet thickness of less than 1.5 mm (0.060 inches or 60 mils) is referred to as thin-gauge thermoforming and is usually used in rigid or disposable packaging applications. At this thickness level, the polyethylene may be transferred to the molding cavity from a roll or pre-cut sheet. Thicker gauges above 1.5 mm (up to about 0.250 inches) are handled as sheets which are pre-cut to fit a particular molding application.

Three different methods (or a combination of methods) are used to form the polyethylene into the mold cavity:

1. *Vacuum method*
   The vacuum is formed between the polyethylene sheet and the mold cavity so that the softened polyethylene is forced against the walls of the cavity to the designed configuration.
2. *Pressure method*
   In this method, a vacuum is applied to one-half of the cavity between the polyethylene sheet and the lower half of the mold, while air-pressure is simultaneously applied to the upper part of the cavity between the polyethylene sheet and the cavity wall.
3. *Mechanical method*
   This method involves forcing the polyethylene sheet into the cavity by using a male/female mold design in which the male portion of the mold pushes the polyethylene into the female-shaped cavity.

Thick-sheet thermoforming can be used to produce hollow parts that are more difficult to fabricate using a blow molding process. In this method, referred to as "twin-sheet" thermoforming, two sheets are fed to the open mold so that the sheets remain separated from each other with a blow-pin placed between the two sheets. The mold is closed and the polyethylene is forced into the mold above and below the sheets with a combination of air pressure introduced through the blow-pin between the sheets, while at the same time drawing a vacuum between the sheets and the molding wall to remove air from the mold above and below the two sheets.

Table 6.20 Commercial grades of polyethylene for thermoforming applications.

Company/Grade	MI ($I_{2.16}$ kg)	HLMI ($I_{21.6}$ kg)	Density (g/cc)	Softening Point (°C)
ChevronPhillips/Marlex K606	0.70	NA	0.964	NA
Dow/DMDA-6400	0.80	57	0.961	NA
Slovnaft/Bralen RB 03-23 LDPE – High Pressure	0.35	NA	0.919	97
LyondellBasell/LR 776031	0.70	NA	0.960+	129

6.11.2   Grades of Polyethylene for Thermoforming

Both LDPE and HDPE are typically used in thermoforming process. High-density polyethylene with a density of about 0.96 g/cc, Melt Index (I2.16kg) value of about 0.7–0.8 and a relatively broad molecular weight distribution provides a grade of polyethylene for a rigid (stiff) end-use product, while the relatively high polyethylene molecular weight provides good melt strength needed during the thermoforming process. Low-density polyethylene (high-pressure polyethylene) is used in flexible packaging applications. Table 6.20 shows a few examples of polyethylene resins used in the thermoforming process.

References

1. T.J. Kresser, *Polyethylene*, Reinhold Publishing Corp., New York, pp. 71-72, 1957.
2. *Modern Plastics*, Vol. 32, No. 5, p. 83, 1955.
3. R.A.V. Raff, and K.W. Doak, *Crystalline Olefin Polymers, Part 1*; High Polymers Series, Vol. XX, Interscience Publishers, p. 831, 1965.
4. J.A. Christiansen, *J. Phys. Chem.*, Vol. 28, p. 145, 1924.
5. H.L.J. Backstrom, *J. Am. Chem. Soc.*, Vol. 49, p. 1460, 1927.
6. H.L.J. Backstrom, *Z. Physik.Chem.*, Vol. B25, p. 99, 1934.
7. K. Ziegler, and L. Ewald, *Ann. Chem.*, Vol. 504, p. 162, 1933.
8. E.H. Farmer, et al., *Trans. Faraday Soc.*, Vol. 38, p. 348, 1942.
9. C. Moureu, and C. Dufraisse, *Chem. Revs.*, Vol. 3, p. 113, 1926.
10. L. Bateman, *Quart. Revs. (London)*, Vol. 8, p. 147, 1954.
11. J.L. Bolland, *Quart. Revs. (London)*, Vol. 3, p. 1, 1949.
12. C. Decker, et al., *J. Poly. Sci., Polymer Chem. Ed.*, Vol. 11, p. 2879, 1973.
13. J.M. Ginhac, et al., *J. Makromol. Chem.*, Vol. 182, p. 1017, 1981.
14. J. Lacoste, et al., *Polymer Degradation and Stability*, Vol 34, pp. 309-323, 1991.
15. N.C. Billingham, and D.M. Wiles, Eds., *Polymer Stabilization Mechanisms and Applications*, Elsevier Applied Science, 1991.
16. F. Gugumus, *Die Angewandte Makromol. Chemie*, Vol. 158/159, p. 151, 1988.
17. F. Gugumus, *J. Polymer Deg. Stab.*, Vol. 27, p. 19, 1990.
18. D.M. Wiles, et al., *J. Polymer Deg. Stab.*, Vol. 19, p. 195, 1987.
19. R.T. Johnson, and E.J. Morrison, "Thermal scission and cross-linking during polyethylene melt processing, polymer durability," in: *Polymer Durability: Degradation, Stabilization, and Lifetime Prediction*, R.L. Clough, N.C. Billingham, and K.T. Gillen, Eds., American Chemical Society, Advances in Chemistry Series, Vol. 249, Chap. 39, p. 651, 1996.
20. Michael Tolinski, *Additives for PolyOlefins*, Elsevier, 2009; ISBN 9780815520511.

21. Internet data from F.A. Ruiz Heritage Plastics, Inc., Picayune, MS, USA; accessed March, 2010.

22. F.T. Kitchel, et al., U.S. Patent 4,784,822, issued November 15, 1988, and assigned to Enron Chemical Co.

23. B.J. Edwards et al., J. of Non-Newtonian Fluid Mechanics, Vol. 152, 168 (2008).

24. T.O.J. Kresser, *Polyethylene*, Reinhold Publishing Corp., New York, pp. 110-111, 1957.

25. K.C.H. Yi, and W.J. Michie, Jr., "UNIPOL II: The continuing revolution," in: International Business Forum on Speciality Polyolefins Conference, 1993.

26. *Modern Plastics*, Vol. 29, No. 6, p. 110, 1952.

27. www.borealisgroup.com.

28. N.C. Lee, *Understanding Blow Molding*, Hanser Publishing, Munich, Germany, 2000.

29. N.C. Lee, *Understanding Blow Molding*, Hanser Publishing, Munich, Germany, pp. 8-9, 2000.

30. T.O.J. Kresser, *Polyethylene*, Reinhold Plastics Applications Series, p. 26, Table 2-6, 1957.

31. T.O.J. Kresser, *Polyethylene*, Reinhold Plastics Applications Series, p. 186, 1957.

32. A. Lustiger, Understanding environmental stress cracking in polyethylene, *Medical Plastics and Biomaterials Magazine*, published July, 1996; accessed from www.devicelink.com.

33. R.A.V. Raff, and J.B. Allison, *Polyethylene*, Interscience Publishers, Inc., New York, pp. 374 and 390; reference cited was: J.B. DeCoste, et.al., *Ind. Eng. Chem.*, Vol. 43, pp. 117-121, 1951.

34. M.P. Mack, and M.A. Page, U.S. Patent 4,603,173, issued July 29, 1986, and assigned to DuPont Co.

35. J.P. Bladt, and P.P. Shirodkar, U.S. Patent 5,589,551, issued Dec. 31, 1996, and assigned to Mobil Oil Corp.

36. A. Poloso, et al., U.S. Patent 7,285,717, issued October 23, 2007, and assigned to Exxon Mobil Chemical Co.

37. H.V. Baann, and A.K. Lindahl, U.S. Patent 6,806,338, issued October 19, 2004, and assigned to Borealis Technology Oy, Porvoo Finland.

38. A. Lustiger, et al., U.S. Patent 7,022,770, issued April 4, 2006, and assigned to Exxon Mobil Chemical Company, Houston, Texas.

39. A.D. Ferguson,*British Plastics*, Vol. 16, pp. 430-432, 1944].

40. H. Rees, *Injection Molding Technology*, Hanser Publishers, 1994.

41. Rotational Molding, www.wikipedia.org and references cited therein.

42. LyondellBasell, Equistar Division,*A Guide to Rotational Molding*, (concise description of rotational molding process), www.lyondellbasell.com.

7

Experimental Methods for Polyethylene Research Program

7.1 Introduction

The growth in the polyethylene industry since the 1940s has been due to the development of new technology primarily in the area of new catalysts and the polymerization processes necessary to manufacture the polyethylene prepared from these catalysts. These innovations were responsible for the introduction of a wide variety of polyethylene structures that were developed for new applications.

It was evident in the beginning of the polyethylene industry in the mid-1930s that the scientists in the research laboratories at Imperial Chemicals Industries (ICI) relied on a unified effort from scientists with a wide variety of different skills to commercialize polyethylene. Such cooperation between scientists and engineers has always been a key component necessary for growth in the polyethylene industry.

As new catalyst technology emerged over the past 65 years for the low-pressure manufacture of polyethylene, scientists that specialize in different areas of the polyethylene industry continued to utilize coordinated teams to commercialize new polyethylene technology. This approach is often seen by the relatively large number of people cited as co-inventors on patents that are issued to protect the intellectual property related to polyethylene areas.

New technology advances through three developmental stages are summarized in Table 7.1.

Technology teams that involve representatives from different research areas such as catalyst development, polymerization process engineer, and product and fabrication specialist are necessary in order to move new technology along the path to possible commercialization. Such teams often work together at the same technology center as a new product proceeds from the laboratory stage through the pilot plant stage, where sufficient amounts of new grades of polyethylene are evaluated using commercial fabrication equipment in order to evaluate new products in the final end-use applications.

Once a new catalyst/product has been identified for evaluation at the manufacturing facility for a commercial trial, the technology teams are expanded to include technical representatives from the polyethylene manufacturing center. These new members will be responsible for implementing the new technology at the commercial scale. For example, a new catalyst needs to be produced in large-scale catalyst preparation facilities in order to produce sufficient catalyst for a commercial trial. This commercial quantity of catalyst is then tested in pilot plant operations to produce polyethylene samples for fabrication into finished products in order to verify that the expected product grades have been produced by the catalyst that will be evaluated in a commercial reactor.

The evaluation of new catalysts in a commercial polyethylene manufacturing process is a complex procedure. When a commercial reactor is taken off-line in order to evaluate a new catalyst and product, the polyethylene business undergoes a loss in revenue from production.

There are primarily three polymerization processes used in the low-pressure manufacture of polyethylene involving a transition metal active

Table 7.1 Innovation steps required to commercialize new products.

Stage	Polyethylene Scale
Laboratory	0.1–1.0 pound
Pilot Plant	50–1,000 pounds
Commercial	50–500 tons

Note: The scale is the amount of polyethylene provided from a developmental program. A pilot plant trial may last 3–7 days and provide sufficient material for fabrication into the end-use product. A commercial trial affords sufficient material for customer evaluation on commercial-scale equipment.

site. These are the solution process, slurry process and gas-phase process. Many large petrochemical companies have commercial reactors involving two of the processes with scientists and engineers dedicated to each type of process. The minimum amount of time needed for a commercial trial is 1–3 days, at least for an initial commercial trial, with 50–200 tons of polyethylene usually needed, as some product may be evaluated by potential customers in their fabrication facility.

This chapter will discuss the equipment necessary to set up a catalyst development laboratory and the analytical instrumentation required to characterize the polyethylene produced with experimental catalysts. The introduction of high throughput research methods since the 1990s has improved the potential productivity of such a laboratory. In addition, utilization of designed experiments in which the effect of several variables on the properties under investigation can be screened in a more efficient manner is also necessary to discover new technology for commercialization.

7.1.1   High Throughput Laboratory Equipment

High throughput laboratory equipment may be used to rapidly screen new ethylene polymerization catalysts to determine both qualitative and quantitative data. The qualitative, high throughput catalyst-screening process may be carried out in simplified equipment that is only screening to determine if any particular catalyst composition is an ethylene polymerization system. However, a more complex high throughput apparatus is necessary to determine catalyst polymerization kinetics and the effect of manufacturing process variables on polyethylene composition. Richard Hoogenboom at the Eindhoven University of Technology and Ulrich S. Schubert at the Dutch Polymer Institute have discussed high throughput equipment for polymer research [1–3].

Since the late 1990s, Symyx Technologies, Santa Clara, CA, USA, has offered commercial, high throughput catalyst screening equipment primarily designed for the polyethylene industry [4].

7.2   Experimental Process

Developing a new grade of polyethylene or a polyethylene with improved properties can be viewed as a three step process involving:

1. Catalyst preparation
2. Polymer synthesis
3. Polymer characterization

7.2.1   Catalyst Preparation

Olefin polymerization catalysts are prepared under an inert atmosphere usually using a nitrogen gas environment in which traces of oxygen and moisture have been removed. Equipment such as a glove box operated under an inert atmosphere, glassware and glassware systems specifically designed for handling air-sensitive chemical reagents, and other items such as serum bottles and syringes for transferring air-sensitive liquids used as catalyst preparation reagents are used in preparing experimental catalysts.

It is very important that detailed training of laboratory personnel takes place in teaching safe methods of handling air-sensitive chemical reagents, as accidental spills of these types of reagents will result in violent chemical reactions taking place, often accompanied by fire. Personal protective equipment such as fire-resistant laboratory coats, safety shields and eye protection goggles are examples of necessary equipment. The chemical companies that supply these types of air-sensitive reagents and equipment used to handle air-sensitive reagents are an excellent source for obtaining detailed instructions on safely handling these reagents.

7.2.1.1   Some Catalyst Preparation Operation Guidelines

1. Glove Box Design: Two entry ports are recommended for the transfer of large and small equipment into the glove box. The second smaller entry port is useful to significantly reduce the time required to transfer small items into the glove box and therefore improves laboratory efficiency. This smaller port should be approximately 4–6 inches in diameter. Maintaining a high quality inert atmosphere in which traces of moisture and oxygen are eliminated is also necessary. Scientists at Mobil Chemical Company would test the inert quality of the glove box daily by exposing a dilute solution of diethylzinc (15–25 wt% in hexane) to the glove box environment. Usually a large spatula would be dipped into the diethylzinc (DEZ) solution and then the spatula containing a few drops of DEZ would be moved around the box to determine if a white "smoke" was created. Any trace of white cloud would indicate that the quality of the inert atmosphere was poor and that the gas purification columns that continuously scrub the internal atmosphere would need to be regenerated. Although oxygen and moisture meters may be purchased as part of a glove box design, the DEZ quality

test was found to be very sensitive in identifying traces of moisture and oxygen, and therefore the cost of the moisture and oxygen meters could be eliminated from the glove box design.

2. Serum Bottle Technique: Handling air-sensitive liquids with syringes is sometimes necessary in both the operation of polymerization reactors and in catalyst preparation. Air-sensitive solutions need to be purchased as dilute solutions (usually up to about 20 wt%) in hydrocarbons such as hexane or heptane. Transfer of these solutions to serum bottles is carried out in an inert atmosphere glove box and serum bottles are only filled to about 60–70% to allow for the safe removal of the contents with a syringe. Serum bottles are capped with the septum that comes with each type of bottle. Only the metal crimping tool supplied for capping the serum bottle properly is used. From personal experience, a second septum, designed for small ground-glass joints was placed over metal serum caps and secured with a nylon cable tie around the neck of the serum bottle. This second cover extended the shelf-life of the contents of the serum bottle because a septum punctured several times with a syringe needle begins to degrade.

3. Syringe Technique: Glass syringes are recommended for training laboratory personnel in the safe handling of

Figure 7.1 Example of two-port stainless steel glove box suitable for a catalyst development laboratory. Source: www.vac-atm.com. From: Vacuum Atmospheres Company, 458 Boston Street, Topsfield, MA. 978-561-1763.

air-sensitive liquids using syringes; however, experienced laboratory personnel may use disposable plastic syringes if certain safety guidelines are followed. For example, a particular syringe should never be filled to more than about 50–70% of capacity to eliminate the possibility of accidently pulling the syringe plunger out of the syringe base and spilling the liquid. Fire-resistant, properly-fitted, laboratory coats, safety goggles and safety gloves are always worn while transferring liquids. Before a liquid is extracted from a serum bottle using a syringe, the syringe is purged several times with dry nitrogen and then the syringe is filled with dry nitrogen equal to or slightly greater than the volume of liquid that needs to be withdrawn from the serum bottle.

The syringe-purging operation is carried out by allowing dry nitrogen to flow through a rubber hose with the needle of the syringe extending several inches into the hose. The nitrogen that remains in the purged syringe is then injected into the inverted serum bottle to the volume level which equals the amount of liquid that is to be withdrawn from the serum bottle. This nitrogen volume is injected into the inverted serum bottle to create a slight internal pressure within the serum bottle and is available to replace the amount of liquid that will be withdrawn. Then the syringe needle is positioned within the liquid and the necessary volume is removed from the inverted serum bottle. The syringe contents are then immediately transferred to the desired container. It is strongly recommended that air-sensitive chemicals be transferred to a syringe behind a shield. The glass sliding panels in a laboratory hood provide a good shield in order to extract liquids from serum bottles.

7.2.2 Catalyst Evaluation Process

Catalysts may be rapidly evaluated in specially designed high throughput equipment available commercially [1]. This type of equipment is capable of preparing milligram quantities of catalysts, starting with various starting reagents normally involving a transition metal source and a series of ligands. Ethylene polymerization characteristics can be examined in the same vessel immediately following the catalyst preparation procedure or catalysts can be examined in dedicated equipment in which multiple polymerization autoclaves (4–16 reactors) are operated simultaneously in a high throughput mode. Ulrich Schubert and coworkers [1–3] have described the high throughput screening process.

However, once a catalyst is identified for additional studies, usually a 1–10 gram quantity of catalyst is required and these preparations are carried out in dedicated catalyst preparation apparatus and are then evaluated in a 1–4 liter polymerization autoclave designed to produce 40–400 grams of polyethylene, which is required for a more detailed evaluation of polymer properties. Catalysts that provide polymer with improved properties are then scaled up for a pilot plant evaluation which is usually similar to commercial equipment already in operation. Approximately 100–1,000 grams of catalyst is required for a pilot plant study, which provides 10–100 lbs/hour of polyethylene. A pilot plant trial will usually take place in a continuous mode lasting 1–5 days and producing a total of 200–1,000 lbs of polyethylene.

Evaluation of an experimental catalyst for possible commercial use involves a complex range of catalyst performance characteristics. A summary of some of these characteristics is shown below.

7.2.3 Catalyst Performance Characteristics

I. Catalyst process attributes
   a. Activity (g PE/g cat/hr)
   b. Polymerization kinetics
   c. Reactivity with higher 1-olefins such as 1-butene, 1-hexene and 1-octene
   d. Polymer molecular weight control using hydrogen and temperature
   e. Operability of catalyst in commercial reactor
      - Polymer fines
      - Resin bulk density
      - Catalyst start-up behavior
      - Catalyst feedability/transfer
      - Catalyst cost
      - Batch time to prepare the catalyst
      - Catalyst reproducibility (lot-to-lot variations)
II. Product attributes
   a. Polymer characterization
      - Polymer sample melt blending and stabilization
      - Melt Index (MI) $I_{2\cdot 16kg}$
      - Flow Index (FI) $I_{21.6kg}$
      - Polymer density (g/cc)
      - Comonomer content mol% or wt%
      - DSC melting point data – branching distribution/crystallinity

    - Rheology – behavior of polymer in liquid (melted) condition
- Shear thinning
- Strain hardening
- Melt strength
- Die-swell
- Melt relaxation times

b. Polymer mechanical properties after fabrication into a product
    - Dart impact for films
    - Film tear strength (machine and transverse direction)
    - Film clarity/gels
    - ESCR
    - Shrinkage/warpage
    - Fabrication rates

7.3   Important Considerations for Laboratory Slurry (Suspension) Polymerization Reactors

7.3.1   Background Information

Experimental equipment that is useful for the rapid screening of catalysts in support of the global polyethylene business must meet two critical requirements: (1) The polymerization reactor needs to be properly designed so that an experiment can be carried out under steady-state polymerization conditions for a minimum of about 20 minutes in order to provide important catalyst activity data and sufficient polymer for complete characterization. (2) A process model is needed in order to quantitatively determine important kinetic parameters of an experimental catalyst.

7.3.2   Basic Laboratory Polymerization Reactor Design

Some key features of a laboratory polymerization system designed for the preparation of approximately 50–100 g of polyethylene are as follows:

Location: The complete polymerization system should be constructed in a walk-in type hood with a very good ventilation system. All glass reservoirs that contain liquids that will be transferred to the polymerization reactor need to be located within the hood.

Reactor: A stainless steel autoclave with a reactor diameter of about 20 cm is recommended for easy cleaning, with a total capacity of 1.0–2.0

liters. An autoclave without bolts to close the reactor is preferred for easy opening. These systems are referred to as a "zipper-clave" design. An air-powered hydraulic lift for easy lowering/raising of the polymerization reactor to facilitate cleaning is also recommended. A stirring system capable of up to 1,000 rpm stirring rate is required. The top of the reactor requires at least 3–4 vent openings for addition of polymerization reagents and for venting the reactor as needed.

Reagents: Heptane or isobutane are recommended as the slurry polymerization solvent. Hexane should be avoided because of toxicity. Heptane is preferred for the preparation of LLDPE samples where the reactor needs to be opened in order to clean the interior of the reactor when a significant amount of copolymer is dissolved in the solvent. Isobutane is preferred for the preparation of HDPE or LLDPE where insignificant amounts of copolymer are soluble in the solvent. This type of system may be cleaned without opening the reactor and using a solvent wash to clean the interior of the reactor. The comonomer used is usually 1-butene, 1-hexene or 1-octene, depending on the type of polyethylene product under investigation. 1-Butene or 1-hexene is recommended as the comonomer for evaluating most ethylene/1-olefin copolymers.

Each of these reagents should pass through a molecular sieve column capable of removing oxygen and water and other polar impurities before the reagent is added to a calibrated glass reservoir above the reactor and stored under about 10–20 psi of dry nitrogen. Each reagent should be added to the reactor while the reactor is maintained at about 25–40°C with the reactor under a slow nitrogen purge. The total volume of the slurry solvent and comonomer should not exceed about 50–60% of the total reactor volume. This amount of head space above the liquid phase is needed for the polyethylene that will be prepared during the experiment and will reduce the possibility of the reactor fouling the top portion of the polymerization reactor. Ethylene is added to the reactor through a mass flow meter that is calibrated to display ethylene flow rate in g/minute and which is also able to record the total amount of ethylene that was transferred into the reactor until the reactor reaches the setpoint total pressure. Ethylene flow into the reactor once the polymerization process begins is recorded on a stripchart recorder during the entire polymerization experiment.

If a cocatalyst such as an aluminum alkyl is required in the experiment, this reagent may be added to the reactor through an addition port that is fitted with a ballvalve and under a slow nitrogen purge. The aluminum alkyl may be added through the addition port using a syringe. The addition port may then be rinsed with a small quantity (5–10 ml) of heptane, and then the ballvalve is closed.

Once the liquid reagents have been added to the reactor, the nitrogen purge is stopped and the reactor vent lines are immediately closed. Next, if hydrogen is needed as a chain transfer agent, then hydrogen is added to the reactor from a small calibrated Hoke bomb that has been pressurized with hydrogen. The amount of hydrogen added to the reactor is determined by lowering the pressure in the Hoke bomb by the desired amount. Therefore, the amount of hydrogen added to the reactor can be determined from the ideal gas law. Finally, the closed reactor is heated to a temperature about 5°C below the desired polymerization temperature while stirring the contents at a stirring rate of at least 600 rpm. Once the reactor temperature has reached this preset temperature, then ethylene is added to the reactor to allow the reactor to reach the final polymerization pressure required for a particular experiment. Total reactor pressure for most laboratory polymerization experiments is usually about 100–500 psig, depending on the slurry solvent employed.

Catalyst Addition Process: It is highly recommended that the catalyst under investigation is added to the polymerization reactor while the reactor is under the steady-state conditions for any particular experiment. Consequently, the last step required once the reactor has reached the necessary polymerization conditions is the injection of the solid catalyst into the reactor.

Several different methods for injecting the catalyst into the reactor have been developed by different research groups. One catalyst addition method is carried out by adding a predetermined amount of catalyst to a Y-shaped addition port with the bottom line leading directly into the top of the reactor and each line fitted with a ballvalve. One side of the entry port is used to add a predetermined amount of catalyst so that the catalyst sits on the closed ballvalve leading directly to the reactor. The other side of the entry port is connected to a 100 cc Hoke bomb which has been fitted with a solvent (heptane or isobutane) feed line, an ethylene feed line, pressure gauge and vent line. Prior to injecting the catalyst into the reactor, the 100cc Hoke bomb is filled with about 50 cc of solvent and pressurized with ethylene to about 50 psig higher than the pressure contained in the reactor. The catalyst is injected into the reactor by opening the ball valve at the bottom of the Hoke bomb and then opening the ballvalve holding the solid catalyst. The ethylene overpressure and the liquid heptane contained in the Hoke bomb wash the catalyst into the reactor. After this step, the ballvalve that held the catalyst is immediately closed.

7.3.3   Polymerization Rate/Total Polymer Yield

In order to eliminate nonsteady-state polymerization conditions, the maximum polymerization rate should not exceed about 1–2 g/minute/liter of reactor volume, and the total polyethylene yield should not exceed about 50–75 g/liter of reactor volume in about 45–60 minutes. Nonsteady-state polymerization conditions usually take place with Ziegler-type catalysts while producing LLDPE. Many Ziegler-type catalysts exhibit a high initial rate of polymerization with a relatively rapid decay in polymerization rate, with the rate often decaying 50% in 10–15 minutes. Note that the polymerization kinetics of any particular catalyst strongly influence the polymer particle morphology in a slurry or gas-phase process. A catalyst that displays a high initial rate of polymerization and rapid decay may be a concern in a commercial plant as it may result in poor particle morphology and relatively low catalyst productivity.

7.3.4   Isolation of Polyethylene Product

Once a polymerization experiment is completed, the reactor is cooled to room temperature and the reactor is opened and lowered using the hydrolytic lowering mechanism. The reactor contents are emptied into a large 3-liter Erlenmeyer flask fitted with flexible hosing and stoppered so that a suction technique may be used to siphon the polyethylene slurry out of the reactor and into the flask by applying a vacuum. Next, the product is transferred to a drying tray and the heptane is allowed to evaporate to isolate the dry polyethylene material. Once dry, the polyethylene yield from the experiment may be determined. If heptane is used as the slurry solvent, then the drying process will result in a non-homogeneous mixture of the polyethylene, with relatively low molecular weight and branched polyethylene material laying on the top of the dried polyethylene, which was the polymer fraction that was soluble in the heptane. This entire product needs to be melt-homogenized in order to get a uniform sample. The melt-mixing may be carried out using a two-roll mill in which an antioxidant has been added to the polyethylene product prior to the roll-milling operation. The sample is now suitable for characterization steps that are necessary. Usually, Melt Index values, MI and HLMI, comonomer content, DSC melting point and sample density are routine tests performed on an experimental sample. In addition, catalyst activity is recorded based on the amount of catalyst used in the experiment and the polyethylene yield in one hour. Catalyst activity of > about 1,500 g PE/g catalyst/hour are acceptable.

7.3.5   Steady-State Polymerization Conditions

Soon after the discovery of the first generation Ziegler catalysts and the chromium-based Phillips catalysts in the mid-1950s, the need developed for properly designed ethylene polymerization reactors in which the polyethylene was prepared under constant polymerization conditions in which ethylene concentration and reactor temperature were constant during the polymerization process. Steady-state polymerization conditions are required in order to accurately and precisely determine the important polyethylene structural features obtained from the polyethylene prepared with an experimental catalyst. The structural features include:

- Molecular weight
- Molecular weight distribution
- Comonomer content
- Comonomer distribution

In addition, steady-state polymerization conditions are required in order to determine meaningful kinetic data from an experimental catalyst such as catalyst activity, catalyst decay rates and catalyst reactivity ratios for comonomers such as 1-butene, 1-hexene and 1-octene.

In a classic 1978 paper [5,6], L.L. Bohm reported on the experimental parameters needed to establish steady-state polymerization conditions in order to eliminate monomer transport phenomena from the experimental results. As pointed out by Bohm, suspension or slurry polymerization takes place if the polymerization temperature is lower than the polyethylene solubility temperature and, therefore, the semicrystalline polymer precipitates from the suspension medium as the polymerization proceeds. The important physical process is the mass transfer of ethylene, comonomer and hydrogen (chain transfer reagent used to control polymer molecular weight) from the gas phase through the suspension medium and into the growing polymer particle to the active site. In order to obtain correct kinetic results, concentration gradients and temperature gradients within the polymer particle need to be removed from the polymerization process to achieve the necessary steady-state polymerization conditions.

Bohm postulates that steady-state polymerization conditions may not have been achieved in many earlier (ca. 1950s–1970s) investigations as insufficient stirring rates and/or inappropriate polymerization rates, i.e., producing the polymer sample at too high a polymerization rate, led to nonsteady-state polymerization conditions. Such conditions would affect the structure of the polyethylene produced. Indeed, in some of the very early years after the initial discovery by Hogan and Banks and Karl Ziegler, polyethylene was produced in pressurized glass polymerization bottles, where additional ethylene was not fed to the container. Such conditions were most certainly nonsteady-state, but nonetheless, did provide important data to these early scientists. However, detailed characterization of the polyethylene produced under such conditions had limited value.

7.3.5.1   Determination of Steady-State Conditions During Polymerization

The first goal in achieving steady-state polymerization conditions is to determine the stirring rate that eliminates the stirring rate of the suspension medium (usually a C_4-C_{12} saturated hydrocarbon) from affecting the polymerization rate. These conditions are determined by evaluating the polymerization activity over a range of catalyst concentration, stirring rate and ethylene pressure and characterizing the polymer produced under these various polymerization conditions.

One method of determining steady-state polymerization conditions for any particular reactor design is to increase stirring rate after the first few minutes of a polymerization experiment and determine if the polymerization rate immediately increases. Usually a stirring rate of at least 600 rpm is necessary to remove mass transfer effects from affecting the polymerization rate. Lowering the stirring rate to <200 rpm during a polymerization experiment will most likely decrease the rate of polymerization. Bohm's data showed that the polymerization rate was independent of stirring rate over the stirring rate range of 500–1500 rpm for any experiment containing the proper amount of catalyst and operating under the guidelines suggested.

Bohm's data also showed the effect of stirring rate on the polyethylene molecular weight if the polymerization reactor is operated under non-steady-state conditions. This data is illustrated in Figure 7.2. A similar curve was also found for polymerization rate vs stirring rate.

Examination of Figure 7.2 shows that nonsteady-state conditions exist below a stirring rate of about 500 rpm, where polyethylene molecular weight decreases with decreasing stirring rate. However, polymer molecular weight is constant with a stirring rate between 500–1500 rpm. Therefore, ethylene flow from the gaseous phase through the suspension medium to the growing polymer particle does not influence the polymer molecular weight at stirring rates above 500 rpm in which the process is under steady-state conditions. Obviously, using a stirring rate significantly higher than 500 rpm for routine experiments is recommended for this particular reactor.

For a one-liter reactor size as outlined by Bohm, additional data reported by him found that the polymerization rate needs to be limited to about 0.5–1.5 gram PE/minute for about a one hour polymerization time, which

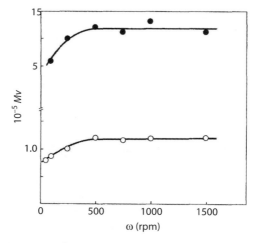

Figure 7.2 Polymerization conditions were as follows: 1.0 liter autoclave; 85°C; 0.6 liter diesel oil as diluent; ethylene pressure 6 bar; with (○) and without (●) hydrogen; AlR_3/Ti = 200. Reprinted from [5,6] with permission from Elsevier Publishing.

corresponds to about 30–90 grams of polyethylene produced for a one-liter reactor. This polymerization rate range will insure that steady-state polymerization conditions are retained over the entire course of the polymerization. Note that the total polymer yield of about 90 grams is also important to insure that the polymer particles remain in suspension over the entire polymerization period so that the steady-state conditions are maintained for the entire experiment. Note that Bohm's reactor was only filled with 60% of the liquid medium in order to provide adequate volume for the polymer particle to occupy during the course of the polymerization. Care must be taken not to produce too much polymer (total polymer yield) over the polymerization period.

In experiments where an ethylene/1-olefin copolymer is produced with hydrogen as a chain transfer agent, the amount of comonomer incorporated in the polymer and the amount of hydrogen consumed as a chain transfer agent also needs to be monitored. The change in the concentration of both the comonomer and the hydrogen should be limited to a decrease of about 10% in order that the polymer structure and the kinetic data are not affected over the course of the polymerization due to large changes in the comonomer and/or hydrogen concentrations during the polymerization experiment.

Ethylene mass-flow meter is an important part in the design of a polymerization reactor. The meter provides instant feedback on the polymerization rate and is connected to a strip chart recorder. In addition, the meter may be used to develop a process model for a particular reactor design.

7.3.5.2   Operation Guidelines for a Slurry Polymerization Reactor under Steady-State Condition

- Slurry solvent and other liquids such as comonomer should occupy 50–60% of the reactor volume.
- Polymerization rate should not exceed 2 grams PE/minute/ liter of reactor volume.
- Polymerization time should be 20–60 minutes.
- Polymer yield should be limited so that the polymerization vessel is not filled to greater than 70–75%.
- Changes in comonomer or chain transfer reagents should be less than 10% of initial value.
- Stirring rate needs to be determined experimentally, however, 600–900 rpm is usually sufficient.

7.3.6   Polymer Characterization

After a polymerization experiment is complete, the polymer yield is isolated by evaporation of the suspension medium. Light hydrocarbons in the range of C_4–C_7 (avoid hexane due to toxicity) are generally preferred to expedite the evaporation process. However, during the evaporation process the polymer yield will be fractionated depending on the amount of polyethylene that was soluble in the polymerization medium. Consequently, as a general guideline, it is recommended that the total polymer yield be stabilized with sufficient antioxidant and the entire polymer yield melt-homogenized using a two-roll mill or some other method that provides a uniform blend of all polymer components. Due to the need to melt blend the total polymer isolated from a particular polymerization experiment, it is usually necessary to produce a minimum of 10 grams of polyethylene, with a total yield of 30–90 grams of polyethylene a suitable range for evaluating an experimental catalyst in a laboratory setting.

7.3.6.1   Laboratory Characterization Equipment

Equipment needed for the characterization of an experimental catalyst and polyethylene is listed in Table 7.2. This subject has also been discussed by Ser van der Ven [7].

7.3.6.2   Melt Index and Density Data

Once a polymer sample has been melt homogenized, the Melt Index ($I_{2.16Kg}$) value and Flow Index ($I_{21.6Kg}$) value of the polymer sample are usually the

Table 7.2 Polymer/catalyst characterization equipment.

Catalyst	Inductive Coupled Plasma (ICP) Metals content of catalyst
	Surface Area and Pore Volume
	Particle size and particle size distribution
	Polymerization Reactors
Polymer	Melt Index Apparatus (Provides relative polymer molecular weight I2.16 (Melt Index) and relative data on polymer molecular weight distribution by determining the Melt Flow Ratio (MFR), which is the ratio of $I_{21·6}$ (Flow Index) value divided by the Melt Index.)
	Two-roll mill or twin-screw batch mixer to melt blend polymer sample
	Infrared Spectrometer –Branching content (comonomer content)
	Differential Scanning Calorimeter (DSC) –Branching distribution and crystallinity
	Gel Permeation Chromatography (GPC) – Number and weight average molecular weight, and molecular weight distribution.
	Temperature Rising Elution Fractionation (TREF)
	Carbon-13 Nuclear Magnetic Resonance (^{13}C NMR)
	Rheology Equipment

first characterization tool utilized in the catalyst laboratory to determine important preliminary data on the type of polyethylene produced by an experimental catalyst. This data can be collected in approximately 30 minutes and provide the scientist with immediate information on the relative molecular weight and molecular weight distribution of the polymer sample.

- Density: ASTM D-1505; A sample plaque is made from the experimental sample according to the method guidelines. The plaque is conditioned for one hour at 100°C to approach equilibrium crystallinity. Measurement for density is then made in a density gradient column and reported as g/cc.
- (MI) or Melt Index ($I_{2.16Kg}$): ASTM D-1238; 190°C with 2,160 gram weight and reported as grams per 10 minutes.
- (FI) or Flow Index ($I_{21.6Kg}$) or HLMI: ASTM D-1238; 190°C with 21,600 gram weight and reported as grams per 10 minutes, where HLMI is designated as high load melt index.
- (MFR) or Melt Flow Ratio: FI/MI correlates with polyethylene MWD.

These values determine the type of end-use application that this particular sample may be utilized for in a commercial application. The Melt Index (MI) value is a relative measure of the polyethylene molecular weight, while the Flow Index (FI) value is a relative measure of the processability of a sample and the Melt Flow Ratio (MFR) value is a relative measure of the shear-thinning behavior of the polyethylene sample which correlates with the MWD of the sample. A relative increase in the shear-thinning characteristics denotes an increase in the MWD of the polymer.

Commercial polyethylene manufactured with a low-pressure process can be classified in terms of molecular weight distribution (MWD) and falls into two categories: polyethylene with a relatively narrow MWD as indicated by Melt Flow Ratio values of 15–30 and polyethylene with a relatively broad MWD as indicated by MFR values of 80–130.

Polyethylene with a narrow or very narrow MWD is used in film, injection molding and rotational molding applications. Polyethylene with a broad MWD is used in HDPE film, sheets, blow molding and pipe applications.

It is important to note that polyethylene molecular weight distribution is a function of the sample molecular weight (MI) or the sample Flow Index (FI). Therefore, the evaluation of a new polyethylene material is carried out by producing a variety of samples over a range of Melt Index and Flow Index values.

An illustration of the amount of information that may be provided with an experimental catalyst in which polyethylene was produced over a range of Melt Index and Flow Index values is shown in Table 7.4. With a Ti-based catalyst, the polymerization temperature may be lowered to prepare polyethylene with a relatively broader MWD and hydrogen may be used as a chain transfer agent to control polymer Flow Index. This type of catalyst provides a greater degree of flexibility compared to a Cr-based catalyst in which hydrogen does not act as a chain transfer agent. This catalyst is based on silica, dibutylmagnesium and $TiCl_4$ and provides polyethylene with a relatively broad MWD, depending on polymerization temperature and catalyst composition [28].

Table 7.3 Catalyst types and corresponding MFR range.

Catalyst Type (Metal)	MFR Range of Commercial Polyethylene
Single-Site (Zr or Ti)	15–18 – Very narrow MWD
Ziegler (Ti/Mg)	24–35 – Narrow MWD
Phillips (Cr)	80–130 – Broad MWD

Table 7.4 Effect of polymerization temperature on MWD.

Temp. (°C)[a]	Flow Index[b]	MFR (FI/MI)
85	10	120
85	80	85
95	10	95
95	80	70

[a]Lowering polymerization temperature usually broadens MWD with Ziegler type catalysts.
[b]At lower temperatures, higher levels of hydrogen are required to hold FI values constant.

Examination of Table 7.4 shows that this particular Ziegler (Ti-based) catalyst provides polyethylene with a relatively broad molecular weight distribution as indicated by MFR values of approximately 70–95 with a polymerization temperature of 95°C, while decreasing polymerization temperature to 85°C significantly increases the molecular weight distribution of the polyethylene as indicated by MFR values of 85–120. In addition, this particular catalyst provides a wide range of polyethylene molecular weight as shown by Flow Index values of 10 to 80. Based on this type of data, the catalyst may have commercial applications in the following areas:

- Blow Molding: Household and industrial containers (HIC); milk bottle and food packaging applications.
- Pipe Markets: Low- and high-pressure applications.
- HDPE: Film markets from high molecular weight (HMW) to medium weight MW.

7.3.6.3   Infrared Method

Infrared spectroscopy was used by Fox and Martin in 1940 [8] to examine the methyl branch content of polyethylene produced by the high-pressure process, and found a higher content of methyl branches than could be accounted for by only polymer end groups. Until this study, scientists thought that the structure of high-pressure polyethylene was much more linear than what was later found. This linearity was suspected because of the crystallinity found in the polyethylene by X-ray data.

Ethylene copolymers with a comonomer such as 1-hexene may be analyzed for comonomer content utilizing an infrared spectrophotometer. The absorbance at 1377 cm^{-1} and 1368 cm^{-1} is determined and the ratio A_{1377}/A_{1368} is determined. An example of the infrared data calibration curve for a series of ethylene/1-hexene copolymers from a high activity Ti/Mg Ziegler catalyst is illustrated in Figure 7.3 [9].

The correlation between polymer density and 1-hexene content (mol%) for this same catalyst is illustrated in Figure 7.4 [9].

The density (corrected to a 1.0 melt index value) vs comonomer content (mol%) relationship is dependent on variables such as branching distribution and branch length. For example, a single-site catalyst that provides a homogeneous branching distribution requires significantly less comonomer

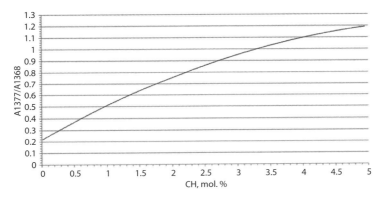

Figure 7.3 Calibration curves for copolymer composition measurement by the infrared method [10].

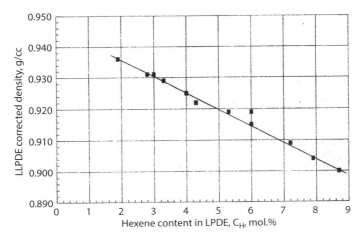

Figure 7.4 Dependence between density of ethylene-hexene copolymers and hexane content [10].

content to produce a LLDPE resin than a typical titanium/magnesium-containing Ziegler catalyst that produces a heterogeneous branching distribution. Moreover, various high-activity Ti/Mg-type catalysts will also exhibit a slightly different density/comonomer content relationship.

7.3.6.4 Differential Scanning Calorimetry

Differential scanning calorimetry (DSC) is a useful method for evaluating the melting characteristics of polyethylene and determining the degree of crystallinity for a particular sample. This method provides information on the intermolecular branching distribution and crystallinity for experimental samples of polyethylene. In general terms, scientists have identified two types of branching distribution in ethylene/1-olefin copolymers (LLDPE), which have been designated as heterogeneous and homogeneous branching distribution, and these are illustrated in Figure 7.5.

Both types of branching were identified soon after the initial discovery in the mid-1950s of the Cr-based Phillips catalyst and Ziegler-type catalysts based on titanium or vanadium. Ethylene/1-olefin copolymers prepared with early versions of catalysts based on titanium and chromium provided heterogeneous (non-uniform) branching distribution due to multiple types of active centers, each of which displayed a different reactivity towards the 1-olefin. However, the homogeneous vanadium-based Ziegler catalysts investigated by Carrick and coworkers [10] produced a homogeneous (uniform) branching distribution due to the presence of only a single type of active center.

Chapter 4 of this book, on single-site catalysts, discusses a wide variety of single-site catalysts that were discovered since about 1980, many

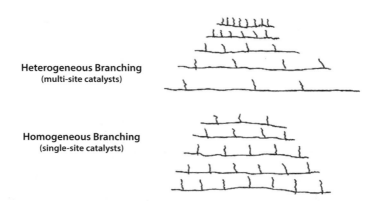

Figure 7.5 Two types of branching distribution for LLDPE.

of which are now commercial catalysts used to prepare new types of polyethylene-based resins that have found new markets and applications for the global polyethylene business.

Differential scanning calorimetry is an extremely useful tool in quickly identifying the type of branching produced by an experimental catalyst. (Note, however, that the polyethylene sample preparation before the melting point endotherm is recorded is an important feature. Generally, the polymer sample needs to be preconditioned by melting the sample and then slowly cooling the melt to room temperature to anneal the sample. This results in better data precision so that melting point differences between various samples may be compared.) The melting endotherm shown in Figure 7.6 was obtained from an ethylene/1-hexene copolymer- which contains approximately 3.5 mol% 1-hexene, and was produced with a highly active Ti/Mg Ziegler catalyst. The shape of the DSC melting endotherm suggests that this particular catalyst provides polyethylene with a heterogeneous branching distribution [12, 13].

Examination of the DSC endotherm shows that the region of the melting endotherm from about 120–125°C is due to the relatively high molecular weight polymer component with a relatively low branch frequency, while the very broad region of the endotherm from about 40–115°C is due to the relatively lower molecular weight highly-branched polymer component. Kissin [9] and coworkers have shown that this type of multi-site Ti-based Ziegler catalyst contains active centers with an r_1 (ethylene/1-hexene

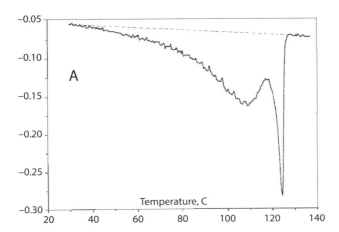

Figure 7.6 DSC melting point scan for LLDPE (containing ca. 3.5 mol% 1-hexene) prepared with high-activity Ti/Mg catalyst. Note: The higher melting point crystallites contain relatively less 1-hexene due to heterogeneous branching distribution [12].

reactivity ratio) value ranging from 8–220. Consequently, this multi-site polymerization catalyst contains active sites that react extremely well with 1-hexene and other active sites that react very poorly with 1-hexene or differ in 1-hexene reactivity by a factor of about 30.

On the other hand, single-site catalysts produce polyethylene with a homogeneous branching distribution that provides a DSC melting endotherm curve with a single melting point that melts over a relatively narrow range, with the melting point a function of the comonomer content of the ethylene/1-olefin copolymer. This melting point data for ethylene/1-hexene copolymers prepared with a zirconocene compound and activated with methylalumoxane is illustrated in Figure 7.7 [13].

For example, examination of the above data for a copolymer that contains 3.5 mol% 1-hexene and a homogeneous branching distribution prepared with this single-site catalyst displays a single maximum melting point peak at about 112°C, as compared to the more complex DSC melting endotherm produced by a similar copolymer using a multi-site catalyst that produces a heterogeneous branching distribution.

Figure 7.8 compares the DSC melting endotherms for each type of branching. The homogeneous sample contained 2.5 mol% 1-hexene, while the heterogeneous branching sample contained 2.8 mol% 1-hexene [13, 14].

7.3.6.5 Gel Permeation Chromatography

Gel permeation chromatography (GPC) is another important tool in the polyethylene characterization process. Although the Melt Flow Ratio (MFR) value obtained with an experimental sample of polyethylene

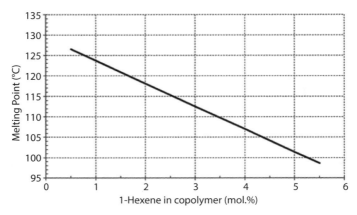

Figure 7.7 DSC determined melting point data for polyethylene prepared with single-site zirconocene/methylalumoxane catalyst [12].

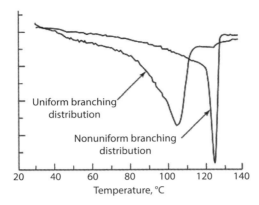

Figure 7.8 DSC melting curves of two ethylene/1-hexene copolymers. The homogeneous (uniform) branching distribution polymer contains 2.5 mol% 1-hexene and the heterogeneous (non-uniform) branching distribution polymer contains 2.8 mol% 1-hexene. Reprinted from Kirk-Othmer Encyclopedia of Chemical Technology, :Polyethylene, Linear Low Density," by Yury V. Kissin (1996). with permission from John Wiley and Sons.

provides information on the relative molecular weight distribution of the polymer sample, the GPC chromatogram provides a detailed examination on the amount of polyethylene with a certain molecular weight. In addition, the GPC provides the number average molecular weight value (Mn) and the weight average molecular weight value (Mw) for the polymer sample. The ratio (Mw/Mn) is designated as the polydispersity index and is also a quantitative measure of the molecular weight distribution. A single-site catalyst has a polydispersity value of 2.0, while polyethylene produced with multi-site catalysts provide a polydispersity index from about 3–6, indicating a relatively narrow MWD and other (e.g., Cr Phillips catalyst) multi-site catalysts provide polyethylene with a polydispersity value of 8–30, indicating a relatively broad MWD.

Three examples of the importance of GPC data in understanding the changes in polyethylene molecular structure due to modifications of ethylene polymerization catalysts will be discussed.

7.3.6.5.1   Titanium-Based Catalyst
The importance of the GPC chromatogram in understanding the complex molecular structure that may be produced with an experimental Ti-based Ziegler catalyst based on silica, dibutylmagnesium, $SiCl_4$, $Si(OBu)_4$ and titanium tetrachloride [15] is shown in the GPC data in Figure 7.9.

This particular experimental catalyst produced a trimodal MWD with molecular weight ranging from 10^2–10^7, while a modification of this

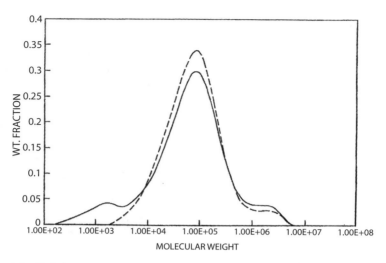

Figure 7.9 GPC chromatogram shows the complex molecular structure of experimental samples of polyethylene [15].

particular catalyst eliminated the very low molecular weight polymer component from the polyethylene and provided a bimodal MWD with only a very high molecular weight polymer component. The GPC data identified the multi-modality of the MWD.

7.3.6.5.2   Cr-Based Phillips Catalyst

Extensive use of GPC to determine very significant changes in the polyethylene molecular structure by modifying the Phillips Cr-based polymerization catalyst was also made by M. P. McDaniel [16]. In one set of experiments, McDaniel prepared five Cr-based catalysts on five aluminophosphate supports using $Cr(DMPD)_2$ where DMPD was bis(dimethylpentadienyl), as the chromium compound. The supports were prepared with a range of P/Al values from 0 to 0.9. The GPC data in Figure 7.10 illustrates the very significant change in the polyethylene molecular structure obtained from these five finished Cr-based catalysts.

7.3.6.5.3   Bimetallic Catalyst Based on Ti/Zr Catalyst

A catalyst based on a silica-supported catalyst that contained both a Ti-based Ziegler catalyst and a Zr-based single-site catalyst [17] provided a high-density polyethylene that contained a bimodal MWD, with the titanium catalyst component producing a relatively high molecular weight polymer component and the Zr-based catalyst component producing a relatively low molecular weight polymer component. The GPC data on this type of catalyst is able to quantify the molecular weight distribution

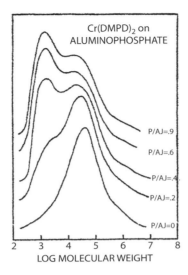

Figure 7.10 GPC data of polyethylene prepared with Cr-based catalysts. Reprinted from [16] with permission from the American Chemical Society.

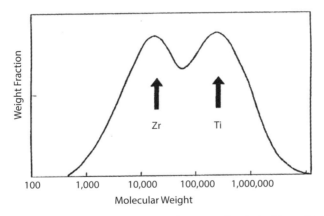

Figure 7.11 GPC curve of a HMW-HDPE sample prepared with the bimetallic (Ti/Zr) catalyst with the polyethylene component provided by each of the two metals identified [17].

and amount of each of the two polymer components. An example of this GPC data is shown in Figure 7.11. This type of polyethylene is useful in applications such as high molecular weight HDPE film, pressure pipe for distribution of natural gas, or blow molding markets where improved environmental stress crack resistance is necessary.

7.3.6.6   Temperature Rising Elution Fractionation

Branching in polyethylene may be either short-chain branching as provided by incorporating higher 1-olefins into the polymer backbone which is utilized in the low-pressure process to manufacture LLDPE; or the branching may be much more complex and consist of both long-chain and short-chain branching as found in LDPE produced by the high-pressure free-radical ethylene polymerization process.

Temperature rising elution fractionation (TREF) is especially useful in probing the branching content in low-density (LDPE) and linear low-density polyethylene (LLDPE). The concept of elution fractionation was described as early as 1950 when Desreux and Spiegel [18] deposited polyethylene sample onto a Celite packed column and obtained the fractions by eluting the packed column with toluene at successively higher temperatures.

This technique was more fully discussed by Wild and coworkers in 1982 [19], where the instrumentation required for the TREF process was illustrated. Briefly, the method requires a solvent reservoir fitted with a degasser, a pump to move the solvent through a fractionating column, a recirculating oil bath that controls the temperature of the column, a detector to monitor the polymer components as they are removed from the column and a strip chart recorder. A diagram of the TREF apparatus used by Wild, which displayed a continuous polymer concentration detection system, is shown in Figure 7.12.

Temperature rising elution fractionation(TREF) is a solvent fractionation technique in which detailed polymer structure information may be obtained by separating molecular species according to their branching content, usually given as number of branches/1000 carbon atoms in the polymer backbone.

The fractionation is achieved by the differences in crystallizability of polyethylene molecules containing different levels of branching. Initially, the polyethylene sample is dissolved in trichlorobenzene at about 140°C, and the fractionation takes place on a stainless steel column packed with Chromosorb P as the polyethylene sample solution is very slowly cooled to room temperature over ca. 3 days at about 1.5°C/hr. The polymer molecules containing the least amount of branching are deposited first onto the column as the solution temperature is lowered. After the entire polyethylene sample has been crystallized onto the support, the column is slowly heated at about 20°C/hr to elude the various branched molecular species from the column to a detector system, which quantifies the amount of polymer as a function of branching frequency. The highly-branched polymer molecules

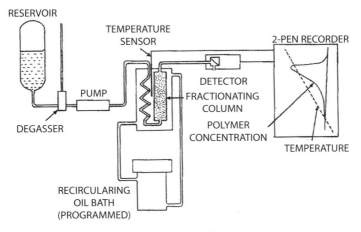

Figure 7.12 Analytical temperature-rising elution fractionation system. Reprinted from [19] with permission from John Wiley and Sons.

which were the last to be crystallized onto the support column are the first molecules that are removed from the support.

Wild and coworkers reported [19] the wide distribution of branching in two LLDPE resins of commercial importance at the time. Both samples had a density of 0.919 g/cc and similar MI values of 1.0 and 1.2 but were produced with two different low-pressure processes. As expected, polyethylene molecular fractions were found with a very heterogeneous branching distribution from about nil to 44 methyl groups/1000 carbons. The slight differences in the TREF data of the two similar LLDPE samples were attributed to the different process used to produce the sample.

In a later paper, F. M. Mirabella and E. A. Ford [20] combined the TREF method with other analytical methods such as DSC, ^{13}C NMR and GPC to more fully characterize the molecular structure of the various polymer fractions provided by TREF from a commercial LLDPE resin. To illustrate the importance of this experimental approach for characterization of polyethylene, some of their data is shown in Table 7.5.

This data show that the heterogeneous branching distribution of LLDPE as illustrated in the Figure 7.13 is an accurate representation of the molecular structure. The relatively higher branched fractions exhibit a relatively broader MWD and decreasing molecular weight. An additional report on the fractionation and thermal behavior of LLDPE using TREF was also reported by another group in 1987 [29].

Table 7.5 Data from Mirabella and Ford illustrating their experimental approach for characterization of polyethylene.

TREF Fraction (elution Temp.)	wt% 1-Butene	SCB/1000 C	$T_{melting}$	MW	Mw/Mn
25–54	14.8	39.9	91.1	57,000	4.8
54–81	6.1	15.6	108.0	75,000	4.1
81–91	3.5	8.8	118.9	93,000	2.9
91–120	0.8	2.1	128.7	124,000	3.2

Note: LLDPE commercial sample with Melt Index of 1.0; density 0.918 g/cc; 7.0 wt% 1-butene; Mw/Mn = 3.8 [20].

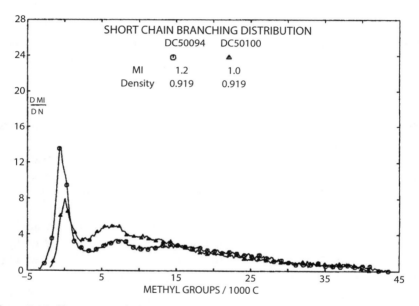

Figure 7.13 Comparison of two LLDPE samples using the TREF technique. Reprinted from [19] with permission from John Wiley and Sons.

7.3.6.7   CRYSTAF Method

As pointed out by Monrabal and coworkers, the TREF is an operational complex method requiring more than one day to perform. The CRYSTAF method is also a separation method that fractionates samples of differing crystallizability by slowly cooling a polymer solution in a single crystallization cycle by monitoring the decrease in solution concentration as the temperature is lowered. Because of this one-step approach, analysis time is reduced to around 6 hours for five simultaneous samples [23].

7.3.6.8 Carbon-13 Nuclear Magnetic Resonance

Carbon-13 nuclear magnetic resonance (^{13}C NMR) is important in understanding more detailed structural information in the backbone of the polyethylene sample. For example, LLDPE is produced commercially with either 1-butene, 1-hexene or 1-octene as the comonomer. Copolymers with a low content of 1-olefin contain only isolated branches, however copolymers containing higher levels of comonomer (e.g., 2–20 mol%) contain a wide variety of complex sequence distributions making ^{13}C NMR a particularly important characterization tool. Information into the sequence distribution of the comonomer in ethylene/1-olefin pentads provides data to determine the reactivity constants K_{ee}, k_{eh}, k_{he} and k_{hh}, where e represents ethylene and h represents 1-hexene.

Some possible Pentad sequences involving ethylene (e) and 1-hexene (h) that may be determined from ^{13}C-NMR data are shown below:

- eehee
- ehehe
- ehhee
- ehhhe

Such sequence structures are required to reduce the crystallinity of the polyethylene sample as the comonomer n-butyl branch from incorporation of 1-hexene into the polymer backbone interrupts the chain-folding mechanism responsible for the crystallinity in polyethylene. Long units of ethylene along the polymer backbone are able to crystallize the polymer sample and are undesirable in polyethylene products that require a large degree of elasticity for particular applications.

Details of the examination of ethylene/1-butene [24, 25]; ethylene/1-hexene [26]; or ethylene/1-octene copolymers [27] by ^{13}C NMR to determine various sequence distributions have been reported.

As discussed previously, the development of a new commercial catalyst requires extensive research on the part of catalyst development scientists, product development specialists and process engineers. Some important catalyst characteristics that relate to the polymerization process are discussed.

7.3.7 Catalyst Process Attributes

7.3.7.1 Catalyst Activity (g PE/g cat) and Polymerization Kinetics

The activity of an experimental catalyst is an important property of a potential commercial catalyst. Because many commercial polyethylene catalysts are supported on a high surface area support such as silica, the catalyst residual inorganic material (ash content) is usually determined after complete removal of the polymer by oxidation. An inorganic residue (ash content) of less than 500 ppm is usually required. However, acceptable catalyst activity of an experimental catalyst in the research laboratory may not translate to acceptable catalyst productivity in a commercial reactor under the reaction conditions required to manufacture a commercial polyethylene resin.

As a general rule, the Cr-based Phillips catalyst used worldwide to manufacture HDPE with a relatively broad MWD, exhibits high catalyst productivity and a steady or increasing rate of polymerization so that polymer morphology is excellent.

However, catalysts that undergo rapid activity decay, rather than a relatively stable polymerization rate, may be problematic in a commercial reactor for slurry and gas-phase polymerization processes. Polymerization kinetics for high-activity Ti-based Ziegler catalyst for the manufacture of HDPE and LLDPE have been discussed in detail [30] and it has been found that the polymerization kinetics depend on both catalyst type and process conditions such as comonomer type, comonomer concentration and chain transfer concentration. Polymerization kinetics in the presence of low levels of comonomer and hydrogen to produce medium- and high-density polyethylene often exhibit a steady rate of polymerization, which is favorable for particle morphology. However, Ti-based catalysts in the presence of high levels of hydrogen required for the manufacture of high Melt Index resins (MI of about 10–80) undergo a significant decrease in activity, as hydrogen reduces polymerization kinetics.

This is not the case for Ti-based catalysts in the presence of high levels of comonomer required for the manufacture of LLDPE. This type of catalyst exhibits a kinetic profile in which the initial rate of polymerization is very high, followed by a rapid decay in kinetics, which may cause operational problems in a commercial reactor. These operational problems are due to the initial rapid rate of polymerization, which affects the manner in which the catalyst particle fragments and the manner in which the polymer particle grows. Other concerns would be partial polymer particle melting due to poor removal of the heat of polymerization, which in turn may cause reactor fouling.

Granular resin bulk density is an important physical property of finished polyethylene material. Polymer particle morphology determines the amount of polymer that may be contained within the reactor and determines the conveying rates of granular material as the polymer is transferred as it exits the polymerization reactor. The initial activity of a catalyst particle during the first five minutes after the particle enters the polymerization reactor determines the catalyst particle fragmentation pattern that affects the finished resin particle morphology. Particles with relatively high resin particle density and a spherical smooth surface are required. Ethylene partial pressure within the reactor needs to be maintained at a level that produces acceptable resin particle morphology, while at the same time does not reduce polyethylene production rates.

7.3.7.2   Reactivity with Higher 1-Olefins such as 1-Butene, 1-Hexene and 1-Octene

The reactivity of experimental catalysts with higher 1-olefins is an important characteristic of a polymerization catalyst. Relative reactivity decreases as the olefin chain increases so that 1-hexene and 1-butene are used only in gas-phase reactors to produce commercial LLDPE resins. Because of lower reactivity and lower volatility 1-octene is not usually used in gas-phase reactors.

7.3.8   Additional Features of Commercial Catalysts

Catalyst costs can range from $20–200/lb depending on the source of the catalyst. Polyethylene manufacturing companies that have the capability of preparing ethylene polymerization catalysts on site have a significant cost advantage over other polyethylene companies that are limited by licensing agreements or other factors that prohibit the manufacture of catalysts for internal use. For example, a catalyst cost of $100/lb with a commercial productivity of 4,000 lb polyethylene/lb of catalyst translates to $0.025/lb polyethylene, which is 2.5% of the cost of polyethylene that has a market price of $1.00 (USD)/lb polyethylene.

Other features of an experimental catalyst that are important considerations in evaluating the commercial potential of a new catalyst are:

- Catalyst startup behavior as a catalyst is added to the polymerization process;
- Catalyst feedability/transfer properties;
- Batch time to prepare the catalyst and catalyst reproducibility (lot-to-lot variations);
- Catalyst raw material costs;
- Catalyst shelf life.

7.4 Polymerization Reactor Design for High-Throughput Methods

Until about the mid-1990s, polyethylene research laboratories usually possessed several polymerization systems depending on the size of the research program. It was common procedure that each investigator had a dedicated polymerization reactor for individual research. These single-batch reactors were about 1–5 liter in capacity and ran in either a slurry or solution mode. Often these individual reactors would be used to carry out a designed experiment in which two or three process variables would be investigated at two levels each. This required 4–8 separate experiments which would take about one week to carry out in the case of three variables at two levels, or 8 individual experiments.

However, in the mid-1990s many polyethylene research groups began to use multi-reactor designs in which 4–16 dedicated reactors could be operated simultaneously to greatly reduce the time required to carry out designed experiments.

In a research laboratory designed to evaluate the effect of process and catalyst variables on the structure of the polyethylene produced, it is necessary to utilize an experimental design in which as many as 16 polymerizations may be carried out simultaneously in a high-through-put mode.

In an experimental program, a minimum of 10 grams of PE is required in order to melt homogenize, stabilize (antioxidants) and characterize polyethylene. For rapid catalyst screening, a series of 16 reactors would provide data to investigate up to four polymerization variables (2^4 experiments) during a single experiment. Key polymerization variables include polymerization temperature, comonomer concentration, and in the case of Ziegler catalysts, hydrogen concentration and co-catalyst type or amount of cocatalyst (aluminum alkyl). A designed experiment (DE) with two levels of three variables would utilize 8 reactors, while investigating two levels of four variables would utilize 16 reactors.

A 200 cm^3 reactor with a zipper-clave closure system is recommended as a suitable reactor size for 8 or 16 polymerization reactors set up to operate simultaneously under high-throughput mode. This size reactor will allow sufficient polymer to be prepared in order to fully characterize the polymer and for complete evaluation of an experimental catalyst. Properly designed, a series of 16 reactors running in a simultaneous system would provide in one day the experimental data that would have taken eight days to collect with only one autoclave.

(a) To completely characterize polymer and an experimental catalyst, a polymerization time of 20–40 minutes is required at a polymerization rate of 0.5–1.0 g/minute. This amount of time is needed in order to determine catalyst kinetic data. The activity profile of a polymerization catalyst provides data which is useful in predicting trends in polymer granular bulk density. In general terms, polymerization catalysts with a very high initial activity provide lower granular bulk density than catalysts with a lower initial rate of polymerization. In addition, a 10 to 40 minute polymerization time will provide the necessary amount of 10–40 grams of polyethylene.

(b) Turnaround time is minimized with a 200 cm³ reactor. The time required to clean and prepare a reactor for another experiment increases significantly as the reactor size increases without providing any additional data.

(c) Laboratory set-up costs and the amount of space necessary to construct this 16 reactor system is greatly reduced with a 200 cm³ reactor size. Reactors need to be placed in a high-ventilation walk-in hood area in order to meet safety requirements.

7.5   Polymer Characterization

A summary of the information obtained for the characterization of the polyethylene produced with an experimental catalyst is listed below.

- Sample melt blending with addition of stabilizer package
- Melt Index (I 2.16)
- Flow Index (I21.6)
- Melt Flow Ratio (MFR); melt index/flow index
- Comonomer content (IR method)
- Density (g/cc) ASTM method
- DSC melting point data (relates to density and branching distribution)
- Rheology (behavior of polymer in molten state)
- Shear thinning
- Strain hardening
- Melt strength
- Die-swell
- Melt relaxation times

Table 7.6 Summarization of data provided by a 550 cc polymerization reactor process model.

Reagent	Liquid Concentration (Molar)
Ethylene	0.403
1-Hexene	2.424
Hydrogen	0.0144

7.6 Process Models

A process model is required for the operation of a laboratory experimental reactor. These models provide the concentration of ethylene, comonomer, and hydrogen for any particular experiment. Petrochemical companies that manufacture polyethylene with either a solution or slurry process have process models for commercial reactors that can be used in laboratory equipment. On the other hand, companies that operate a gas-phase reactor process, or other scientists in an academic laboratory, often carry out laboratory research using a slurry process, so that a laboratory process model needs to be developed.

A process model may be determined experimentally by using an ethylene mass flow meter as part of the reactor design and gas chromatography equipment to analyze the vapor-phase content under a variety of polymerization conditions.

An example of the data provided by a process model for a 550 cc polymerization reactor operating at 80°C, 145 cc heptane, 75 cc of 1-hexene and 0.05 g of hydrogen gas added through a calibrated hoke bomb, and 125 psig total pressure is summarized in Table 7.6:

Under these conditions, the 1-hexene/ethylene vapor ratio is 0.126 and the hydrogen/ethylene vapor ratio is 0.366. This information is sufficient for determining kinetic data, comonomer reactivity ratios and the effect of hydrogen on polyethylene molecular weight.

References

1. R. Hoogenboom, and U. Schubert, *Review of Scientific Instruments*, Vol. 76, p. 062202, 2005; see www.symyx.com or www.chemspeed.comfor example of commercial equipment specifically designed for high throughput experimentation.
2. U.S. Schubert, M. Meier, and R. Hoogenboom, *Macromolecular Rapid Comm.*, Vol. 25, pp. 21-33, 2004.

3. S. Schmatloch, and U.S. Schubert, *Macromolecular Rapid Comm.*, Vol. 25, pp. 69-76, 2004.

4. G.A. Diamond, et al., U.S. Patent 7,078,164, issued July 18, 2006, and assigned to Symyx Technologies, Inc., Santa Clara, CA (USA).

5. L.L Böhm, *Polymer*, Vol. 19, pp. 553-560, 1978.

6. L.L Böhm, *Polymer*, Vol. 19, pp. 545 and 562, 1978.

7. S. van der Ven, *Polypropylene and Other Polyolefins*, Elsevier Publishing, pp. 447-489, 1990.

8. J.J. Fox, and A.E. Martin, *Trans. Faraday Soc.*, Vol. 36, p. 897, 1940.

9. T.E. Nowlin, Y.V. Kissin, and K.P. Wagner, *J. Polymer Science, Part A: Polymer Chemistry*, Vol. 26, pp. 755-764, 1988).

10. W.L. Carrick, et al., *J. Am. Chem. Soc.*, Vol. 82, p. 1502, 1960.

11. Y.V. Kissin, R.I. Mink, A.J. Brandolini, and T.E. Nowlin, *J. Polymer Science, Part A: Polymer Chemistry*, Vol. 47, p. 3271, 2009.

12. Y.V. Kissin, et al., *J. Polymer Sci., Part A: Polymer Chemistry*, Vol. 26, p. 755, 1988.

13. Y.V. Kissin, "Polyethylene, linear low density," in: *Kirk-Othmer Encyclopedia of Chemical Technology*, John Wiley & Sons, Figure 1, 1996.

14. Y.V. Kissin, et al., U.S. Patent 5,258,345, issued Nov. 2, 1993, assigned to Mobil Oil Corp.

15. Y.V. Kissin, U.S. Patent 5,258,345 issued to Mobil Oil Corporation, Nov. 2, 1993.

16. M.P. McDaniel, Controling polymer properties with the phillips chromium catalysts, *Ind. Eng. Chem. Res.*, Vol. 27, No. 9, p. 1559, 1988.

17. T.E. Nowlin, et al., U.S. Patent 5,332,706, issued July 26, 1994, assigned to Mobil Oil Corporation. See also; T.E. Nowlin, et al., U.S. Patent 5,473,028, issued Dec. 5, 1995, assigned to Mobil Oil Corp.; T.E. Nowlin, et al., U.S. Patent 5,539,076, issued July 23, 1996, assigned to Mobil Oil Corp.; T.E. Nowlin, et al., U.S. Patent 5,525,678, issued June 11, 1996, assigned to Mobil Oil Corp.; T.E. Nowlin, et al., U.S. Patent 5,614,456, issued March 25, 1997, assigned to Mobil Oil Corp.; and T.E. Nowlin et al., U.S. Patent 6,740,617, issued May 25, 2004, assigned to Univation Technologies, LLC (Houston, TX). The last patent in the group issued to Univation Technologies, LLC, Houston, TX. T.E. Nowlin, et al., Premium piple resins, 2002, U.S. Patent 6,403,181, issued June 11, 2002, assigned to Mobil Oil Corp. (Fairfax, VA) relates to polyethylene pipe.

18. V. Desreux, and M.C. Spiegel, *Bull. Soc. Chim. Belges.*, Vol. 59, p. 476, 1950.

19. L. Wild, et al., *Journal Polymer Sci., Polymer Physics Ed.*, Vol. 20, p. 441, 1982.

20. F.M. Mirabella, Jr., and E.A. Ford, *J. Polymer Science, Part B: Polymer Physics*, Vol. 25, p. 777, 1987.

21. Y.V. Kissin, R.I. Mink, and T.E. Nowlin, *J. Polymer Science, Part A: Polymer Chemistry*, Vol. 37, p. 4255, 1999.

22. P. Schouterden, G. Groeninckx, B. Van der Heijden, and F. Jansen, *Polymer*, Vol. 28, p. 2099, 1987.

23. B. Monrabal, J. Blanco, J. Nieto, and J.B.P. Soares, *J. Polymer Science, Part A, Polymer Chemistry*, Vol. 37, pp. 89-93, 1999.
24. E.T. Hsieh, and J.C. Randall, *Macromolecules*, Vol. 15, p. 353, 1982;
25. G.J. Ray, et al., *Macromolecules*, Vol. 14, p. 1323, 1981.
26. E.T. Hsieh, and J.C. Randall, *Macromolecules*, Vol. 15, p. 1402, 1982.
27. K. Kimura, et al., *Polymer*, Vol. 25, p. 441, 1984.
28. T.E. Nowlin, et al., *J. Polymer Science, Part A, Polymer Chemistry*, Vol. 29, p. 1167, 1991.
29. P.Schouterden,G.Groeninckx,B.VanderHeijden,and F.Jansen,*Polymer*,Vol.28, p. 2099, November 1987.
30. T.E. Nowlin, R.I. Mink, and Y.V. Kissin, *Handbook of Transition Metal Polymerization Catalysts*, Ray Hoff and R.T. Mathews, Editors, Chap. 6, John Wiley and Sons, 2010.

Index

Advanced Sclairtech Process, 208, 215
AkzoNobel, 247, 340
Albizzati, E., 66
Algae, 44
Alkanox 240, 320
Allison, J.B., 229
American Chemistry Council, 35
Ammonium fluoroborate, 197
Anderson, A.W., 297
Anti-block agents, 325
Antioxidants, 317, 319, 320
Anti-static, 318
Armstrong, S., 341
Arndt, M., 180, 222
Atiqullah, M., 222
Atwood, J.L., 182
Aufbau process, 48
Autocatalytic oxidation, 311
Autoclave reactor, 250, 251, 368

B. F. Goodrich, 317
Backbone scission, 313
Backstrom, H.L.J., 311
Bade, O.M., 160
Baker, L., 114
Banks, R.L., 9, 50, 70, 109, 256
Barron, Andrew R., 181–183
Basell Polyolefine GmbH, 273
BASF, 208, 234, 278, 279, 280, 319
BASF gas phase reactor, 279, 280, 282
Bateman, L., 311

Battenfield Gloucester Engineering
    Co. Inc., 330
Battle of Britain, 224
Baugh, Lisa S., xviii
Bayer AG, 35
Beerman, C., 58, 59
Bent strip test, 344
Bernhardt, T.M., 40
Best, S.A., 79
Bestian, H., 58
Beta hydride elimination, 115
Biaxial rotation, 355
Bimetallic catalyst, 216
Bimodal MWD, 215, 217, 334
Bis(n-butylcyclopentadienyl)
    zirconium dichloride, 193
Bis(triphenylsilyl)chromate
    catalyst, 127
Bis-phosphinimine ligand, 209
Blow molding, 341
Blowing agents, 324
Blown Film, 328-332, 334
Blow-up ratio, 336
Bohm, L.L., 92, 372–374
Boitsfort, F., 63
Bolland, J.L., 311
Boor, John Jr., 50, 71, 107
Bordon Dairy, 343
Borealis, 208, 294, 335, 340, 350
BorPEX, 340
Borstar, 215

Branching distribution, 189, 296
Breslow, David, 55, 58, 172, 173
Breuer, F.W., 107
Briel, H., 49
Britovsek, G.J.P., 222
Brown, Stephen J., 208
Buderi, Robert, 7
Bullis, Kevin, 44
Burnett, G.M., 244
Burns and McDonnell, 38

CAB-O-SIL, 88
Cabot Corp. Cab-O-Sil, 325
Caiman Energy, 35
Calcium carbonate, 318, 323
Calcium stearate, 322, 324
Camille Dreyfus Laboratory, 226
Canich, Jo Ann M., xviii
Cann, K., 130
Carbon black, 317, 323
Carbowax 4600, 325
Carpenter, W.B., 341
Carrick, W., 58, 114, 380
Cast film, 337-339
Catalyst addition process, 370
Catalyst poison, 272
Cationic Chromium (III)
    Complex, 142
Cellophane, 238, 239
Celluloid Manufacturing
    Company, 341
Chadwick, John C., xviii
Chain scission, 313, 315, 316
Chain termination, 61
Chanzy, H.D., 71
Chemical Market Resources, 299
Chemical Marketing Associates, 14
Chemplex Company, 77, 78
Chien, J.C.W., 55, 179
Chromocene catalyst, 132
Ciba-Geigy, 319
Cis- 1,4 polyisoprene, 50
Commercial Loop reactor -
    first process, 265

Constrained Geometry Catalyst
    (CGC), 202
Continuous extrusion blow
    molding, 343
Cossee, P., 57, 59, 61
Coupling agents, 323
Coville, N.J., 222
Cracking Agent, IGEPAL CO-630, 344
Crosslink Findland Oy, 341
Crosslinking, 315, 316
Crude Oil, 31
CRYSTAF method, 388
Cyclic olefin copolymer, 211

Dart Industries, Inc., 279
DEAC/DBM mixtures, 187
Decomposition reaction, high
    pressure process, 232
DeCoste, J.B., 344
Delbouille, A., 63
Desreux, V., 386
Dewar-Chatt-Duncanson, 60
Diatomaceous earth, 325
Dibutylmagnesium, 82
Die swell, 345
Diethyl zinc (DEZ), 364
Differential Scanning Calorimetry,
    (DSC), 380
Dobreva, D., 75
Dombro, R.A., 78
Dow Chemical, 63, 202–206,
    212, 213, 298, 299, 314
Dowlex, 299
Draft-tube design, slurry process, 262
Dreiling, M.J., 125
Drusco, G., 284
Dufraisse, C., 311
DuPont Corp., 8, 231, 233,
    295, 298, 308
DuPont of Canada, 174
Dutch State Mines (DSM), 298
Dye, R.F., 277, 278
Dym, J.B., 351

Edwards, B.J., 327
Eindhoven University of
      Technology and Dutch
      Polymer Institute, 363
Eisch, J.J., 184, 185
Elston, C.T., 174, 296
Energy Information Administration
      (EIA), 36
Engineering Industries, Inc., 354
ESCR (environmental stress crack
      resistance), 258, 343, 350
Esteruelas, M.A., 149
Ethane, 28, 29, 38
Ethylene mass flow meter, 394
Ethylene vinyl acetate copolymer, 233
Ethylene vinyl alcohol, 330
Ethylene/1-butene copolymers
      (HDPE), 257
Ethylene/styrene copolymer, 212, 213
European Union (EU), 21
Ewald, L., 311
Ewen, J.A., 192, 222
Extrusion Methods, 304
Exxon Chemical Co., 69, 208
ExxonMobil, 33, 36, 44, 242

Farmer, E.H., 311
Fawcett, E.W., 8, 230, 351
Federal District Court (Delaware), 112
Ferguson, A.D., 351
Fink, G., 185
Flame retardants, 317, 324
Fletcher, K.L., 75, 93
Flory, P.J., 243
Ford, E.A., 253, 387, 388
Fox, J.J., 378
Fu, S., 139, 140

G-20 Countries, 21
Galli, P., 88, 89
Gas-phase reactor /skin
      temperatures, 288, 289
Gas-phase reactor Union
      Carbide, 282, 283

Gas-phase, fluidized bed reactor, 282
Gas-phase, horizontal stirred
      system, 281
Gaylord, N.G., 50, 107
Geipel, L.E., 107
General Dyestuff Corp., 344
General Films Inc., 330
Georgia Institue Technology, 40
Gertz, G., 23
Giannini, U., 66
Gibson, R.O., 8, 230, 351
Gibson, V.C., 145-149, 222
Glove box, 364, 365
Glove box, design, 364
Goeke, G.L., 67
Goins, Robert R., 274, 276
Goodrich Gulf, 50
Gramann, Paul, 351
Greco, A., 91
Grignard reagents, 75, 94
Groppo, E., 116, 126
Gugumus, F., 313

Habgood, B.J., 317
Hackmann, M., 222
HALS (hindered amine light
      stabilizers), 321
Ham, George E., 225
Hancock, T., 317
Happoldt, W.B., 317
Haward, R.N., 75, 93
Hemperly, W.F., 337, 338
Hercules Research Center, 50, 55
Herwig, J., 178, 179
Heterogeneous branching, 175
High Pressure Process, 224
High through put laboratory
      equipment, 363
High-pressure pump, 229
High-stalk extrusion HMW-
      HDPE, 336
High-turbulent flow, 264
Hill, W.N., 351
Hoechst, AG, 64, 92, 266

Hoff, Ray, xviii, 77, 121
Hogan, J.P., 9, 50, 70, 109, 118, 255, 256
Holmes, Kim R., 25
Holzkamp, E., 49
Homogeneous branching, 174
Homolytic cleavage, 313
Hoogenboom, R., 363
Horton, A.D., 222
Household Industrial chemicals
    (HIC), 346
Howard, J.B., 343
Hyatt, John, 351

Immergut, E.H., 226
Imperial Chemical Industries,
    6, 8, 230, 306, 308
INEOS, 274, 291, 292
INEOS- Innovene, 291, 292
Infrared method, 378
Injection molding, 348-350
Inorganic fillers, 321
Intramolecular back-biting
    mechanism, 245

Jaffe, Amy M., 33
Jezl, James L., 279
Johnston, R.T., 316
Jolly, P.W., 155–157
Jordan, R.F., 186, 187
Joseph, Stephen, 42
Joule Unlimited, 44

Kamfjord, T., 196
Kaminsky, W., 12, 168, 176–178,
    196, 211, 222
Kanoh, N., 65
Karapinka, G.L., 132
Karol, F.J., 67, 133
Keim, W., 159
Kharas, Homi, 23
Kimble, H.S., 264
Kissin, Y.V., xix, 95, 99, 187, 381, 383
Koppers Company, Inc., 225, 226, 234
Kresser, T., 225, 249
Kyrtcheva, R., 75

Lacoste, J., 312
L'Alleund, B., 63
Landman, Uzi, 40
Lanning, W.C., 275
Lee, D.H., 196
Lee, Norman C., 341, 344
Lesnikova, N.P., 130, 131
Liquidfied Natural Gas
    (LNG), 31, 33, 37
Lo, F.Y., 194
Lockhart, M.W., 38
Loebel, A.B., 107
Long chain branching, 201, 204
Long, W.P., 185
Loop reactor, improved design, 273
Lubricants, 318, 324
Ludlum, D.B., 53
Lunsford, J., 134, 139, 140
Luperox, 250
Lupotech T Process, 242
Lustiger, A., 343
LyondellBasell, 210, 242, 291, 292

Magnesium alkyls, 74, 75
Malpass, D.B., xviii, 175, 255, 297
Marcellus Shale, 35
Mark, H.F., 50, 107, 243
Martin, A.E., 378
Marwil, S.J., 262
Mathers, Robert T., xviii
Max Planck Institute, 9, 50
McCarthy, Shawn, 35
McDaniel, M.P., xix, 122-126,
    141, 198, 222, 384
McDaniel, N.D., 222
Melt index instrument, 306
Melt strength, 306
Methylalumoxane, 176, 178, 180-183
Mexico, Coatzacoalcos, 242
Michels, A., 229
Milk bottle resin, 343, 347
Miller, Adam R., 282
Miller, Terry, 25
Mink, Robert I., xix, 80, 81
Mirabella, F.M., 253, 387, 388

Mitsubishi Chemical, 64, 65, 87, 90, 247
Mitsui, 208, 335
Mobil Chemical Co., 69, 215, 216, 335
Mobil Oil Corp., 81
Mohring, P.C., 222
Molding methods, 304
Molecular weight, 305
Molecular weight distribution, 305
Monoi, T., 161
Monrabal, B., 388
Monsanto, 49
Montagna, A.A., 222
Montedison, 66, 87, 90,91,93
Morrison, E.J., 316
Moureu, C., 311
Mulheim process, 49, 53, 57
Multi-layer Films, 330

Naphtha, 28, 29
Natta, G., 50, 112
Natural gas, 32
Natural gas - dry well, 34
Neftochim Research Centre, 75
Neithamer, David R., 202
Nobel Prize, 112
Non-Newtonian flow, 306, 326
Norwood, D., 262, 263, 265
Nova Chemical, 35, 207, 209, 213, 298
Nowlin, Thomas E., 80, 81
Nylon, 330

Occidental Petroleum, 210
Octadecane oxidation, 311
OECD (Organization for Economic Cooperation and Development), 41
Olabisi, O., 222
OPEC (Organiztion of Petroleum Exporting Countries), 4, 30
Organic peroxides, 244, 246, 247
Osgood, W.V., 309
Osswald, T.A., 351
Oxidation products, 314
Oxygen barrier, 330

Oxygen based free radical, 232

Parison, 341,
Particle form, 69, 256
Particle Morphology, 72, 73
Pasadena, Texas, 257
Pasynkiewicz, S., 181, 222
Peacock, Andrew J., xviii
Perrin, M.W., 230
Personal protective equipment, 364
Peters, E.F., 279
Peters, R., 355
Petkov, L., 75
Phillips catalyst, 109, 112
Pipe applications, 339
Pirinoli, F., 66
Plax Corp., 343
Polyethylene Product Space, 307
Polyethylene shrinkage, 349
Polyethylene structure, high-pressure, 252
Polypropylene, 10, 50
Potential Gas Committee, 36
Product Life Cycle, 16
Pullukat, T.J., 77, 120

Radar, 7
Radenkov, P., 75
Raff, R.A., 225, 229
Range Resources, 35
Reactivity ratio, 95, 206
Reactor kill system, 272
Reciprocating screw, 353
Rees, Herbert, 351
Reich, Leo, 107, 226
Renner, F., 196
Return on capital (ROC), 2
Rexall Chemical Company, 226
Reynolds number, 264
Rheology, 326
Richardson, John, 4
Rieger, B., 222
Robertson, E., 38
Roedel, M.J., 254
Rohlfing, R.G., 262, 264

Roper, A.N., 75, 93
Royal Dutch Shell, 35
Rubin, Irvin I., 351
Rutland Plastics Inc., 353
Rytter, E., 196

SABIC (Saudi Arabia Basic
    Industries Corp.), 3, 298
Sailor, H.R., 111, 255
Saudi Arabia, Jubail, 69
Saudi Arabia, Yanbu, 69
Saudi Ethylene and Polyethylene
    Company (SEPC), 242
Scheirs, J., 222
Schindler, A., 107, 226
Schmid, K., 280
Schubert, U.S., 363, 366
Sclair process, 175, 295
Sclairtech process, 299
Scoggin, J.S., 264
Serum bottles, 365
Severn, John R., xviii
Shale gas, 33, 35
Shear-thinning, 306, 327
Shida, M., 77
Shipton, E., 351
Shrinkage/warpage in injection
    molding, 353
Silica, 76
Sinn, H., 177, 222
Small, B.L., 153-155
Sobata, P., 90
Solid acid supports, 197
Solvay and Cie, 62
Specialty Minerals Inc., 325
Spencer Chemical Co., 49, 225, 234
Spherilene- LyondellBasell, 291, 292
Spiegel, M.C., 386
Spitzmesser, S.K., 222
Stamatoff, G.S., 297
Standard Oil, 111, 281
Steady-state polymerization
    conditions, lab. slurry
    reactor, 372

Steam Cracking, 28, 38
Stevens, Jacques, 63
Stevens, James C., 168, 202
Stoppette, 343
Strain, D.E., 309
Strain-hardening, 306
Sunoco Logistics Partners, 35
Symyx Corp., 363
Synthetic Genomics Inc., 44
Syringe technique, 365, 366

Takeuch, S., 284
Talc, 323
Tanaka, T., 65
Tandem reactors, 335
Tax rates, Corporate Global, 41, 42
Tebbe reagent, 58
Technology teams, 362
Temperature Rising Elution
    Fractionation (TREF),
    253, 386, 387
Tetraethyl lead (TEL), 15
Theopold, K.H., 134, 142, 144, 145
Thermoforming, 357
Thermoplastic, 304
Thermoplastic materials, 13
Thixotropic behavior, 183
Tie layer, 330
Titanium dioxide, 323
Tolinski, Michael, 319
TOPAS Advanced Polymers, 211
Totty, Michael, 44
Trigonox, 340
Tubular process, 231, 250
Turner, H.W., 193
Turng, L.S., 351

Uniloy Mahines, 343
Union Carbide Corp., 8, 63,
    126, 231, 308, 337
UNIPEX, 341
UNIPOL, 286, 291
UNIPOL (II) process, 334, 335
Univation, 274, 284

University of Amsterdam, 229
University of Ulm, 40
UV-light Stabilizers, 321

Vacuum Atmospheres Company, 365
Van der Ven, Sar, 375
Vanadium-based catalysts,
        175, 295, 297
Vandenberg, E.J., 50
Vertical pipe-loop reactor, 262, 264
Viscoelasticity, 305
Viscosity, 306
Vollmer, H., 177

W. R. Grace, Davison, 67, 80, 94,
        112, 135,187,194, 198
Wagner, Burkhard E., 67, 72
Wang, B.P., 179
Wang, Q., 213
Wass, D.F., 222

Welborn, H.C., 192
Welch, M.B., 122
Wester, T., 196
Wild, L., 253, 387, 388
Williams, E.G., 309
Witt, D.R., 119
Woldt, R., 177
Wolfensohn Center, 23
Wunderlich, B., 71

Xu, W., 209

Yamaguchi, K., 65

Zarn Company, 343
Ziegler Process, 265
Ziegler, Karl, 9, 48, 112
Zinc stearate, 324
Zipper-clave reactor, 369, 392
Zirconium cation, 183

Also of Interest

Introduction to Industrial Polyethylene: Properties, Catalysts, Processes

By Dennis P. Malpass

ISBN 978-0-470-62598-9

Demystifies the largest volume manmade synthetic polymer by distilling the fundamentals of what polyethylene is, how it is made and processed, and what happens to it after its useful life is over.

"I found this to be a straightforward, easy-to-read, and useful introductory text on polyethylene, which will be helpful for chemists, engineers, and students who need to learn more about this complex topic. The author is a senior polyethylene specialist and I believe we can all benefit from his distillation of knowledge and insight to quickly grasp the key learnings."

R.E. King III; Ciba Corporation (part of the BASF group)